Lecture Notes in Physics

W0232284

Springer-Verlag Berlin Heidelberg GmbH

The Editorial Policy for Proceedings

The series Lecture Notes in Physics reports new developments in physical research and teaching – quickly, informally, and at a high level. The proceedings to be considered for publication in this series should be limited to only a few areas of research, and these should be closely related to each other. The contributions should be of a high standard and should avoid lengthy redraftings of papers already published or about to be published elsewhere. As a whole, the proceedings should aim for a balanced presentation of the theme of the conference including a description of the techniques used and enough motivation for a broad readership. It should not be assumed that the published proceedings must reflect the conference in its entirety. (A listing or abstracts of papers presented at the meeting but not included in the proceedings could be added as an appendix.)

When applying for publication in the series Lecture Notes in Physics the volume's editor(s) should submit sufficient material to enable the series editors and their referees to make a fairly accurate evaluation (e.g. a complete list of speakers and titles of papers to be presented and abstracts). If, based on this information, the proceedings are (tentatively) accepted, the volume's editor(s), whose name(s) will appear on the title pages, should select the papers suitable for publication and have them refereed (as for a journal) when appropriate. As a rule discussions will not be accepted. The series editors and Springer-Verlag will normally not interfere with the detailed editing except in fairly obvious cases or on technical matters.

Final acceptance is expressed by the series editor in charge, in consultation with Springer-Verlag only after receiving the complete manuscript. It might help to send a copy of the authors' manuscripts in advance to the editor in charge to discuss possible revisions with him. As a general rule, the series editor will confirm his tentative acceptance if the final manuscript corresponds to the original concept discussed, if the quality of the contribution meets the requirements of the series, and if the final size of the manuscript does not greatly exceed the number of pages originally agreed upon. The manuscript should be forwarded to Springer-Verlag shortly after the meeting. In cases of extreme delay (more than six months after the conference) the series editors will check once more the timeliness of the papers. Therefore, the volume's editor(s) should establish strict deadlines, or collect the articles during the conference and have them revised on the spot. If a delay is unavoidable, one should encourage the authors to update their contributions if appropriate. The editors of proceedings are strongly advised to inform contributors about these points at an early stage.

The final manuscript should contain a table of contents and an informative introduction accessible also to readers not particularly familiar with the topic of the conference. The contributions should be in English. The volume's editor(s) should check the contributions for the correct use of language. At Springer-Verlag only the prefaces will be checked by a copy-editor for language and style. Grave linguistic or technical shortcomings may lead to the rejection of contributions by the series editors. A conference report should not exceed a total of 500 pages. Keeping the size within this bound should be achieved by a stricter selection of articles and not by imposing an upper limit to the length of the individual papers. Editors receive jointly 30 complimentary copies of their book. They are entitled to purchase further copies of their book at a reduced rate. As a rule no reprints of individual contributions can be supplied. No royalty is paid on Lecture Notes in Physics volumes. Commitment to publish is made by letter of interest rather than by signing a formal contract. Springer-Verlag secures the copyright for each volume.

The Production Process

The books are hardbound, and the publisher will select quality paper appropriate to the needs of the author(s). Publication time is about ten weeks. More than twenty years of experience guarantee authors the best possible service. To reach the goal of rapid publication at a low price the technique of photographic reproduction from a camera-ready manuscript was chosen. This process shifts the main responsibility for the technical quality considerably from the publisher to the authors. We therefore urge all authors and editors of proceedings to observe very carefully the essentials for the preparation of camera-ready manuscripts, which we will supply on request. This applies especially to the quality of figures and halftones submitted for publication. In addition, it might be useful to look at some of the volumes already published. As a special service, we offer free of charge LaTeX and TeX macro packages to format the text according to Springer-Verlag's quality requirements. We strongly recommend that you make use of this offer, since the result will be a book of considerably improved technical quality. To avoid mistakes and time-consuming correspondence during the production period the conference editors should request special instructions from the publisher well before the beginning of the conference. Manuscripts not meeting the technical standard of the series will have to be returned for improvement.

For further information please contact Springer-Verlag, Physics Editorial Department II, Tiergartenstrasse 17, D-69121 Heidelberg, Germany

Jean Cleymans   Hendrik B. Geyer
Frederik G. Scholtz (Eds.)

Hadrons in Dense Matter and Hadrosynthesis

Proceedings of the Eleventh Chris Engelbrecht
Summer School Held in Cape Town, South Africa,
4–13 February 1998

Springer

Editors

Jean Cleymans
Department of Physics
University of Cape Town
7701 ZA Rondebosch, South Africa

Hendrik B. Geyer
Frederik G. Scholtz
Department of Physics
University of Stellenbosch
6400 ZA Stellenbosch, South Africa

Library of Congress Cataloging-in-Publication Data.

Die Deutsche Bibliothek - CIP-Einheitsaufnahme

Hadrons in dense matter and hadrosynthesis : proceedings of the
Eleventh Chris Engelbrecht Summer School, held in Cape Town,
South Africa, 4 - 13 Februrary 1998 / Jean Cleymans ... (ed.).
(Lecture notes in physics ; Vol. 516)
ISBN 978-3-662-14238-7   ISBN 978-3-540-49483-6 (eBook)
DOI 10.1007/978-3-540-49483-6

ISSN 0075-8450
ISBN 978-3-662-14238-7

Originally published by Springer-Verlag Berlin Heidelberg New York in 1999
Softcover reprint of the hardcover 1st edition 1999

Typesetting: Camera-ready by the authors/editors
Cover design: *design & production*, Heidelberg

SPIN: 10644254   55/3144 - 5 4 3 2 1 0 – Printed on acid-free paper

Preface

This volume contains lectures presented at the Eleventh Chris Engelbrecht Summer School held at the University of Cape Town during the first half of February 1998.

The school gave lecturers the opportunity to present their fields of research in great detail with four or five lectures devoted to a single topic.

The topic of the lectures included in this volume is the study of dense hadronic matter in relativistic heavy ion collisions and in astrophysics.

In relativistic heavy ion collisions one can study the properties of highly compressed nuclear matter, test models describing the creation of hadrons, describe the evolution of hot hadronic matter and look for signals for the phase transition from nuclear to quark matter.

The lectures included in this volume provide excellent introductions to the fields of chiral symmetry at finite temperature, the use of light cone variables and the use of statistical methods applied to relativistic heavy ion collisions. The lectures also give a very thorough review of the experimental results at the GSI/SIS accelerator and a detailed presentation of the methods used in astrophysics for the theoretical study of dense stars.

We would like to take this opportunity to thank all the speakers for their efforts and for making the school a most enjoyable experience.

We gratefully acknowledge the financial support of the Foundation for Research Development (FRD, Pretoria) and the University Research Committee of the University of Cape Town.

Cape Town, October 1998   *Jean Cleymans*

Contents

List of Participants

Becattini, Francesco   becattini@avaxfi.fi.infn.it
    INFN Sezione di Firenze,
    Largo Enrico Fermi 2,
    I-50125 Florence, Italy
Boonzaaier, L.   eggers@physics.sun.ac.za
    Department of Physics,
    University of Stellenbosch,
    Stellenbosch 7600, South Africa
Braun-Munziger, Peter   P.Braun-Munzinger@gsi.de
    KPI GSI,
    Planckstr. 1,
    64220 Darmstadt, Germany
Cleymans, Jean   cleymans@physci.uct.ac.za
    Department of Physics,
    University of Cape Town,
    Rondebosch 7701, South Africa
Dadić, Ivan   dadic@thphys.irb.hr
    Ruder Boskovic Institute,
    P.O. Box 1016,
    HR-41001 Zagreb, Croatia
De Wet, Antony   jadew@global.co.za
    P.O. Box 514,
    Plettenberg Bay 6600, South Africa
Dumitru, Adrian   dumitru@th.physik.uni-frankfurt.de
    Inst. f. Theor. Physik,
    JW Goethe University ,
    D-60054 Frankfurt, Germany
Elliott, Duncan   elliott@physci.uct.ac.za
    Department of Physics,
    University of Cape Town,
    Rondebosch 7701, South Africa

Fetea, Mirella · · · · · · · · · · · · · · · · · · mfetea@physci.uct.ac.za
  Department of Physics,
  University of Cape Town,
  Rondebosch 7701, South Africa

Fetea, Remus · fetea@physci.uct.ac.za
  Department of Physics,
  University of Cape Town,
  Rondebosch 7701, South Africa

Geyer, Hendrik · · · · · · · · · · · · · · · · · Hbg@sun-akad2.sun.ac.za
  Department of Physics,
  University of Stellenbosch,
  Stellenbosch 7600, South Africa

Goldstein, Kevin · · · · · · · · · · · · · · · · keving@physci.uct.ac.za
  Department of Physics,
  University of Cape Town,
  Rondebosch 7701, South Africa

Klevansky, Sandi · · · · · · sandi@frodo.tphys.uni-heidelberg.de
  Inst. f. Th. Physik,
  Ruprecht-Karls-Univ.
  Philosophenweg 16,
  D-69120 Heidelberg, Germany

Koll, Matthias · · · · · · · · · · · · · koll@pythia.itkp.uni-bonn.de
  ITKP Bonn
  Rathausstr. 14,
  D-53111 Bonn, Germany

Madsen, Jes · jesm@dfi.aau.dk
  Inst. of Physics, Astron.
  Århus Univ.,
  Langelandsgade,
  DK-8000 ÅRHUS C, Denmark

Marais, Mark · · · · · · · · · · · · · · · · · · · marais@physci.uct.ac.za
  Department of Physics,
  University of Cape Town,
  Rondebosch 7701, South Africa

Munyaneza, Faustin · · · · · · · · · · · · · · munya@physci.uct.ac.za
  Department of Physics,
  University of Cape Town,
  Rondebosch 7701, South Africa

Murugan, Jeff · · · · · · · · · · · · · · · · · · murugan@physci.uct.ac.za
  Department of Physics,
  University of Cape Town,
  Rondebosch 7701, South Africa

Oeschler, Helmut   oeschler@axp602.gsi.de
  GSI ,
  Technische Hochschule,
  Darmstadt, Germany
Piròvano, Luca   pirovano@physci.uct.ac.za
  Department of Physics,
  University of Cape Town,
  Rondebosch 7701, South Africa
Redlich, Krzysztof   redlich@physik.uni-bielefeld.de
  Inst. Fizyki Teoretycznej,
  University of Wroclaw,
  Plac Maxa Borna 9,
  50-204 Wroclaw, Poland
Rischke, Dirk   drischke@nt1.phys.columbia.edu
  RIKEN-BNL Research Center,
  Brookhaven National Laboratory,
  Upton, NY 11973, USA
Scholtz, Frikkie   fgs@sunvax.sun.ac.za
  Department of Physics,
  University of Stellenbosch,
  Stellenbosch 7600, South Africa
Schumann, Marc   schumann@physci.uct.ac.za
  Department of Physics,
  University of Cape Town,
  Rondebosch 7701, South Africa
Sollfrank, Josef   josef.sollfrank@physik.uni-regensburg.de
  Inst. of Theor. Physics,
  University of Regensburg ,
  Postfach 397,
  D-93053 Regensburg, Germany
Suhonen, Esko   esko.suhonen@oulu.fi
  Dept. of Physical Sciences,
  University of Oulu,
  SF-90570 Oulu 57, Finland
Van Biljon, A.J.
  Department of Physics,
  University of Stellenbosch,
  Stellenbosch 7600, South Africa
Van Gend, Carel   vangend@physci.uct.ac.za
  Department of Physics,
  University of Cape Town,
  Rondebosch 7701, South Africa

Venugopalan, Raju   venugopa@alf.nbi.dk
    Niels Bohr Institute ,
    University of Copenhagen ,
    DK-2100 Copenhagen, Denmark
Viollier, Raoul   viollier@physci.uct.ac.za
    Department of Physics,
    University of Cape Town,
    Rondebosch 7701, South Africa
Wyngaardt, S.M.   eggers@sunvax.sun.ac.za
    Department of Physics,
    University of Stellenbosch,
    Stellenbosch 7600, South Africa

Pion and Kaon Production as a Probe for Hot and Dense Nuclear Matter

Helmut Oeschler[1]

Institut für Kernphysik, Technische Universität Darmstadt,
D - 64289 Darmstadt, Germany
for the KaoS Collaboration*

Abstract. The study of particle production in heavy ion reactions represents a valuable tool to extract information on the properties of hot and dense nuclear matter. Pions, kaons and protons were detected in mass-symmetric heavy ion reactions from C+C to Au+Au at incident energies between 0.6 and 2.0 A·GeV with the magnetic spectrometer KaoS installed at SIS, GIS. The study of K^+ mesons is considered to represent an ideal tool to extract information on the nuclear equation of state (EOS). First results indicate a soft EOS. The yield of K^- in Ni+Ni collisions is higher than expected from NN collisions. A possible interpretation of this observation is an in-medium mass modification. The center-of-mass pion spectra deviate from a Boltzmann distribution. The results indicate that high-energy pions are emitted at an early stage of the collision while low-energy pions can be emitted also rather late. This is evidenced using (i) the centrality dependence of the pion yield, (ii) a comparison of oppositely charged pion spectra and (iii) the shielding by spectator matter in peripheral collisions.

1 Introduction

1.1 Heavy Ion Physics

Interactions between heavy ions at various incident energies exhibit specific characteristics. Their study represents different research goals:

- **Around 100 A·MeV** incident energy the interpenetration is rather small leading to an increase in density to less than $1.2 \cdot \rho_0$ (with ρ_0 the normal nuclear density). The reaction mechanism is strongly influenced by the nuclear mean field. This is seen by attractive deflections and the transition to repulsive deflections which is called at higher incident energies "directed flow". Already a significant amount of the incident beam energy can be converted into excitation of the nuclei leading to new and interesting phenomena like e.g. multifragmentation.
- **Around 1 A·GeV** incident energy in central collisions the nuclei are expected to be stopped leading to densities of $(2 - 3) \cdot \rho_0$. Excitation energies of about $\approx 100 - 200$ MeV/nucleon are reached and part of this energy is converted into collective motion, i.e. as radial expansion of the compressed system. The phenomenon of "collective flow" is observed

by studying the kinetic energies of the outgoing fragments which are much higher than expected from Coulomb repulsion alone. The excitation energies are sufficient to create new particles. It can be easily seen that the number and species of produced particles represent a probe to test the energy content in the collision zone. This subject is the main theme of this work.

- **Around 10 A·GeV** incident energy still higher densities can be reached ($\approx 8 \cdot \rho_0$). The observed "flow" phenomena are similar to those at somewhat lower energies, yet many more particles are produced. The energy content (or energy density) is such that a condition might be reached which is close to the expected transition of normal matter to the "quark-gluon-plasma".

- **Around 100 A·GeV** incident energy it is under debate whether the colliding nuclei are stopped. This question refers to the question whether in the collision zone mainly mesons are found (non-stopping) or also baryons (stopping). In any case the energy content in this zone is very high and one expects that conditions are reached in which a phase transition to the "quark-gluon-plasma" might occur. However, a definite experimental prove is still missing. The latter two aspects are covered by the talk of P. Braun-Munzinger.

1.2   Particle Production Around 1 A·GeV

In central collisions of heavy ions at relativistic energies the colliding nuclei are expected to be stopped leading to dense and highly excited nuclear matter in their collision zone. The investigation of particle production is a well established method to explore the properties of this hot and excited dense nuclear matter [1–3].

Around 1 A·GeV only the production of pions is possible in single NN collisions while for other mesons the energy needed for the production has to be accumulated from more than one elementary NN collision. Figure 1 shows as arrows on the abscissa the thresholds in the center-of-mass frame for the production of various mesons and the excitation of baryonic resonances. The solid line gives the conversion into the laboratory system (ordinate) neglecting Fermi motion. The dashed lines represent the inclusion of Fermi motion in the two extremes, one corresponds to the Fermi motion towards the center of mass reducing the threshold and one pointing away from the center of mass increasing the threshold.

Pions are the most abundantly produced particles. They can easily be produced in individual nucleon-nucleon collisions. Pions interact strongly with nuclear matter by forming baryonic resonances e.g. via $\pi + N \rightarrow \Delta$. These resonances decay again mostly by pion emission. Therefore, pions are expected to leave the collision zone in a late stage of the collision when the system has expanded and cooled down [4].

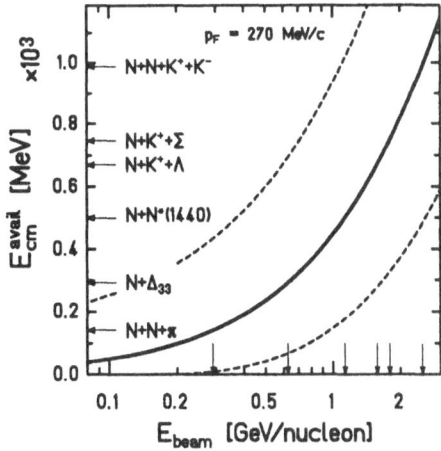

Fig. 1. Thresholds for meson production and excitation of baryonic resonances in the center-of-mass frame (abscissa) and the conversion into the laboratory system via the solid line. The dashed lines represent the extremes when including Fermi motion of the nucleon.

The K^+ **production** well below threshold was measured to occur preferentially in central collisions [5]. This suggests that they are produced in multi-step collisions where in a first step via $NN \rightarrow N\Delta$ baryonic resonances are excited. In a second step via $N\Delta \rightarrow N\Lambda K^+$ (or via $N\pi$ collisions) kaons can be produced. Due to these multi-step mechanisms the K^+ yield is sensitive to the available energy and thus to the nuclear equation of state (EOS). The systematic study of the dependence of the measured cross sections on centrality and beam energy is a promising tool to extract information on the EOS.

K^- **production** is compared to K^+ production at equivalent beam energies (at incident energies with equal values of $\sqrt{s} - \sqrt{s_{thres}}$ which compensates the different production threshold in the NN system). It turns out that the measured ratio K^-/K^+ is much higher for Ni+Ni collisions than for NN collisions [6]. A possible interpretation for this enhanced yield is a mass reduction of K^- in the nuclear medium.

Of special interest is the origin of **high-energy pions**, i.e. pions with a total energy above the available energy in free nucleon-nucleon collisions. They can be called "subthreshold" particles. Their yields are compared to K^+ production, as the total energy needed for the production is similar. Results are presented indicating that high-energy pions are emitted at an early stage of the collision. This is evidenced by studying (i) the centrality dependence of the yield [7,8], (ii) a comparison of π^+ and π^- spectra [9,10] and (iii) the shielding by spectator matter in peripheral collisions [9,11].

2   The Experiment

The experiments are performed with the **Kaon** Spectrometer (KaoS) installed at the heavy ion synchrotron SIS at GSI, Darmstadt.

This spectrometer [12] (Fig. 2) was designed to identify kaons over a wide range of momenta and angles in the presence of a high background of protons and pions. It consists of a quadrupole and a dipole magnet. KaoS combines a compact geometry to minimize the decay in flight, a large solid angle ($\Omega = 15 - 35$ msr) and a broad momentum range ($p_{max}/p_{min} \approx 2$ up to 1.7 GeV/c). The intrinsic momentum resolution without tracking is $\delta p/p \simeq 3\%$. A time start detector (16 scintillator paddles) is located in between the quadrupole and dipole magnet. The stop detector consists of 50 plastic scintillator paddles along the focal plane of the spectrometer. This arrangement allows for a very fast time-of-flight trigger which is indispensable for the efficient detection of rare particles. Three multi-wire proportional chambers – one located between the quadrupole and dipole magnet, one at the exit of the dipole and one close to the focal plane – are used for offline tracking. The pions, kaons, protons and heavier particles are identified by the time-of-flight and momentum information. A tracking analysis is used to suppress background.

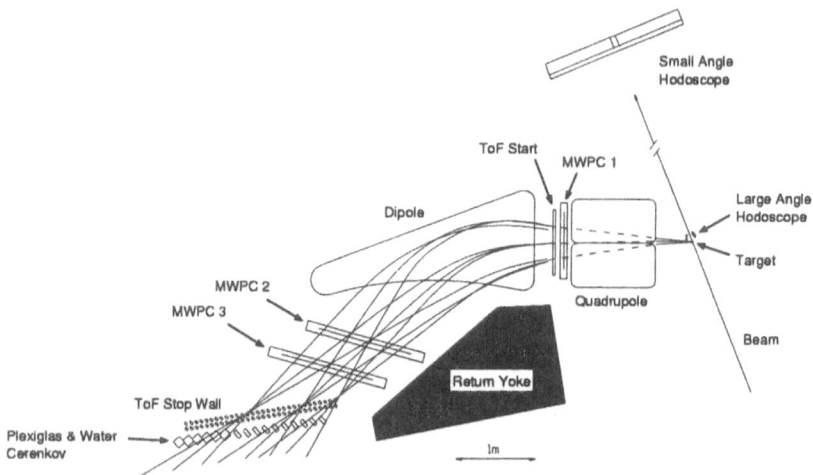

Fig. 2: Layout of the double-focusing Kaon Spectrometer at SIS/GSI together with the detector setup.

In the analysis mesons from central and peripheral collisions are separated by means of the hit multiplicity of charged particles in the Large Angle Hodoscope (LAH). This hodoscope consists of a 84-fold segmented detector close to the target at angles of 12–48 degrees. In this angular range participating protons are the most abundant particles. The impact-parameter selection has been controlled[13] by the correlation of this multiplicity with the summed nuclear charge of projectile fragments observed in the Small Angle Hodoscope (SAH). This 380-fold segmented detector covers polar angles between 0.5 and 11 degrees. It is located 7 m downstream of the target.

3 K^+ Production

Strangeness is produced by the creation of a $\bar{s}s$ pair. The positively charged kaon consists of $\bar{s}$ and u quarks. The u-quark originates from a nucleon and the s-quark is build in the nucleon converting it into a lambda. The corresponding energy needed in the center-of-mass frame is 671 MeV plus the kinetic energy of the K^+ (see Fig. 1). Hence, the beam energy of 1 A·GeV is not sufficient to produce K^+ in NN collisions (neglecting the Fermi motion) and the production is called "subthreshold". This fact causes a great sensitivity of the K^+ yield to collective effects, i.e. on the energetic condition and it allows to extract information on the properties of the nuclear equation of state (EOS)[3,14–17].

A key advantage of studying emitted K^+ is that due to their property "antistrangeness" they hardly interact with nuclear matter. This is illustrated in Fig. 3 by the comparison of K^+ and π^+ interaction with p. The cross section for K^+p are much smaller than for π^+p. Therefore, K^+ constitute a direct probe of the hot collision zone.

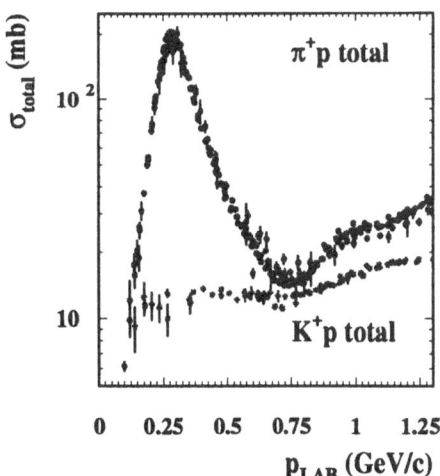

Fig. 3. Elementary cross sections for K^+p and π^+p interactions (from [18]) evidencing the contrast between the two species. Note the logarithmic scale on the ordinate.

The sensitivity on the EOS reduces with increasing incident energy, since close or above threshold K^+ can be produced via NN collisions. The sensitivity is less for light systems than for heavier systems. This is due to the

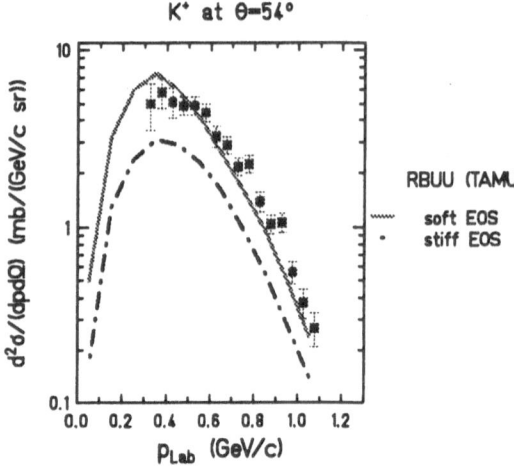

Fig. 4. Measured K^+ spectrum for Au+Au at 1 A·GeV [19] compared with BUU calculations (see text).

different densities reached in the collisions. Higher densities favour multiple collisions and the collective production of K^+. Consequently, the "best" choice to extract information on the EOS is to study the K^+ spectra of a heavy system at energies around or even better below 1 A·GeV. These spectra can then be compared with model calculations using different values for the stiffness of the EOS. This is done in Fig. 4 for Au+Au collisions at 1 A·GeV using the BUU calculations of [20]. This figure indicates a soft EOS ($\kappa = 220$ MeV). I would like to stress, however, that this conclusion is rather premature. Present models contain several uncertainties e.g. elementary cross sections. Therefore, models have to prove that they describe "insensitive" K^+ production properly, like the K^+ production in light collision systems C+C. Furthermore, they have to describe the measured pion spectra, too.

Another quantity, which according to model calculations exhibits a sensitivity on the EOS is the variation of the K^+ multiplicity with A_{part} as shown in Fig. 5 for the system Au+Au at 1 A·GeV. The figure shows an increase of M_{K+}/A_{part} as a function of A_{part}. An emission proportional to the number of nucleons in the collisions zone would yield a constant value. Hence, in central collisions more K^+ per A_{part} are produced than in peripheral reactions. This rise can be parameterized with $M_{K+} \propto A_{part}^{\alpha}$. A soft EOS yields according to [21] higher values of α, in agreement with the measured value of $\alpha = 1.8 \pm 0.2$, than a stiff EOS.

This result is rather similar to the one obtained from studying the inclusive K^+ multiplicity per A for A+A collisions as shown in the upper part of Fig. 6. This figure evidences the contrast to pion production given in the lower part. M_π/A decreases with A+A (see also [8]). Pion emission proportional to the number of nucleons would yield a constant value. The decrease might be due to absorption.

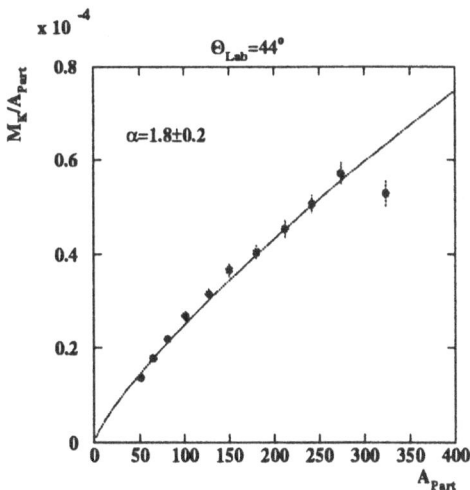

Fig. 5. K^+ multiplicity per number of A_{part} as a function of A_{part} for Au+Au collisions at 1 A·GeV [19].

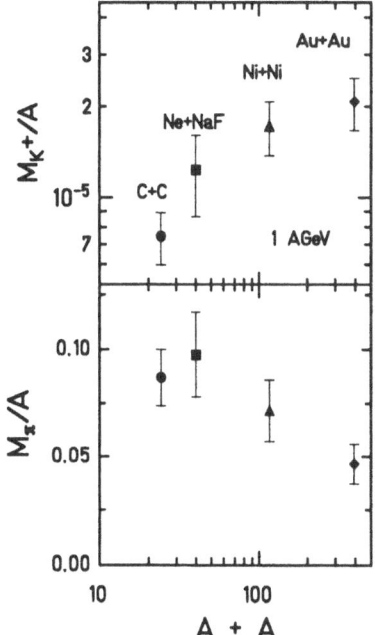

Fig. 6. Pion (π^+, π^0, π^-) and K^+ multiplicities per A as a function of A+A.

4 K^- Production

The information which can be extracted from studying K^- production is very different to that of K^+. The threshold for K^- is higher (987.4 MeV) than for K^+ since a pair of (K^-, K^+) has to be produced. Furthermore, the interaction of K^- with nuclear matter is rather strong as K^- can be "absorbed" by a nucleon forming a Λ.

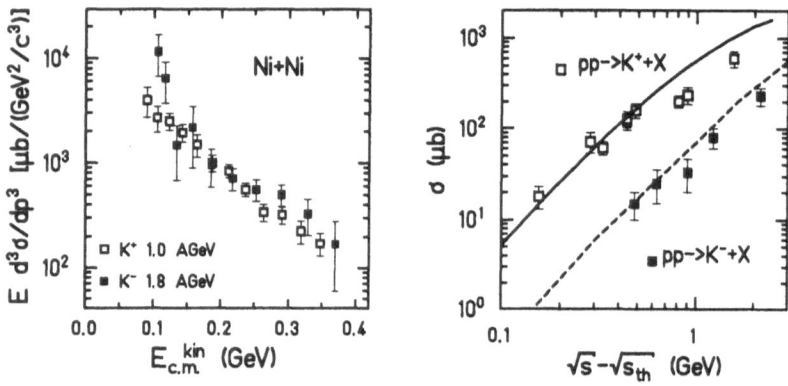

Fig. 7: Cross section for K^+ and K^- production in Ni+Ni (left) [6] and pp collisions (right) [22] at equivalent energies, i.e. same energy relative to the threshold.

Here, we compare K^- and K^+ production using Ni+Ni collisions. To compensate for the different thresholds, the K^- spectrum obtained at 1.8 A·GeV is compared to the K^+ spectrum at 1.0 A·GeV in Fig. 7, left. In both cases the incident energy is 230 MeV below the NN threshold. Nearly identical spectra are obtained. This result does not appear astonishing. However, the elementary cross section for $pp \rightarrow K^+$ and $\rightarrow K^-$ at the same incident energy with respect to the threshold (see Fig.7, right) exhibits a factor 10 difference in favour for K^+ production (for NN collisions the enhancement factor is 7 due to isospin). Furthermore, as already mentioned, K^- are expected to be absorbed strongly in nuclear matter. This should further reduce the yield of K^- in heavy ion collisions. In this respect, the nearly equal yields of K^- and K^+ are astonishing.

Two explanations are discussed at present:

(i) Hadrons in dense matter might change their properties [23,24]. For K^- a significant reduction of the mass is expected when the density of the nuclear environment increases, while for K^+ a slight increase of the mass with density is predicted.

(ii) If a thermal and chemical equilibrium is achieved, the individual cross section do no longer matter. In a first attempt [25] the measured ratio of K^+ to K^- at the incident energy of 1.8 A·GeV could be described within a thermal concept. This approach does not necessarily contradict the mass modifications as discussed in [26].

5 Pion Spectra

Pion spectra up to laboratory momenta of 1400 MeV/c have been measured in mass symmetric systems from A = 12 to A = 197 and at incident energies from 0.6 to 2.0 A·GeV. As a selection, Fig. 8 shows double differential cross sections of positively and negatively charged pions in the Boltzmann representation $1/(pE)\ d^2\sigma/(dEd\Omega)$ for various collision systems and at different incident energies. The spectra are measured at laboratory angles corresponding to a center-of-mass angular range within 90±30 degrees. Note that in this representation thermal distributions exhibit straight lines. All spectra in Fig. 8 exhibit a concave, non-thermal shape. Straight lines (Boltzmann distributions) fitted to the high-energy tail, i.e. to kinetic energies above the corresponding free NN kinematical limit, are shown. The variation of these inverse slope parameters with the mass of the collision system is rather weak. However, proton spectra measured close to midrapidity show a much stronger increase of the inverse slope parameters with system mass. The inverse slope parameters of the high-energy pions increase strongly with incident energy. For details see Ref. [8].

It is of interest to study the variation of the slopes of the high-energy part of the pion spectra with centrality, i.e. as a function of A_{part}. Fig. 9 shows that the inverse slope parameters increase with A_{part}. All collision systems follow a common line. Furthermore, the inverse slope parameters obtained for positively charged kaons agree with this systematics. These findings together with the obtained slope parameters from participating protons fit into a picture of a thermal, radially expanding source. Since the influence of flow increases with the mass of the emitted particle, protons show higher "apparent temperatures". At incident energies around 1 A·GeV pions are either "free" or "bound" in baryonic resonances. At freeze out the pion spectra are then composed of a "thermal" component and another one governed by the decay kinematics of the excited baryonic resonances. Indeed, the measured shapes (Fig. 8) can be qualitatively understood by such a scenario. Recent quantitative examples of such a decomposition are found e.g. in Ref. [27–29].

The arguments presented so far are pointing towards the interpretation within a thermal concept. A recent attempt to understand the particle ratios and spectra is given in Ref. [26]. Next, arguments are given that the assumption of a unique freeze-out time for all particles and even for one particle species of different kinetic energy, here pions, is highly questionable.

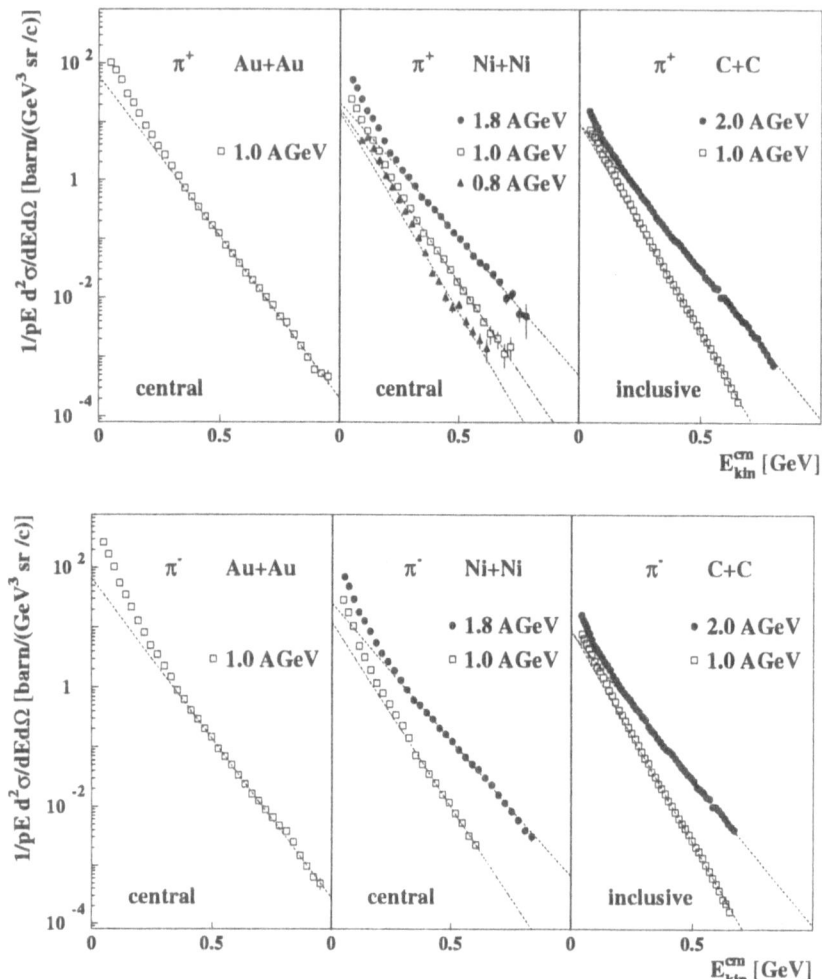

Fig. 8: Spectra of positively (upper part) and negatively (lower part) charged pions in the center-of-mass frame in a "Boltzmann" representation for various reactions (preliminary data).

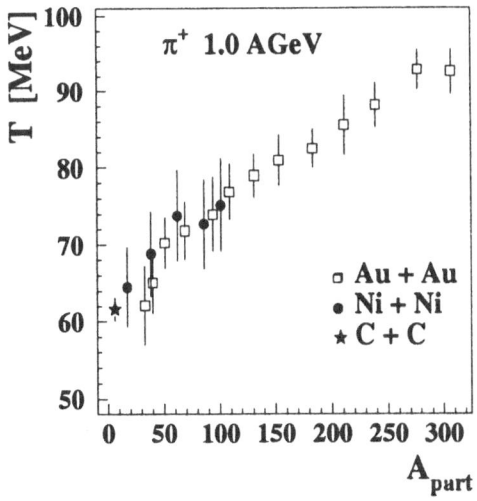

Fig. 9. The inverse slope parameters of positively charged high-energy pion for different collision systems reveal a very similar behaviour if plotted as a function of A_{part}.

6   The Time Evolution of Pion Emission

In this chapter three observations demonstrate that high-energy pions are emitted during an early stage of the collision. The first one is based on a comparison of the centrality dependence of high-energy pions with that of positive kaons [8]. The second argument is based on a comparison of spectra of positively and negatively charged pions. A third independent argument for an early emission of high-energy pions is based on a detailed study of the shapes of pion spectra under different geometrical conditions [11].

6.1   High-Energy Pions

Figure 10 (left top) shows the ratio $(d\sigma_{\pi^+}/d\Omega_{CM})$ / $(d\sigma_p/d\Omega_{CM})$ (labeled π^+/p) as a function of the average number of participating nucleons (experimentally deduced see Ref. [8]) for the heavy mass systems at 0.8 and 1 A·GeV beam energy. Parameterizing $(d\sigma_{\pi^+}/d\Omega_{CM})$ / $(d\sigma_p/d\Omega_{CM}) \propto A_{part}^{\alpha-1}$ (solid lines in Fig. 10), an exponent α of 1.04 ± 0.13 (1.05 ± 0.13) is obtained at 1.0 (0.8) A·GeV incident energy. This result demonstrates that the number of pions at midrapidity, dominated by the low-energy part of the spectra, exhibits a linear increase with the number of participating nucleons as already reported in Refs. [7,8,30,31] using the assumption that the number of high-energy protons emitted close to midrapidity $(d\sigma_p/d\Omega_{CM})$ scales linearly with A_{part}. Absorption is expected to play a minor role in these trends as only the spatial distribution of the mass varies. That is the advantage of studying the A_{part} dependence of one mass system only and not comparing different mass systems.

A different trend in σ_{π^+}/σ_p (Fig. 10, right top) is observed when taking into account the high-energy part of the pion spectra alone. This ratio increases with the size of the reaction zone resulting $\alpha = 1.63 \pm 0.19$ ($\alpha = 1.86 \pm 0.19$) for 1.0 (0.8) A·GeV incident energy (lines in Fig. 10). Here, only those pions are taken into account which have a total energy above 671 MeV in the center-of-mass frame. This value has been chosen to compare directly with the results from positive kaon production and it represents the minimum energy needed for the production of positive kaons ($m_K + (m_\Lambda - m_N)$), see below.

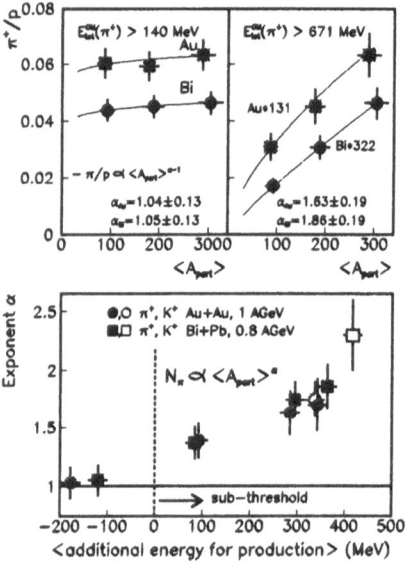

Fig. 10. Upper part: The ratio π^+/p for all (left) and high-energy (right) pions from Au+Au (1 A·GeV) and Bi+Pb (0.8 A·GeV) as a function of the average number of participating nucleons. Lines represent a fit $\propto A^{\alpha-1}$. Lower part: Exponent α for positive pions and kaons as a function of the "average additional production energy" (see text for definition) [8].

In the lower part of Fig. 10 the dependence of pion production on the beam energy and on the kinetic energy of the pions is combined: We study the dependence of the exponent α on the energy which is available to produce a particle in a nucleon-nucleon collision, corrected for the kinematical limit in free nucleon-nucleon collisions ($E_{kin}+E_{threshold}-(\sqrt{s_{NN}}-2m_N)$). The resulting quantity which is called "additional energy needed for production", is defined to be positive for subthreshold production.

Figure 10 (bottom) evidences an exponent $\alpha \approx 1$ for particles produced above threshold and a continuous increase of the exponent α with the energy which is needed to produce a particle with a given kinetic energy. For positive kaons a non-linear dependence of the yield with the number of participating nucleons has been reported, too [5,19,32]. To include these results in Fig. 10 the corresponding full kaon spectra were integrated and the average kinetic energy was determined. To obtain the average energy needed for the production of these kaons, the minimum energy to produce a positive kaon of 671 MeV was added. Within the error bars no significant difference between particle species is observed in Fig. 10. This result evidences the similarity in production characteristics of the positive kaons and the high-energy pions, characterized by the exponent α, and can be understood by the fact that the same total energy is needed for their production. As mentioned above, a key in this representation is that absorption which is present in heavy ion collisions, likely cancels and only the production properties are seen. The observed trend suggests that the more energy is needed to produce a particle the more secondary collisions during the course of the heavy ion reaction have to take place to accumulate the energy needed for their production. For positive kaons such a suggestion has already been made from a theoretical point of view [15,21,33]. Secondary collisions happen more frequently in central collisions of heavy nuclei during the hot and dense stage of the reaction.

Our observations indicate that both kaons as well as high-energy pions are produced and emitted at the same early stage of the reaction. The majority of pions are emitted in a later stage.

6.2   Comparison of Positively and Negatively Charged Pions

In this section, we present a comparison of positively and negatively charged pions emitted in ^{197}Au+^{197}Au collisions at 1.0 A·GeV incident kinetic energy. The observed difference in the π^- and π^+ spectra is attributed to the different isospins and to the oppositely acting Coulomb field. At the incident energy of 1 A·GeV the π production occurs essentially via the formation of the Δ_{33} resonance. Hence, the influence of the isospin can be calculated. Furthermore, only the one-pion decay is relevant. For details see Ref. [9,10].

Figure 11 shows the π^- and π^+ cross section $d^2\sigma/(dE^{kin}_{c.m.}\,d\Omega_{c.m.})$ as a function of the kinetic energy $E^{kin}_{c.m.}$ in the center-of-mass frame for central Au+Au reactions measured at a laboratory angle of (44 ± 4) degrees which corresponds to midrapidity. These central collisions represent (14 ± 4) % of the total reaction cross section. At low kinetic energies of the pions, the π^- yield exceeds the π^+ yield and approaches it at higher kinetic energies.

The energy integrated π^-/π^+ ratio $R^{tot}_{exp} = (d\sigma(\pi^-)/d\Omega)/(d\sigma(\pi^+)/d\Omega)$ is determined by extrapolating the energy distribution to $E_{c.m.} = 0$ describing the spectra with the sum of two Maxwell-Boltzmann distributions (see also [7,8]). The experimental value of 1.94±0.05 agrees well with the ratios derived from an isospin decomposition using the parametrization given in Ref. [34] (1.90) and with the nearly identical values using the isospin formulas corresponding to a formation purely via the Δ_{33} resonance (1.95). This agreement indicates that the π^-/π^+-ratio reflects the N/Z asymmetry of the colliding system and motivates the assumption that the observed energy dependence is caused by the oppositely acting Coulomb field.

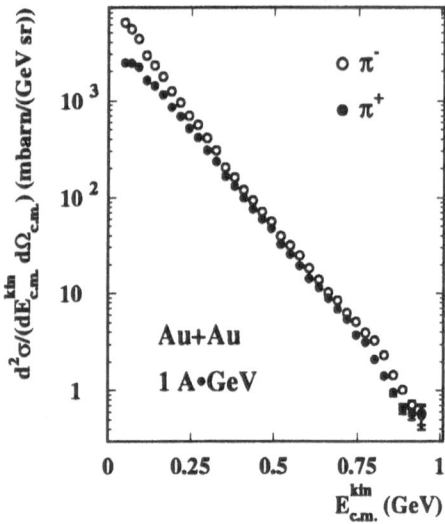

Fig. 11. Double differential cross sections of negatively and positively charged pions from central collisions of the reaction system ^{197}Au+^{197}Au at an incident beam energy of 1 A·GeV and at an emission angle of $\theta_{LAB} = (44 \pm 4)$ degrees.

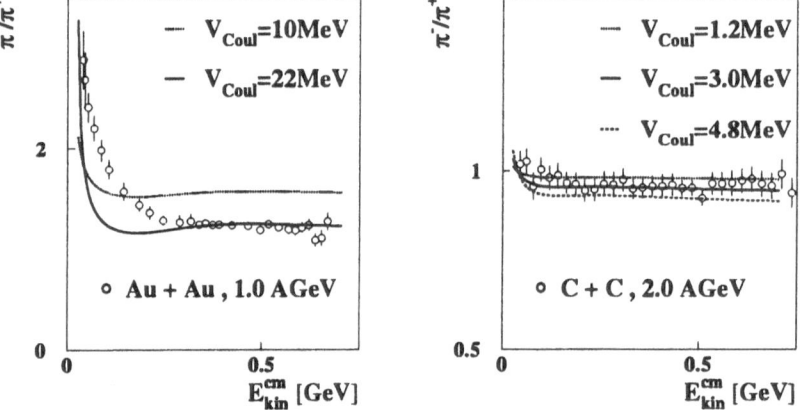

Fig. 12: Ratio of π^-/π^+ as a function of the kinetic energy of pions for Au+Au (left) and C+C collisions (right).

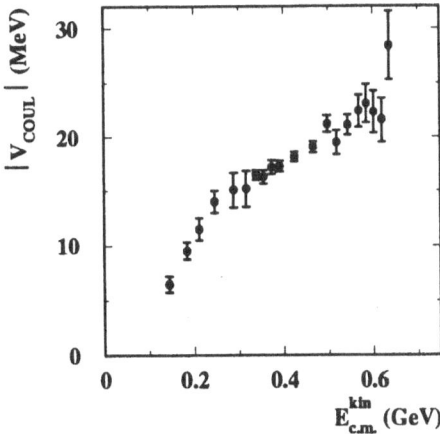

Fig. 13. Variation of $|V_{Coul}|$ with the pion kinetic energy as deduced from the measured energy-dependent π^-/π^+ ratio for central ^{197}Au+^{197}Au collisions.

Figure 12, left, shows the π^-/π^+ ratio for Au+Au collisions as a function of the pion kinetic energy in the c.m.-system. The π^-/π^+ ratio decreases from 2.8 at low pion energies to nearly a constant value of 1.1 for pion energies above 0.4 GeV.

We assume that the π^-/π^+ ratio is independent of the pion energy if Coulomb effects are disregarded. This assumption is supported by transport model calculations [35,36] which find a constant π^-/π^+ ratio of ≈ 1.9 if the Coulomb interaction is switched off. The Coulomb field affects the charged pions directly by the proton phase-space distribution during the collision at the instant of emission. Hence, it is intriguing to extract the strength of the Coulomb force starting from a static approximation for the Coulomb field. According to Ref. [37], the Coulomb force distorts the pion spectra by modifying both the kinetic energies of the particles and the available phase space. The solid line in Fig. 12, left, demonstrates that the π^-/π^+ ratio can only above 0.6 MeV pion kinetic energy be described with a Coulomb potential of 22 MeV. For lower pion energies a lower Coulomb potential is required. The dashed line in Fig. 12 shows the π^-/π^+ ratio for a V_{coul} of 10 MeV. It is clear that a constant Coulomb potential fails to describe the measured results. A Coulomb potential varying with pion energy is needed as shown in Fig. 13, see also [10].

For comparison, Fig. 12, right, shows the π^-/π^+ ratio for C+C at 2 A·GeV [29]. The integrated π^-/π^+ ratio is one as expected for $N = Z$ nuclei. The rather weak influence of the Coulomb field is demonstrated by the three lines. This figure evidences the precision of the data as π^- and π^+ are measured in different magnetic field settings.

The extracted variation of V_{coul} can be interpreted that high-energy pions are emitted early from a compact configuration while pions of lower energy leave the system at a more dilute stage. Hence, it is worthwhile to extract the radii and densities of the pion emitting sources. For central collisions the mean number of participating charges Z_{part} has been measured to 110±8. In a simple assumption of an emission of pions from the surface of a charged sphere, the Coulomb potential ($V_{coul} = Z\alpha\hbar c/r_{eff}$) yields an effective radius of $r_{eff} \approx 7.2$ fm for high-energy pions. This can be converted into an effective density of $\rho_{eff} = (1.1\,^{+0.2}_{-0.3}) \cdot \rho_0$. For pion around 0.2 GeV $V_{coul} = 10$ MeV is extracted. The corresponding density is $\rho_{eff} = (0.1 \pm 0.03) \cdot \rho_0$.

We have observed that the Coulomb field which acts on low-energy pions, is weaker than the field acting on high-energy pions. This corresponds to a more dilute charge distribution at freeze-out for low-energy pions. Similar conclusions were obtained from pion-correlation studies [38,39].

6.3   Shielding by Spectator Matter – The Pion Camera

Direct experimental evidence for the time evolution of pion emission is presented based on the shadowing of spectator matter in certain space-time regions. For this purpose we have chosen peripheral collisions of Au+Au at 1.0 A·GeV incident energy. The moving spectator matter acts like a shutter of a camera shielding the pion, i.e. modifying the pion emission pattern according to the spatial distribution of the spectator matter at the time of the pion freeze out. The motion of the spectator serves as a calibrated clock since the c.m.-velocity is well defined. A preferential emission perpendicular to the reaction plane has already been observed for this collision system [43,44]. Recently, an enhanced in-plane emission of pions was observed [40]. This "antiflow" behaviour is found to be pronounced only in peripheral collisions. In this work we reveal that the effects of "flow" and "antiflow" are resulting from the interplay of the time evolution of pion emission with the shadowing of the surrounding matter.

Figure 14 illustrates the emission in respect to the reaction plane. The orientation of the reaction plane is determined using the recipe given by Danielewicz [42].

In previous works [43,44] the observation of preferential emission of pions perpendicular to the reaction plane has been reported. Here, we subdivide the emission in plane, comparing the number of pions emitted to the same side as the projectile remnants (projectile side) with the one on the opposite side (target side). Assuming that the out-of-plane emission reflects the least disturbed pion spectra, the in-plane spectra are divided by the out-of-plane one for normalization purpose. As an example Fig. 15 shows these ratios obtained at a laboratory angle of 84 degrees ($0.01 \leq y/y_{beam} \leq 0.10$). The most interesting observation is the drastic drop for high-energy pions on the "projectile side", while on the "target side" the ratio is about one. For low-energy pions one observes a slight reduction only on the "target side". The observations for π^- and π^+ are nearly identical, demonstrating that the effect is not caused by the opposite Coulomb force. To illustrate the

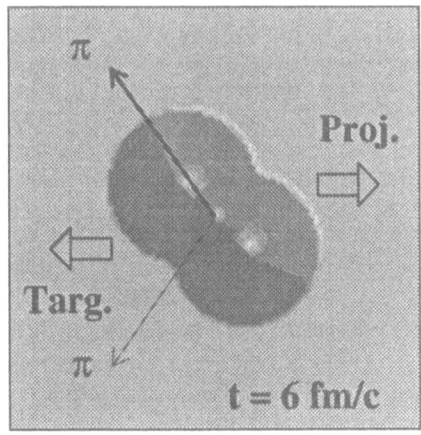

Fig. 14. Illustration of the geometrical situation of two colliding nuclei showing the reaction plane and out-of-plane emission (from [41]).

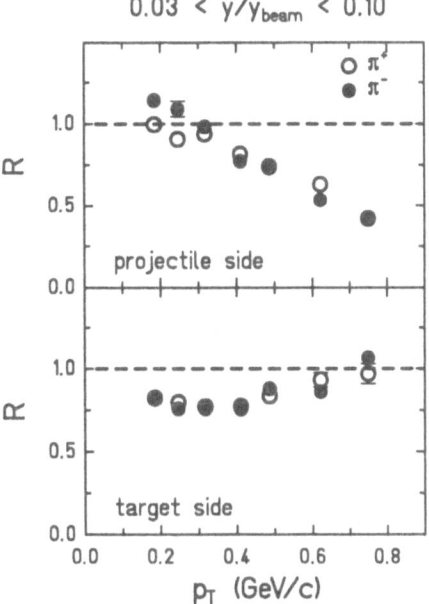

Fig. 15. Ratio of the pion spectra for in-plane emission to the one out of plane in peripheral Au+Au collisions at 1 A·GeV. The upper part refers to the emission to the "projectile side", the lower one to the "target side". Full (open) symbols refer to π^- (π^+) emission.

reduction of high-energy pions on the "projectile side", Fig. 16 shows the geometrical situation just at the beginning of the collision. The projectile spectator is just inbetween the fireball and the detector on the projectile

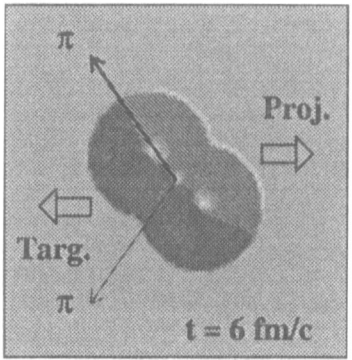

Fig. 16. Sketch of the geometrical situation of the spectators at 6 fm/c after the instant when the nuclei touch for Au+Au collisions at 1 A·GeV.

side, thus shielding pions emitted at this instant. However, at this "early" stage the target spectator is not shielding the emission to the "target side" as can be observed in Fig. 15. Hence, high-energy pions are emitted at this early stage ($\approx$ 6fm/c) of the collision. The emission time interval is estimated to be $\approx$ 10 fm/c from the time of fly-by of the projectile residue. Low-energy pions seem to be emitted during the whole collision as no pronounced suppression is seen. The slight reduction on the target side indicates a preferred emission at later times.

The results of the three methods can be interpreted as follows: The excited nuclear system is expanding and pion emission occurs during the whole collision process. High-energy pions are emitted only during the early phase of the collision. This is evidenced by the shadowing of spectator matter and by the high Coulomb field acting on these pions. Lower pion energies do not exhibit a pronounced shadowing, the repective Coulomb field are rather low. This points to a rather long emission time interval at a late stage of the collision. Microscopic calculation also state that high-energy pions are emitted aerlier than pions of lower kinetic energy [45].

7   Summary

The study of particle production in heavy ion collisions offers a rich field to explore nuclear matter at high densities and at high excitations.

The extraction of information on the nulear equation of state is one of the main topics. It was demonstrated that the study of K^+ production represents a favourite tool for this task. The present results indicate a soft ($\kappa \approx$ 220 MeV) compression modulus. A complete systematics, i.e. an excitation function for K^+ of a light and a heavy collision system are needed [46] to compare with model prediction before this long-outstanding question can be answered.

The yield of K^- production in Ni+Ni is similar to the K^+ cross section when compared at equivalent energies (same c.m.-energy below the repective thresholds). This is astonishing as the elementary cross section exhibits a factor 7 larger yield for K^+ than for K^- at the same energy above threshold. A reduction of the K^- mass in the nuclear medium is discussed as interpretation of these yields.

Pion energy spectra deviate from a pure Boltzmann shape. They show a low energy enhancement which is interpreted as dominated by decaying baryonic resonances. A comparison of the inverse slope parameters of pions, K^+ and p seem to favour an interpretation with a chemical and thermal equilibrium. Also particle ratios seem to favour this concept [26]. In spite of this success, detailed studies of the high-energy part of the pion spectra exhibit that they are emitted only at an early stage of the collision while low-energy pions seem to be emitted during the whole collision process. This was evidenced by three independent analysing techniques.

This work was supported by the German Federal Minister for Research and Technology (BMBF), by the Gesellschaft für Schwerionenforschung (GSI) and by the Polish Committee of Scientific Research.

* The members of the **KaoS Collaboration**:

A. Wagner[1], C. Müntz[2], A. Förster, H. Oeschler, A. Surowiec, C. Sturm, F. Uhlig, (*Institut für Kernphysik, Technische Universität, Darmstadt, D - 64289 Darmstadt, Germany*),

W. Ahner, R. Barth, P. Koczon, F. Laue, M. Mang, P. Senger, (*Gesellschaft für Schwerionenforschung, D-64291 Darmstadt, Germany*),

D. Brill, Y. Shin, H. Ströbele (*Johann Wolfgang Goethe-Universität, D-60325 Frankfurt/Main, Germany*),

I. Böttcher, B. Kohlmeyer, M. Menzel, F. Pühlhofer, J. Speer, (*Philipps-Universität, D-35037 Marburg, Germany*),

M. Debowski, W. Walus (*Jagiellonian University, PL-30-059 Kraków, Poland*),

F. Dohrmann, E. Grosse[3], L. Naumann, W. Scheinast (*Forschungszentrum Rossendorf, D-01314 Dresden, Germany*)

References

1. S. Nagamiya et al., *Phys. Rev.* C **24**, 1981 (971).
2. R. Brockmann et al., *Phys. Rev. Lett* **53**, 1984 (2012).
3. R. Stock, *Phys. Rep.* **135**, 1986 (259).
4. S. Nagamiya, *Phys. Rev. Lett* **49**, 1982 (1383).
5. D. Miśkowiec et al., *Phys. Rev. Lett* **72**, 1994 (3650).
6. R. Barth et al., *Phys. Rev. Lett* **78**, 1997 (4007).

[1] now at NSCL, MSU, USA
[2] now at Univ. Frankfurt
[3] also at Tech. Univ. Dresden, Germany

7. C. Müntz et al., *Z. Phys.* A **352**, 1995 (175).
8. C. Müntz et al., *Z. Phys.* A **357**, 1997 (399).
9. A. Wagner, PhD Thesis (1996), Technische Universität Darmstadt, Germany.
10. A. Wagner et al., *Phys. Lett.* B **420**, 1998 (20).
11. A. Wagner et al., to be published.
12. P. Senger et al., *Nucl. Instrum. Methods* A **327**, 1993 (393).
13. W. Ahner et al., *Z. Phys.* A **341**, 1991 (123).
14. J. Aichelin et al., *Phys. Rev. Lett* **58**, 1987 (1926).
15. W. Cassing et al., *Phys. Rep.* **188**, 1990 (363).
16. J. Aichelin, *Phys. Rep.* **202**, 1991 (233).
17. A. Lang, W. Cassing, U. Mosel, K. Weber, *Nucl. Phys.* A **541**, 1992 (507).
18. Particle Data Group, *Phys. Rev.* D **54**, 1996 (1).
19. M. Mang, PhD Thesis (1997), University of Frankfurt , Germany.
20. X.S. Fang et al., *Nucl. Phys.* A **575**, 1994 (766).
21. C. Hartnack et al., *Nucl. Phys.* A **580**, 1994 (643).
22. CERN HERA Report 84-01 (1984).
23. D.B. Kaplan, A.E. Nelson, *Phys. Lett.* B **175**, 19986 (57).
24. G.E. Brown et al., *Phys. Rev.* C **43**, 1991 (1881).
25. J. Cleymans, D. Elliott, A. Keränen, E. Suhonen, *Phys. Rev.* C **57**, 1998 (3319).
26. J. Cleymans, H. Oeschler, K. Redlich, to be published.
27. B. Hong et al., *Phys. Lett.* B **407**, 1997 (115).
28. W. Weinhold, B. Friman, W. Nörenberg, *Phys. Lett.* B , 1998 ().
29. A. Förster, diploma thesis, Technische Universität Darmstadt, Germany 1998.
30. O. Schwalb et al., *Phys. Lett.* B **321**, 1994 (20).
31. J.W. Harris et al., *Phys. Rev. Lett* **58**, 1987 (463).
32. M. Cieślak , PhD Thesis (1995), Jagiellonian University Cracow, Poland.
33. J. Aichelin et al., *Phys. Rev. Lett* **58**, 1987 (1926).
34. B. J. Ver West, R. A. Arndt, *Phys. Rev.* C **25**, 1982 (1979).
35. S.A. Bass et al., *Phys. Rev.* C **51**, 1995 (3343).
36. S. Teis et al., *Z. Phys.* A **356**, 1997 (421).
37. M. Gyulassy and S.B. Kauffmann, *Nucl. Phys.* A **362**, 1981 (503).
38. D. Beavis et al., *Phys. Rev.* C **34**, 1986 (757).
39. R. Bock et al., *Z. Phys.* A **333**, 1989 (193).
40. J.C. Kintner et al., *Phys. Rev. Lett* **78**, 1997 (4165).
41. J. Konopka, PhD Thesis (1995), University of Frankfurt , Germany.
42. P. Danielewicz, *Phys. Rev.* C **51**, 1995 (716).
43. D. Brill et al., *Phys. Rev. Lett* **71**, 1993 (336).
44. D. Brill et al., *Z. Phys.* A **357**, 1997 (207).
45. S.A. Bass et al., *Phys. Rev.* C **50**, 1994 (2167).
46. C. Sturm, PhD Thesis , Technische Universität Darmstadt, Germany, in preparation.

Fluid Dynamics
for Relativistic Nuclear Collisions

Dirk H. Rischke[1]

RIKEN-BNL Research Center, Brookhaven National Laboratory, Upton, NY 11973, USA

Abstract. I give an introduction to the basic concepts of fluid dynamics as applied to the dynamical description of relativistic nuclear collisions.

1 Introduction and Conclusions

Modelling the dynamic evolution of nuclear collisions in terms of fluid dynamics has a long-standing tradition in heavy-ion physics, for a review see [1–3]. One of the main reasons is that one essentially does not need more information to solve the equations of motion of ideal fluid dynamics than the equilibrium equation of state of matter under consideration. Once the equation of state is known (and an initial condition is specified), the equations of motion uniquely determine the dynamics of the collision. Knowledge about microscopic reaction processes is not required. This becomes especially important when one wants to study the transition from hadron to quark and gluon degrees of freedom, as predicted by lattice simulations of quantum chromodynamics (QCD) [4]. The complicated deconfinement or hadronization processes need not be known in microscopic detail, all that is necessary is the thermodynamic equation of state as computed by e.g. lattice QCD. This fact has renewed interest in fluid dynamics to study the effects of the deconfinement and chiral symmetry restoration transition on the dynamics of relativistic nuclear collisions. Such collisions are presently under intense experimental investigation at CERN's SPS and Brookhaven National Laboratory's AGS and (beginning Fall 1999) RHIC accelerators.

In this set of lectures I give an overview over the basic concepts and notions of relativistic fluid dynamics as applied to the physics of heavy-ion collisions. The aim is not to present a detailed and complete review of this field, but to provide a foundation to understand the literature on current research activities in this field. This has the consequence that the list of references is far from complete, that I will not make any attempt to compare to actual experimental data, and that some interesting, but more applied topics (such as transverse collective flow) will not be discussed here. In Section 2, I discuss the basic concepts of relativistic fluid dynamics. First, I present a derivation of the fluid-dynamical equations of motion. A priori, there are more unknown functions than there are equations, and one has to devise approximation schemes in order to close the set of equations of motion. The most simple is the ideal fluid approximation, which simply discards some of the unknown functions. Another one is the assumption of small deviations from local thermodynamical equilibrium, which leads to the equations of dissipative fluid dynamics. A brief discussion of multi-fluid models concludes this section. In Section 3 I discuss numerical aspects of solution schemes for ideal relativistic fluid dynamics. Section 4 is devoted to a discussion of one-dimensional solutions of ideal fluid dynamics. After presenting a classification of possible wave patterns in one spatial dimension, for both thermodynamicall normal as well as anomalous matter, I discuss the expansion of semi-infinite matter into vacuum. This naturally leads to the discussion of the Landau model for the one-dimensional expansion of a finite slab of matter. The Landau model is historically the first fluid-dynamical model for relativistic nuclear collisions. However, more realistic is, at least for ultrarelativistic collision energies, the so-called the Bjorken model which is subsequently presented. The main result of this section is the time delay in the expansion of the system due to the softening of the equation of state in a phase transition region. This may have potential experimental consequences for nuclear collisions at RHIC energies, where one wants to study the transition from hadron to quark and gluon degrees of freedom. Finally, Section 5 concludes this set of lectures with a discussion on how to decouple particles from the fluid evolution in the so-called "freeze-out" process and compute experimentally observable quantities like single inclusive particle spectra.

Units are $\hbar = c = k_B = 1$. The metric tensor is $g^{\mu\nu} = \text{diag}(+, -, -, -)$. Upper greek indices are contravariant, lower greek indices covariant. The scalar product of two 4-vectors a^μ, b^μ is denoted by $a^\mu g_{\mu\nu} b^\nu \equiv a^\mu b_\mu \equiv a \cdot b$. The latter notation is also used for the scalar product of two 3-vectors a, b, $a \cdot b$.

2   The Basics

In this section, I first derive the conservation equations of relativistic fluid dynamics. If there are n conserved charges in the fluid, there are $4 + n$ conservation equations: 1 for the conservation of energy, 3 for the conservation of 3-momentum, and n for the conservation of the respective charges. In the general case, however, there are $10 + 4n$ independent variables: the 10 independent components of the energy-momentum tensor (which is a symmetric tensor of rank 2), and the 4 independent components of the 4-vectors of the n charge currents. Thus, the system of fluid-dynamical equations is not closed, and one requires an approximation in order to solve it.

The simplest approximation is the ideal fluid assumption which reduces the number of unknown functions to $5 + n$. The equation of state of the fluid then provides the final equation to close the system of conservation equations and to solve it uniquely. Another approximation is the assumption of small deviations from an ideal fluid and leads to the equations of dissipative fluid dynamics. In this approximation one provides an additional set of $6 + 3n$ equations to close the set of equations of motion. Finally, I also briefly discuss multi-fluid-dynamical models.

2.1   The Conservation Equations

Fluid dynamics is equivalent to the conservation of energy, momentum, and net charge number. Consider a single fluid characterized locally in space-time by its energy-momentum tensor $T^{\mu\nu}(x)$ and by the n conserved net charge currents $N_i^\mu(x)$, $i = 1, \ldots, n$. (Conserved charges are for example the electric charge, baryon number, strangeness, charm, etc.) Consider now an arbitrary hypersurface Σ in 4-dimensional space-time. The tangent 4-vector on this surface is $\Sigma^\mu(x)$. The normal vector on a surface element $d\Sigma$ of Σ is denoted by $d\Sigma_\mu(x)$. By definition, $d\Sigma \cdot \Sigma = 0$. The amount of net charge of type i and of energy and momentum flowing through the surface element $d\Sigma$ is given by

$$dN_i \equiv d\Sigma \cdot N_i \; , \quad i = 1, \ldots, n \; , \tag{1}$$

$$dP^\mu \equiv d\Sigma_\nu T^{\mu\nu} \; , \quad \mu = 0, \ldots, 3 \; . \tag{2}$$

Now consider an arbitrary space-time volume V_4 with a *closed* surface Σ. If there are neither sources nor sinks of net charge and energy-momentum inside Σ, one has

$$\oint_\Sigma d\Sigma \cdot N_i \equiv 0 \; , \quad i = 1, \ldots, n \; , \tag{3}$$

$$\oint_\Sigma d\Sigma_\nu T^{\mu\nu} \equiv 0 \; , \quad \mu = 0, \ldots, 3 \; . \tag{4}$$

Gauss theorem then leads immediately to the *global conservation of net charge and energy-momentum*:

$$\int_{V_4} d^4x \, \partial_\mu N_i^\mu \equiv 0 \; , \quad i = 1, \ldots, n \; , \tag{5}$$

$$\int_{V_4} d^4x \, \partial_\nu T^{\mu\nu} \equiv 0 \; , \quad \mu = 0, \ldots, 3 \; . \tag{6}$$

For arbitrary V_4, however, one has to require that the integrands in (5,6) vanish, which leads to *local conservation of net charge and energy-momentum*:

$$\partial_\mu N_i^\mu \equiv 0 \; , \quad i = 1, \ldots, n \; , \tag{7}$$

$$\partial_\mu T^{\mu\nu} \equiv 0 \; , \quad \nu = 0, \ldots, 3 \; . \tag{8}$$

These are the equations of motion of relativistic fluid dynamics [5]. Note that there are $4 + n$ equations, but $10 + 4n$ independent unknown functions $T^{\mu\nu}(x)$, $N_i^\mu(x)$. ($T^{\mu\nu}$ is a symmetric tensor of rank 2 and therefore has 10 independent components, the N_i^μ are 4-vectors with 4 independent components.) Therefore, the system of fluid-dynamical equations is a priori not closed and cannot be solved in complete generality. One requires additional assumptions to close the set of equations. One possibility is to reduce the number of unknown functions, another is to provide $6 + 3n$ additional equations of motion which determine all unknown functions uniquely. Both possibilities will be discussed in the following subsections.

2.2   Tensor Decomposition and Choice of Frame

Before discussing approximations to close the system of conservation equations, it is convenient to perform a tensor decomposition of $N_i^\mu, T^{\mu\nu}$ with respect to an *arbitrary, time-like, normalized* 4-vector u^μ, $u \cdot u = 1$. The projector onto the 3-space orthogonal to u^μ is denoted by

$$\Delta^{\mu\nu} \equiv g^{\mu\nu} - u^\mu u^\nu , \quad \Delta^{\mu\nu} u_\nu = 0, \quad \Delta^{\mu\alpha} \Delta^\nu_\alpha = \Delta^{\mu\nu} . \tag{9}$$

Then the tensor decomposition reads:

$$N_i^\mu = n_i u^\mu + \nu_i^\mu , \tag{10}$$

$$T^{\mu\nu} = \epsilon u^\mu u^\nu - p \Delta^{\mu\nu} + q^\mu u^\nu + q^\nu u^\mu + \pi^{\mu\nu} , \tag{11}$$

where

$$n_i \equiv N_i \cdot u \tag{12}$$

is the *net density* of charge of type i in the frame where $u^\mu = (1, 0)$ (subsequently denoted as the *local rest frame*, LRF),

$$\nu_i^\mu \equiv \Delta^\mu_\nu N_i^\nu \tag{13}$$

is the *net flow* of charge of type i in the LRF,

$$\epsilon \equiv u_\mu T^{\mu\nu} u_\nu \tag{14}$$

is the *energy density* in the LRF,

$$p \equiv -\frac{1}{3} T^{\mu\nu} \Delta_{\mu\nu} \tag{15}$$

is the *isotropic pressure* in the LRF,

$$q^\mu \equiv \Delta^{\mu\alpha} T_{\alpha\beta} u^\beta \tag{16}$$

is the *flow of energy* or *heat flow* in the LRF, and

$$\pi^{\mu\nu} \equiv \left[\frac{1}{2} \left(\Delta^\mu_\alpha \Delta^\nu_\beta + \Delta^\mu_\beta \Delta^\nu_\alpha \right) - \frac{1}{3} \Delta^{\mu\nu} \Delta_{\alpha\beta} \right] T^{\alpha\beta} \tag{17}$$

is the *stress tensor* in the LRF. Note that the particular projection (17) is *trace-free*. (The trace of the projection $\Delta_\alpha^\mu T^{\alpha\beta} \Delta_\beta^\nu$ is absorbed in the definition of p.) The tensor decomposition replaces the original $10 + 4n$ unknown functions by an equal number of new unknown functions n_i (n), ν_i^μ $(3n)$, ϵ (1), p (1), q^μ (3), and $\pi^{\mu\nu}$ (5).

So far, u^μ is arbitrary. However, one can give it a physical meaning by choosing it either to be

$$u_E^\mu \equiv \frac{N_i^\mu}{\sqrt{N_i \cdot N_i}} \, , \tag{18}$$

or (which is an implicit definition)

$$u_L^\mu \equiv \frac{T_\nu^\mu u_L^\nu}{\sqrt{u_L^\alpha T_\alpha^\beta T_{\beta\gamma} u_L^\beta}} \, . \tag{19}$$

The first choice means that u_E^μ is the physical 4-velocity of the *flow* of net charge i. The LRF is then the *local rest frame of the flow of net charge i*, i.e., the frame where $N_i^\mu = (N_i^0, 0)$. In this frame, there is obviously no flow of charge i, $\nu_i^\mu \equiv 0$, and $N_i^0 \equiv n_i$. This LRF is called *Eckart frame*. Note, however, that not all net charges need to flow with the same velocity, ν_j^μ might be $\neq 0$ for $j \neq i$. The number of unknown functions is still $10 + 4n$, since the 3 previously unknown functions ν_i^μ have been merely replaced by the 3 independent components of u_E^μ ($u_E \cdot u_E = 1!$), which now have to be determined dynamically from the conservation equation for N_i^μ.

The second choice means that u_L^μ is the physical 4-velocity of the *energy flow*. The LRF is the *local rest frame of the energy flow*. It is obvious that in this frame $q^\mu \equiv 0$. This frame is called *Landau frame*. The number of unknown functions is still $10 + 4n$, since the 3 previously unknown functions q^μ have been merely replaced by the 3 independent components of u_L^μ ($u_L \cdot u_L = 1!$), which now have to be determined dynamically from the conservation equation for $T^{\mu\nu}$. Other choices of rest frames are also possible, for a discussion, see [6].

2.3   Ideal Fluid Dynamics

Consider an ideal gas in *local thermodynamical equilibrium*. The single-particle phase space distribution for fermions or bosons then reads

$$f_0(k, x) = \frac{g}{(2\pi)^3} \frac{1}{\exp{(k \cdot u(x) - \mu(x))/T(x)} \pm 1} \, , \tag{20}$$

where $u^\mu(x)$ is the *local* average 4-velocity of the particles, $\mu(x)$ and $T(x)$ are *local* chemical potential and temperature, and g counts internal degrees of freedom (spin, isospin, color, etc.) of the particles. The chemical potential of the particles is defined as $\mu \equiv \sum_{i=1}^n q_i \mu_i$, where μ_i are the chemical potentials which control the net number of charge of type i, and q_i is the individual charge of type i carried by a particle. The chemical potential for antiparticles is $\bar{\mu} = -\mu$ (in thermodynamical equilibrium). Let us define the single-particle phase space distribution for antiparticles by $\bar{f}_0(\bar{\mu}) = f_0(-\mu)$.

The kinetic definitions of the net current of charge of type i and of the energy-momentum tensor are [6]

$$N_i^\mu(x) \equiv q_i \int \frac{d^3k}{E} k^\mu \left[f_0(k,x) - \bar{f}_0(k,x) \right] , \tag{21}$$

$$T^{\mu\nu}(x) \equiv \int \frac{d^3k}{E} k^\mu k^\nu \left[f_0(k,x) + \bar{f}_0(k,x) \right] , \tag{22}$$

where $E \equiv \sqrt{k^2 + m^2}$ is the on-shell energy of the particles and m their rest mass. Inserting (20) one computes

$$N_i^\mu = n_i u^\mu , \tag{23}$$
$$T^{\mu\nu} = \epsilon u^\mu u^\nu - p \Delta^{\mu\nu} , \tag{24}$$

where

$$n_i \equiv g q_i \int \frac{d^3k}{(2\pi)^3} [n(E) - \bar{n}(E)] \tag{25}$$

is the *thermodynamic* net number density of charge of type i of an *ideal gas*, and the Fermi–Dirac or Bose–Einstein distribution was denoted by $n(E) \equiv 1/(\exp[(E-\mu)/T]\pm 1)$, $\bar{n}(E) \equiv 1/(\exp[(E+\mu)/T]\pm 1)$. Furthermore,

$$\epsilon \equiv g \int \frac{d^3k}{(2\pi)^3} E [n(E) + \bar{n}(E)] \tag{26}$$

is the *thermodynamic ideal gas* energy density, and

$$p \equiv g \int \frac{d^3k}{(2\pi)^3} \frac{k^2}{3E} [n(E) + \bar{n}(E)] \tag{27}$$

is the *thermodynamic ideal gas* pressure. The form (23,24) implies that for an ideal gas in local thermodynamical equilibrium the functions $\nu_i^\mu = q^\mu = \pi^{\mu\nu} = 0$, i.e., there is no flow of charge or heat with respect to the particle flow velocity u^μ, and there are no stress forces. This implies furthermore (and can be confirmed by an explicit calculation) that for an ideal gas in local thermodynamical equilibrium $u_E^\mu \equiv u_L^\mu \equiv u^\mu$, i.e., Eckart's and Landau's choice of frame coincide with the local rest frame of particle flow.

This consideration of an ideal gas in local thermodynamical equilibrium serves as a motivation for the so-called *ideal fluid approximation*. In this approximation, one starts on the *macroscopic* level of fluid variables N_i^μ, $T^{\mu\nu}$ and *a priori* takes them to be of the form (23) and (24). The corresponding fluid is referred to as an *ideal fluid*. Without any further assumption, however, the corresponding system of $4 + n$ equations of motion contains $5 + n$ unknown functions, ϵ, p, u^μ, and n_i, $i = 1, \ldots, n$. One therefore has to specify an *equation of state* for the fluid, for instance (and most commonly taken) of the form $p(\epsilon, n_1, \ldots, n_n)$. This closes the system of equations of motion.

The equation of state is the *only* place where information enters about the nature of the particles in the fluid and the *microscopic* interactions between them. Usually, the equation of state for the fluid is taken to be the *thermodynamic* equation of state, as computed for a system in *thermodynamical equilibrium*. The process of closing the system of equations of motions by assuming a thermodynamic equation of state therefore involves the *implicit assumption* that the *fluid is in local thermodynamical equilibrium*. It is important to note, however, that the *explicit form* of the equation of state is *completely unrestricted*, for instance it can have anomalies like phase transitions.

The ideal fluid approximation therefore allows to consider a wider class of systems than just an ideal gas in local thermodynamical equilibrium, which served as a motivation for this approximation. An ideal gas has a very specific equation of state without any anomalies and is given by (27) which defines $p(T, \mu_1, \ldots, \mu_n)$ (which in turn allows to determine all other thermodynamic functions from the first law and the fundamental relation of thermodynamics, and thus to specify $p(\epsilon, n_1, \ldots, n_n)$, see the following remarks).

I close this subsection with three remarks. The first concerns the notion of an equation of state which is *complete* in the thermodynamic sense. Such an equation of state allows (by definition) to determine, for given values of the independent thermodynamic variables, all other thermodynamic functions from the first law of thermodynamics (or one of its Legendre transforms)

$$\mathrm{d}s = \frac{1}{T}\,\mathrm{d}\epsilon - \sum_{i=1}^{n} \frac{\mu_i}{T}\, n_i \ , \tag{28}$$

s being the entropy density, and from the fundamental relation of thermodynamics

$$\epsilon + p = T\,s + \sum_{i=1}^{n} \mu_i\, n_i \ . \tag{29}$$

Obviously, for independent thermodynamic variables $\epsilon, n_1, \ldots, n_n$, an equation of state of the form $s(\epsilon, n_1, \ldots, n_n)$ is complete in this sense, since partial differentiation of this function yields, from (28), the functions $1/T, \mu_1/T, \ldots, \mu_n/T$. Then, the fundamental relation (29) yields the last unknown thermodynamic function, p.

Another example of a complete equation of state is $p(T, \mu_1, \ldots, \mu_n)$, since the (multiple) Legendre transform of (28) reads

$$\mathrm{d}p = s\,\mathrm{d}T + \sum_{i=1}^{n} n_i\, \mathrm{d}\mu_i \tag{30}$$

(which is also known as the Gibbs–Duhem relation), such that the thermodynamic functions $s, n_1, \ldots, n_n$ can be determined from partial differentiation of p. The last unknown thermodynamic function, ϵ, can then be determined from (29).

The equation of state $p(\epsilon, n_1, \ldots, n_n)$ is, however, *not a complete* equation of state in the thermodynamic sense. Partial differentiation of this function yields thermodynamic functions $\partial p/\partial \epsilon$, $\partial p/\partial n_i$, $i = 1, \ldots, n$, which in general do *not allow* to infer the values of T, s, and μ_i, $i = 1, \ldots, n$.

The second remark concerns the assumption of local thermodynamical equilibrium. In order to achieve local thermodynamical equilibrium, spatio-temporal variations of the macroscopic fluid fields have to be small compared to microscopic reaction rates which drive the system (locally) towards thermodynamical equilibrium. A quantity that characterizes spatio-temporal variations of the macroscopic fields is the so-called *expansion scalar* $\theta \equiv \partial \cdot u$. It determines the (local) rate of expansion of the fluid. Microscopic reaction rates are essentially given by the product of cross section and local particle density, $\Gamma \simeq \sigma n$. The criterion for local thermodynamical equilibrium then reads

$$\Gamma \gg \theta, \text{ or } \sigma \gg \theta/n . \tag{31}$$

The third remark concerns entropy production. In ideal fluid dynamics, the entropy current is defined as

$$S^\mu \equiv s\, u^\mu . \tag{32}$$

Taking the projection of energy-momentum conservation in the direction of u_ν one derives

$$0 = u_\nu\, \partial_\mu T^{\mu\nu} = \dot{\epsilon} + (\epsilon + p)\, \theta , \tag{33}$$

where $\dot{a} \equiv u \cdot \partial a$ is a comoving time derivative and where use has been made of the fact that u^μ is normalized, i.e., $\partial_\mu(u \cdot u) = 0$. With the first law of thermodynamics (28) and the fundamental relation of thermodynamics (29) one rewrites this as

$$T\,(\dot{s} + s\,\theta) + \sum_{i=1}^n \mu_i\,(\dot{n}_i + n_i\,\theta) = 0 . \tag{34}$$

Finally, employing net charge conservation $\partial \cdot N_i \equiv \dot{n}_i + n_i\,\theta = 0$ yields

$$\dot{s} + s\,\theta \equiv \partial \cdot S = 0 , \tag{35}$$

i.e., the entropy current is conserved in ideal fluid dynamics. As we shall see in one of the following section, however, this proof only holds where the partial derivatives in these equations are well-defined, i.e., for continuous solutions of ideal fluid dynamics. Discontinuous solutions will in fact be shown to produce entropy.

2.4   Dissipative Fluid Dynamics

In dissipative fluid dynamics one does not set ν_i^μ, q^μ, $\pi^{\mu\nu}$ a priori to zero, but specifies them through additional equations. There are two ways to obtain the latter. The first is phenomenological and starts from the second law of thermodynamics, i.e., the principle of non-decreasing entropy,

$$\partial \cdot S \geq 0 \ . \tag{36}$$

The second way resorts to kinetic theory to derive the respective equations. In principle, both ways require the additional assumption that deviations from local thermodynamical equilibrium are small. To make this statement more concise, let us introduce the *equilibrium pressure* $p_{eq} = p_{eq}(\epsilon, n_1, \ldots, n_n)$, i.e., it is the pressure as computed from the equation of state for given values of ϵ, $n_1, \ldots, n_n$. In a general non-equilibrium (dissipative) situation, however, p_{eq} is different from the isotropic pressure p defined through (15). Denote the difference by $\Pi \equiv p_{eq} - p$. Then, the requirement that deviations from local thermodynamical equilibrium are small is equivalent to requiring ν_i^μ, q^μ, $\pi^{\mu\nu}$, and Π to be small compared to ϵ, p_{eq}, and n_i.

I first outline the phenomenological approach to derive the equations of dissipative fluid dynamics. For the sake of definiteness, in the remainder of this subsection let us consider a system of one particle species only and let us assume that the total *particle number* of this species is conserved (implying that no annihilation or creation processes take place, i.e., we do not consider the corresponding antiparticles). The particle number current then replaces the net charge current. We shall also work in the Eckart frame, where $\nu^\mu \equiv 0$. Let us make an Ansatz for the entropy 4-current S^μ. In the limit of vanishing q^μ, $\pi^{\mu\nu}$, and Π, the entropy 4-current should reduce to the one of ideal fluid dynamics, $S^\mu \to s u^\mu$. The only non-vanishing 4-vector which can be formed from the available tensors u^μ, q^μ, and $\pi^{\mu\nu}$ is βq^μ, where β is an arbitrary coefficient (remember $\pi^{\mu\nu} u_\nu = 0$). Therefore,

$$S^\mu = s u^\mu + \beta q^\mu \ . \tag{37}$$

With this Ansatz one computes with the help of $u_\nu \partial_\mu T^{\mu\nu} = 0$ and $\partial \cdot N = \dot{n} + n\theta = 0$:

$$T\partial \cdot S = (T\beta - 1)\partial \cdot q + q \cdot (\dot{u} + T\partial\beta) + \pi^{\mu\nu}\partial_\mu u_\nu + \Pi\,\theta \geq 0 \ . \tag{38}$$

The simplest way to ensure this inequality is to choose

$$\beta \equiv 1/T \ , \tag{39}$$

$$\Pi \equiv \zeta\,\theta \ , \tag{40}$$

$$q^\mu \equiv \kappa T\,\Delta^{\mu\nu}\left(\partial_\nu \ln T - \dot{u}_\nu\right) \ , \tag{41}$$

$$\pi^{\mu\nu} \equiv 2\eta\left[\frac{1}{2}\left(\Delta_\alpha^\mu \Delta_\beta^\nu + \Delta_\beta^\mu \Delta_\alpha^\nu\right) - \frac{1}{3}\Delta^{\mu\nu}\Delta_{\alpha\beta}\right]\partial^\alpha u^\beta \ , \tag{42}$$

where ζ, η, and κ are the (positive) *bulk viscosity*, *shear viscosity* and *thermal conductivity* coefficients. Note that these equations define the dissipative corrections as *algebraic* functions of gradients of the flow velocity u^μ and the equilibrium temperature T. With these choices,

$$\partial \cdot S = \frac{\Pi^2}{\zeta T} - \frac{q \cdot q}{\kappa T^2} + \frac{\pi^{\mu\nu} \pi_{\mu\nu}}{2\eta T} \; , \tag{43}$$

which is obviously larger or equal to zero (remember that $q \cdot q < 0$, which can be most easily proven from $q \cdot u = 0$ in the frame where $u^\mu = (1, \mathbf{0})$). While this ensures the second law of thermodynamics, it was shown [7] that the resulting equations of motion are *unstable* under perturbations and support *acausal*, i.e., *superluminous* propagation of information. They are therefore not suitable as candidates for a *relativistic* theory of dissipative fluid dynamics.

A solution to this dilemma was presented by Müller [8], and Israel and Stewart [9]. They observed that the Ansatz (37) for the entropy current should not only contain first order terms in the dissipative corrections, but also second order terms:

$$S^\mu = s \, u^\mu + \beta \, q^\mu + Q^\mu \; , \tag{44}$$

where

$$Q^\mu \equiv \alpha_0 \, \Pi \, q^\mu + \alpha_1 \, \pi^{\mu\nu} \, q_\nu + u^\mu \left(\beta_0 \, \Pi^2 + \beta_1 \, q \cdot q + \beta_2 \, \pi^{\nu\lambda} \pi_{\nu\lambda} \right) \tag{45}$$

is second order in the dissipative quantities Π, q^μ, and $\pi^{\mu\nu}$. Inserting this into $\partial \cdot S \geq 0$ leads to *differential equations* for Π, q^μ, and $\pi^{\mu\nu}$ which involve the coefficients ζ, η, κ, α_0, α_1, β_0, β_1, β_2. It can be shown that, for reasonable values of these coefficents, the resulting 14 equations of motion (the 9 equations that determine Π, q^μ, and $\pi^{\mu\nu}$ and the 5 conservation equations for N^μ, $T^{\mu\nu}$) are stable and causal.

In the phenomenological approach, the values of these coefficients are not determined. In the second approach, however, based on kinetic theory, they can be explicitly computed along with deriving the additional 9 equations of motion for Π, q^μ, and $\pi^{\mu\nu}$. This will be outlined in the following.

Let us start by writing the single-particle phase space distribution in local equilibrium (20) as

$$f_0(k, x) = \frac{g}{(2\pi)^3} \left[\exp\{y_0(k, x)\} \pm 1 \right]^{-1} \; , \tag{46}$$

where $y_0(k, x) \equiv [k \cdot u(x) - \mu(x)]/T(x)$. Now assume that the *non-equilibrium* phase space distribution, written in the form

$$f(k, x) \equiv \frac{g}{(2\pi)^3} \left[\exp\{y(k, x)\} \pm 1 \right]^{-1} \; , \tag{47}$$

deviates only slightly from the equilibrium distribution function $f_0(k, x)$, or in other words:

$$y(k, x) \simeq y_0(k, x) + \varepsilon_1(x) + k \cdot \varepsilon_2(x) + k_\mu k_\nu \, \varepsilon_3^{\mu\nu}(x) \; , \qquad (48)$$

where $\varepsilon_1(x)$, $\varepsilon_2^\mu(x)$, and $\varepsilon_3^{\mu\nu}$ are small compared to $T(x)$, $\mu(x)$. Then one can expand $f(k, x)$ around $f_0(k, x)$ to first order in these small quantities:

$$f(k, x) \simeq f_0(k, x) \left(1 + \left[1 \mp \frac{(2\pi)^3}{g} f_0(k, x) \right] [y(k, x) - y_0(k, x)] \right) \; . \qquad (49)$$

Note that $f(k, x)$ depends on the 14 variables $\mu/T - \varepsilon_1$, $u^\mu/T + \varepsilon_2^\mu$, and $\varepsilon_3^{\mu\nu}$. ($\varepsilon_3^{\mu\nu}$ is a symmetric tensor of rank 2, and therefore naively has 10 independent components. However, its trace can be absorbed in a redefinition of the first variable $\mu/T - \varepsilon_1$, therefore it actually has only 9 independent components.) Inserting $f(k, x)$ into the kinetic theory definition of N^μ and $T^{\mu\nu}$, (21) and (22), (with f_0 replaced by f and, since we do not consider antiparticles, discarding $\bar{f}_0$), one can establish relations between the 14 unknown *macroscopic* functions (in the Eckart frame) ϵ, n, u^μ, Π, q^μ, $\pi^{\mu\nu}$ and the 14 variables $\mu/T - \varepsilon_1$, $u^\mu/T + \varepsilon_2^\mu$, $\varepsilon_3^{\mu\nu}$. This uniquely determines the non-equilibrium single-particle phase space distribution $f(k, x)$ in terms of the macroscopic, i.e., fluid-dynamical variables. This identification involves one subtlety: as in ideal fluid dynamics one still has to know the value of the (equilibrium) pressure p_{eq} to determine all unknown quantities. The equilibrium pressure p_{eq} is, however, only known as a function of the *equilibrium* energy density ϵ_0 and the *equilibrium* particle number density n_0, but not as function of the actual energy density ϵ and particle number density n. Two additional assumptions are required, namely that

$$\epsilon \equiv u_\mu T^{\mu\nu} u_\nu = \epsilon_0 \equiv u_\mu T_0^{\mu\nu} u_\nu \; , \qquad (50)$$

$$n \equiv u \cdot N = n_0 \equiv u \cdot N_0 \; , \qquad (51)$$

where $T_0^{\mu\nu}$ and N_0^μ are the (kinetic) energy-momentum tensor and particle number current computed with the *equilibrium* phase space distribution $f_0(k, x)$. Then $p_{\text{eq}}(\epsilon, n) \equiv p_{\text{eq}}(\epsilon_0, n_0)$ and the value of the equilibrium pressure p_{eq} is also determined. On close inspection, these additional assumptions do not pose any further restriction on the set of 14 unknown functions, but merely serve as definitions of (equilibrium) temperature T and chemical potential μ corresponding to a given energy density ϵ and particle number density n. Another way to say this is that the assumptions (50), (51) determine a local equilibrium phase space distribution $f_0(k, x)$. However, in a non-equilibrium context this distribution has no actual dynamical meaning, and one is therefore free to choose it in a way which fulfills (50) and (51).

The next step consists of deriving the equations of motion for the 14 unknown functions ϵ, n, u^μ, Π, q^μ, $\pi^{\mu\nu}$. To this end, one takes the first *three* moments of the Boltzmann equation for $f(k, x)$,

$$k \cdot \partial f(k, x) = \mathcal{C}[f] \; . \qquad (52)$$

This results in

$$\int \frac{\mathrm{d}^3 k}{E} \, k \cdot \partial \, f(k, x) \equiv \partial \cdot N = \int \frac{\mathrm{d}^3 k}{E} \, C[f] \equiv 0 \ , \tag{53}$$

$$\int \frac{\mathrm{d}^3 k}{E} \, k^\mu k^\nu \, \partial_\mu \, f(k, x) \equiv \partial_\mu T^{\mu\nu} = \int \frac{\mathrm{d}^3 k}{E} \, k^\nu \, C[f] \equiv 0 \ , \tag{54}$$

$$\int \frac{\mathrm{d}^3 k}{E} \, k^\mu k^\nu k^\lambda \, \partial_\mu \, f(k, x) \equiv \partial_\mu S^{\mu\nu\lambda} = \int \frac{\mathrm{d}^3 k}{E} \, k^\nu k^\lambda \, C[f] \equiv X^{\nu\lambda} \ . \tag{55}$$

Note that conservation of particle number, energy, and momentum leads to vanishing right-hand sides for eqs. (53) and (54). The structure of the microscopic collision term C is such that these requirements are fulfilled (particle number and energy-momentum conservation in microscopic collisions between particles) [6]. On the other hand, the right-hand side of Eq. (55) does not vanish, since there is no corresponding microscopic conservation law. Note that the trace of (55) is equivalent to m^2 times Eq. (53), such that $X^\nu_\nu \equiv 0$. Therefore, only 9 out of the set of 10 equations (55) are independent. Together with the 5 equations (53) and (54), these 9 equations determine the set of 14 unknown functions of dissipative fluid dynamics. The 9 independent equations (55) are equivalent to the 9 equations derived from $\partial \cdot S \geq 0$ in the phenomenological approach. The unknown phenomenological coefficents ζ, κ, η, α_0, α_1, β_0, β_1, and β_2 can now be explicitly identified from suitable projections of $X^{\nu\lambda}$. Israel and Stewart have shown [9] that the resulting equations fulfill the requirements of hyperbolicity and causality.

This concludes the brief survey of dissipative fluid dynamics. So far, no serious attempt has been made to apply relativistic dissipative fluid dynamics towards the description of heavy-ion collisions. First steps were done by Mornas and Ornik [10] who investigated the broadening of collisional shock waves through dissipative effects in a simple one-dimensional geometry. Also, Prakash et al. generalized the Israel–Stewart theory to a mixture of several particle species [11].

2.5 Multi-fluid Dynamics

In multi-fluid dynamics one considers not a single, but several fluids $j = 1, \ldots, M$, characterized by the net charge currents N^μ_{ij} (the net current of conserved charge i in fluid j) and energy-momentum tensors $T^{\mu\nu}_j$. There is overall net charge and energy-momentum conservation,

$$\partial \cdot N_i = 0 \ , \quad N^\mu_i \equiv \sum_{j=1}^{M} N^\mu_{ij} \ , \tag{56}$$

$$\partial_\mu T^{\mu\nu} = 0 \ , \quad T^{\mu\nu} \equiv \sum_{j=1}^{M} T^{\mu\nu}_j \ , \tag{57}$$

but not for each fluid separately,

$$\partial \cdot N_{ij} = S_{ij} \ , \quad \partial_\mu T_j^{\mu\nu} = S_j^\nu \ . \tag{58}$$

The right-hand sides define the so-called *source terms* which according to (56), (57) obey

$$\sum_{j=1}^{M} S_{ij} = 0 \ , \quad \sum_{j=1}^{M} S_j^\nu = 0 \ . \tag{59}$$

The source terms are parameters of a particular model and have to be specified e.g. from kinetic theory. Let us consider the Boltzmann equation for particles from fluid j:

$$k \cdot \partial f_j(k,x) = \sum_{klm} \left[C_{lm}^{jk} - C_{jk}^{lm} \right] \ . \tag{60}$$

The right-hand side involves the collision terms for the microscopic 2-particle reactions $lm \to jk$ (the *gain term* C_{lm}^{jk}) where particles from fluid l and fluid m (l and m not necessarily different) collide to produce particles of fluid j and k (again, j and k not necessarily different), and $jk \to lm$ (the *loss term* C_{jk}^{lm}) where particles from fluid j and k collide to produce particles of fluid l and m. Taking the zeroth and first moment of this equation yields

$$\partial \cdot N_{ij} \equiv q_i \int \frac{d^3k}{E} k \cdot \partial f_j(k,x) = q_i \sum_{klm} \int \frac{d^3k}{E} \left[C_{lm}^{jk} - C_{jk}^{lm} \right] \equiv S_{ij} \ , \tag{61}$$

$$\partial_\mu T_i^{\mu\nu} \equiv \int \frac{d^3k}{E} k^\mu k^\nu \partial_\mu f_j(k,x) = \sum_{klm} \int \frac{d^3k}{E} k^\nu \left[C_{lm}^{jk} - C_{jk}^{lm} \right] \equiv S_j^\nu \tag{62}$$

This defines the source terms through the microscopic collision rates.

Results of any specific multi-fluid model will not be discussed here, I instead refer the reader to the literature on this subject [12]. I close with two remarks: (a) a *single* fluid may consist of several *different* particle species (for instance, π, K, N, Λ etc.), as long as it is reasonable to assume that they stay in local thermodynamical equilibrium among each other. Then, the only place where information enters about these different particle species is the equation of state $p(\epsilon, n_1, \ldots, n_n)$. (b) *Different* fluids may consist of the *same* particle species (with the *same* equation of state $p(\epsilon, n_1, \ldots, n_n)$). This situation occurs for instance in the initial stage of relativistic heavy-ion collisions, where the single-particle phase space distributions of target and projectile nucleons, while overlapping in space-time, are still well separated in momentum space due to the high initial relative velocity between them. This is a situation where there is local thermodynamical equilibrium in target and projectile separately, but not between them. It therefore is reasonable to treat target and projectile, although consisting of the same particle species, as two separate fluids.

3   Numerical Aspects

In this section, I discuss basic aspects of numerical solution schemes for relativistic ideal fluid dynamics. For the sake of simplicity, let us consider the case of one conserved charge only. Define

$$R \equiv N^0 = n u^0 = n\gamma \ , \tag{63}$$

$$E \equiv T^{00} = (\epsilon + p)\gamma^2 - p \ , \tag{64}$$

$$M \equiv \{T^{0i}\}_{i=x,y,z} = (\epsilon + p)\gamma^2 v \ , \tag{65}$$

where $u^\mu = \gamma(1, v)$ is the fluid 4-velocity, $\gamma = (1 - v^2)^{-1/2}$. With these definitions, the conservation laws (7), (8) take the form

$$\partial \cdot N \equiv \partial_t R + \nabla \cdot (Rv) = 0 \ , \tag{66}$$

$$\partial_\mu T^{\mu 0} \equiv \partial_t E + \nabla \cdot [(E + p)v] = 0 \ , \tag{67}$$

$$\partial_\mu T^{\mu i} \equiv \partial_t M^i + \nabla \cdot (M^i v) + \partial_i p = 0 \ . \tag{68}$$

In this form, the conservation equations can be solved numerically with any scheme that also solves the non-relativistic conservation equations. There is, however, one fundamental difference between the non-relativistic equations and the relativistic ones. In order to solve the latter for R, E, M, the net charge density, energy density, and momentum density in the *calculational frame*, one has to know the equation of state $p(\epsilon, n)$ and v. The equation of state, however, depends on n, ϵ, the net charge density and energy density *in the rest frame of the fluid*. One therefore has to locally transform from the calculational frame to the rest frame of the fluid in order to extract n, ϵ, v from R, E, M. In the non-relativistic limit, there is no difference between n and R, or ϵ and E and the equation of state can be employed directly in the conservation equations. Also, the momentum density of the fluid is related to the fluid velocity by a simple expression. The transformation between rest frame and calculational frame quantities is described explicitly in the next subsection.

3.1   Transformation Between Calculation Frame and Fluid Rest Frame

In principle, the transformation is explicitly given by equations (63) – (65), i.e., by finding the roots of a set of 5 nonlinear equations (the non-linearity enters through the equation of state $p(\epsilon, n)$). In numerical applications, however, this transformation has to be done several times in each time step and each cell. It is therefore advisable to reduce the complexity of the transformation problem. This is done as follows [13].

First note that M and v are parallel, thus

$$M \cdot v \equiv M v = (\epsilon + p)\gamma^2 v^2 = (\epsilon + p)(\gamma^2 - 1) = E - \epsilon \ , \tag{69}$$

where $M \equiv |\boldsymbol{M}|$, $v \equiv |\boldsymbol{v}|$. Therefore,

$$\epsilon = E - M v \ , \quad n = R\sqrt{1 - v^2} \ , \tag{70}$$

where the second equation is a simple consequence of (63). With these equations ϵ and n can be expressed in terms of R, E, M and v. The 5-dimensional root search is therefore reduced to finding the modulus of v for given R, E, and M, which is a simple one-dimensional problem. To solve this, use the definition of M,

$$M = (\epsilon + p)\gamma^2 v = (E + p)v \ . \tag{71}$$

This equation can be rewritten as a fixed point equation for v for given R, E, M:

$$v = \frac{M}{E + p\left(E - M v, R\sqrt{1 - v^2}\right)} \ . \tag{72}$$

The fixed point yields the modulus of the fluid velocity, from which one can reconstruct $\boldsymbol{v} = v\,\boldsymbol{M}/M$, and find ϵ and n via (70). The equation of state $p(\epsilon, n)$ then yields the final unknown variable, the pressure p.

3.2   Operator Splitting Method

In general, to model a heavy-ion collision with ideal fluid dynamics requires to solve the 5 conservation equations in three space dimensions. Since this is in general a formidable numerical task, one usually resorts to the so-called *operator splitting method*, i.e., the full 3-dimensional solution is constructed by solving sequentially three one-dimensional problems. More explicitly, all conservation equations are of the type

$$\partial_t U + \sum_{i=x,y,z} \partial_i F_i(U) = 0 \ , \tag{73}$$

U being R, E, or M^i. Such an equation is numerically solved on a space-time grid, and time and space derivatives are replaced by finite differences:

$$U_{ijk}^{n+1} = U_{ijk}^n - \Delta t\, G\left[U_{ijk}^n\right] \ , \tag{74}$$

where i, j, k are cell indices (the cell number in x, y, and z direction) and n denotes the time step. Δt is the time step width. $G\left[U_{ijk}^n\right]$ is a suitable finite difference form of the 3-divergence in (73).

It can be shown that in the continuum limit instead of solving (74) it is equivalent to solve the following set of *predictor-corrector* equations

$$U_{ijk}^{(1)\,n+1} = U_{ijk}^n - \Delta t\, G_x\left[U_{ijk}^n\right] \ ,$$

$$U_{ijk}^{(2)\,n+1} = U_{ijk}^{(1)\,n+1} - \Delta t\, G_y\left[U_{ijk}^{(1)\,n+1}\right] \ , \tag{75}$$

$$U_{ijk}^{n+1} = U_{ijk}^{(2)\,n+1} - \Delta t\, G_z\left[U_{ijk}^{(2)\,n+1}\right] \ , \tag{76}$$

and that the solution converges towards the solution of (73). Here, the $G_i[U]$, $i = x, y, z$, are finite difference forms of the partial derivatives $\partial_i F_i(U)$ (no summation over i) in x, y, or z direction. $U_{ijk}^{(1)\,n+1}$ is the first *prediction* for the full solution U_{ijk}^{n+1}. It is generated by solving a finite difference form of the one-dimensional equation

$$\partial_t U + \partial_i F_i(U) = 0 \ , \tag{77}$$

where $i = x$. Subsequently, the first prediction $U_{ijk}^{(1)\,n+1}$ is used to solve a finite difference form of (77), where now $i = y$, to obtain the second *prediction* $U_{ijk}^{(2)\,n+1}$ for the full solution. ($U_{ijk}^{(1)\,n+1}$ has been *corrected* to $U_{ijk}^{(2)\,n+1}$.) Finally, the full solution U_{ijk}^{n+1} is obtained by using $U_{ijk}^{(2)\,n+1}$ to solve a finite difference form of (77) with $i = z$. ($U_{ijk}^{(2)\,n+1}$ has been *corrected* to U_{ijk}^{n+1}.)

In other words, the solution to the partial differential equation (73) in three space dimensions is obtained by solving a sequence of partial differential equations (77) in one space dimension. The 3-divergence operator in (73) was *split* into a sequence of three partial derivative operators. Physically speaking, in a given time step one first propagates the fields in x direction, then in y direction, and then in z direction. In actual numerical applications it is advisable to permutate the order xyz to minimize systematical errors.

The advantage of the operator splitting method is that there exists a variety of numerical algorithms which solve evolution equations of the type (77) in one space dimension (see, for instance, [14] and refs. therein). One of them is discussed in the following subsection.

3.3   The Relativistic Harten–Lax–van Leer–Einfeldt Algorithm

The relativistic Harten–Lax–van Leer–Einfeldt (HLLE) algorithm [14,15] solves equations of the type

$$\partial_t U + \partial_x F(U) = 0 \ , \tag{78}$$

i.e., propagation of a field U in one space dimension. For ideal relativistic fluid dynamics, $U = R, E$, or M and $F(U) = Rv, (E + p)v$, or $Mv + p$. (For one-dimensional propagation, it is sufficient to consider only the components of the momentum density M and the fluid velocity v in the direction of propagation. They are here denoted by M and v, respectively.)

The idea behind the relativistic HLLE scheme is the following. Consider the initial distribution of the density U on a numerical grid. U is assumed to be constant inside each cell, but different from cell to cell, i.e., the initial distribution consists of a sequence of constant flow fields inside the cells separated by discontinuities at the cell boundaries, cf. Fig. 1.

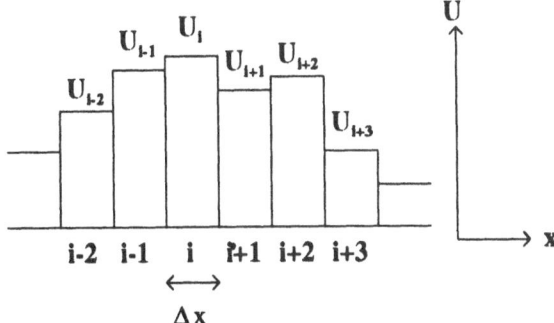

Fig. 1: The initial distribution of the density U on the numerical grid.

In the further time evolution these discontinuities will decay, resulting in the transport of U across the grid. The decay of a discontinuity between two regions of constant flow is, however, a well-known problem in fluid dynamics, the so-called *Riemann problem.* For simple equations of state it is even analytically solvable. Consider the discontinuity to be located at $x = 0$. Denote the density in the region of constant flow to the left of the discontinuity by U_1, and that to the right by U_r. The initial condition at time $t = 0$ then reads

$$U(x,0) = \begin{cases} U_1 , & x < 0 \\ U_r , & x \geq 0 \end{cases} , \tag{79}$$

cf. Fig. 2 (a). For the sake of definiteness, consider $U_1 > U_r$. For $t > 0$, the solution looks qualitatively as in Fig. 2 (b). There is a rarefaction fan propagating into the region of higher density (in this case to the left), and a shock front into the region of lower density (in this case to the right). Between fan and shock wave there are two regions of constant flow separated by a contact discontinuity (a discontinuity where the pressure is equal on both sides). It is evident that a numerical algorithm can be constructed which solves the fluid dynamical equations simply by solving a sequence of Riemann problems for the discontinuities at all cell boundaries in a given time step. Such algorithms are called *Godunov algorithms* [16].

The relativistic HLLE is a so-called *Godunov-type* algorithm [16], i.e., it does not employ the full solution of the Riemann problem but approximates it by a region of constant flow between U_l and U_r, cf. Fig. 2 (c):

$$U(x,t) = \begin{cases} U_1 , & x < b_1 t \\ U_{lr} , & b_1 t \leq x < b_r t \\ U_r , & x \geq b_r t \end{cases} . \tag{80}$$

Here, $b_1 < 0$ and $b_r > 0$ are the so-called *signal velocities.* They characterize the velocities with which information about the decay of the discontinuity

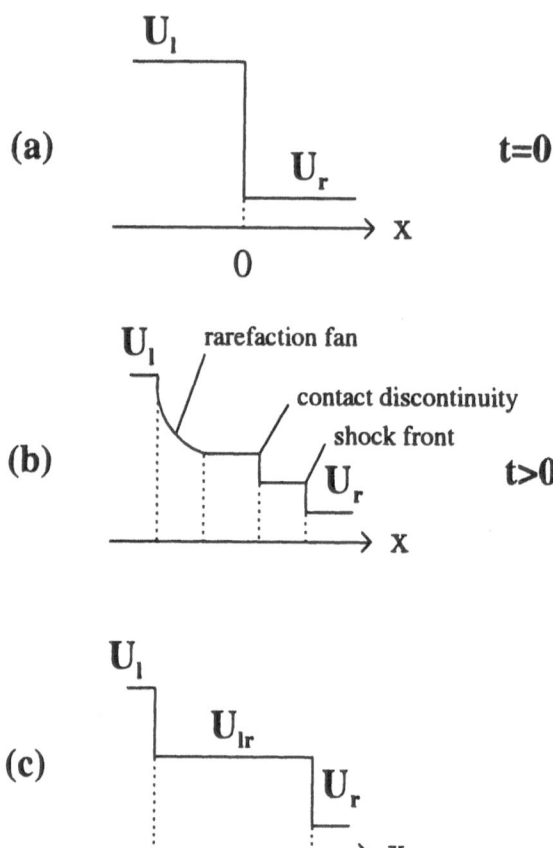

Fig. 2: (a) The initial condition of the Riemann problem at $t = 0$. (b) The solution of the Riemann problem at $t > 0$. (c) The approximate solution of a Godunov-type algorithm.

travels to the left and right into the regions of constant flow. The value U_{lr} in the region of constant flow between U_l and U_r is determined in accordance with the conservation laws. To this end, integrate (78) over a fixed interval $[x_{min}, x_{max}]$, $x_{min} < b_l t$, $x_{max} > b_r t$. One obtains:

$$U_{lr} = \frac{b_r U_r - b_l U_l + F(U_l) - F(U_r)}{b_r - b_l} . \tag{81}$$

The value of the flux $F(U_{lr})$ corresponding to the density U_{lr} is determined by integrating (78) over the fixed interval $[0, x_{max}]$ or $[x_{min}, 0]$:

$$F(U_{lr}) = \frac{b_r F(U_l) - b_l F(U_r) + b_l b_r (U_r - U_l)}{b_r - b_l} . \tag{82}$$

Upon discretization, the differential operator $\partial_x F(U)$ in the evolution equation for the density U_i in cell i assumes the form $[F(U_{i+1/2}) - F(U_{i-1/2})]/\Delta x$

where Δx is the cell size (grid spacing) and $U_{i\pm1/2}$ are values of the density at the position of the right and left boundary of cell i. These values are taken *after* the decay of the respective discontinuities at the cell boundaries, i.e., they are the corresponding values U_{lr} given by (81) and the respective $F(U_{i\pm1/2})$ are the corresponding values $F(U_{\text{lr}})$ given by (82). This yields the following explicit expressions for the relativistic HLLE scheme

$$U_i^{n+1} = U_i^n - \frac{\Delta t}{\Delta x}\left[F\left(U_{i+1/2}^n\right) - F\left(U_{i-1/2}^n\right)\right] , \tag{83}$$

$$F\left(U_{i+1/2}^n\right) = \frac{b_{\text{r}}\,F\left(U_i^n\right) - b_{\text{l}}\,F\left(U_{i+1}^n\right) + b_{\text{r}}\,b_{\text{l}}\left(U_{i+1}^n - U_i^n\right)}{b_{\text{r}} - b_{\text{l}}} . \tag{84}$$

A reasonable estimate for the signal velocities is to take them as the relativistic addition (subtraction) of flow velocities and sound velocities in the respective cells adjacent to the cell boundary:

$$b_{\text{r}} = \max\left\{0, \frac{v_{i+1}^n + c_{\text{s},i+1}^n}{1 + v_{i+1}^n\,c_{\text{s},i+1}^n}\right\} , \tag{85}$$

$$b_{\text{l}} = \min\left\{0, \frac{v_i^n - c_{\text{s},i}^n}{1 - v_i^n\,c_{\text{s},i}^n}\right\} . \tag{86}$$

As described above, this scheme is accurate to first order in time. A scheme which is accurate to second order can be obtained using half-step updated values $F\left(U_{i\pm1/2}^{n+1/2}\right)$, for more details see [17].

4 One-Dimensional Solutions

In this section I discuss solutions of ideal relativistic fluid dynamics in one space dimension. I first introduce the notion of characteristic curves. Then, I discuss possible one-dimensional wave patterns for thermodynamically normal and anomalous media. Choosing a representative equation of state which features both thermodynamically normal and anomalous regions I then discuss the expansion of semi-infinite matter into the vacuum. The emerging wave patterns will help us to understand the possible solutions of the Landau model, which was historically the first fluid-dynamical model for relativistic heavy-ion collisions. Finally, also the Bjorken model for ultrarelativistic heavy-ion collisions is discussed.

4.1 One-Dimensional Wave Patterns

For flow in one spatial dimension (say, in x direction) the two conservation equations for energy and for momentum read:

$$\partial_t T^{00} + \partial_x T^{x0} = 0 , \quad \partial_t T^{0x} + \partial_x T^{xx} = 0 . \tag{87}$$

A suitable linear combination of these equations leads to the equivalent set of equations

$$\left(\partial_t + \frac{v \pm c_s}{1 \pm v c_s}\, \partial_x\right) \mathcal{R}_\pm = 0 \ , \tag{88}$$

where $c_s^2 \equiv \partial p/\partial \epsilon|_{s/n}$ is the velocity of sound squared (s/n is the specific entropy) and

$$\mathcal{R}_\pm \equiv y - y_0 \pm \int_{\epsilon_0}^{\epsilon} d\epsilon' \, \frac{c_s(\epsilon')}{\epsilon' + p(\epsilon')} \tag{89}$$

are the so-called *Riemann invariants*, $y \equiv \mathrm{Artanh}v$ is the fluid rapidity. Equation (88) has the obvious interpretation that the Riemann invariants $\mathcal{R}_\pm$ are constant along world lines $x_\pm(t)$ defined by

$$\frac{dx_\pm(t)}{dt} \equiv w_\pm = \frac{v \pm c_s}{1 \pm v c_s} \ . \tag{90}$$

These world lines are the so-called *characteristic curves* or *characteristics* $C_\pm(x, t)$. It is also obvious that these curves are the world lines of *sonic perturbations* or *sound waves* on top of the fluid-dynamical wave pattern. $C_+(x, t)$ characterizes sound waves moving to the right (in positive x direction) while $C_-(x, t)$ characterizes those moving to the left (in negative x direction). For the simple example of constant flow, the characteristic curves are shown in Fig. 3.

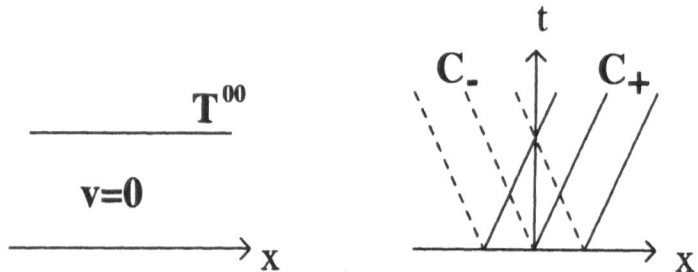

Fig. 3: The characteristic curves for a constant flow pattern.

Let us now consider a so-called *simple rarefaction wave* moving to the right, cf. Fig. 4. (For the definition of a simple wave, see [18], for our purposes it is sufficient to remark that in one spatial dimension a simple wave is the only possible wave that can connect two regions of constant flow. A rarefaction wave denotes a wave where the energy density decreases in the direction of propagation.) Then, one can prove that $\mathcal{R}_+ = $ const. everywhere (for the proof, see [18]; analogously, for simple waves moving to the left, $\mathcal{R}_- = $ const.).

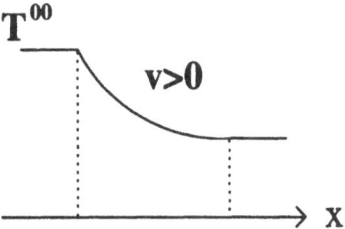

Fig. 4: A continuous simple wave between two regions of constant flow, moving to the right.

It is therefore sufficient to consider the equation for the $\mathcal{R}_-$ invariants, or the $\mathcal{C}_-$ characteristics, respectively. Let us consider how the slope of the $\mathcal{C}_-$ characteristics changes with x at constant t:

$$\left.\frac{\partial w_-}{\partial x}\right|_t \equiv w'_- = \frac{v'(1-c_s^2) - c'_s(1-v^2)}{(1-v\,c_s)^2} \; . \tag{91}$$

From $\mathcal{R}_+ = $ const. everywhere one infers

$$v' = -(1-v^2)\frac{c_s}{\epsilon+p}\epsilon' \; , \tag{92}$$

while

$$c'_s = \left.\frac{1}{2\,c_s}\frac{\partial^2 p}{\partial\epsilon^2}\right|_{s/n}\epsilon' \; . \tag{93}$$

Therefore,

$$w'_- = -\frac{1-w_-^2}{2\,c_s(1-c_s^2)}\Sigma\epsilon' \; , \tag{94}$$

where

$$\Sigma \equiv \left.\frac{\partial^2 p}{\partial\epsilon^2}\right|_{s/n} + 2\,c_s^2\frac{1-c_s^2}{\epsilon+p} \; . \tag{95}$$

Equation (94) is an important qualitative result: Since the first factor is always positive (w_- as well as c_s are causal), and since the energy density decreases with x for the rarefaction wave considered here, $\epsilon' < 0$, the sign of w'_- is solely determined by the sign of Σ. The quantity Σ, however, is solely determined by the equation of state of matter under consideration, i.e., its sign (and absolute value) is an intrinsic property of the fluid. Matter with $\Sigma > 0$ is called *thermodynamically normal*, while matter with $\Sigma < 0$ is *thermodynamically anomalous*. More specifically, if $\Sigma > 0$, then $w'_- > 0$, and if $\Sigma < 0$, then $w'_- < 0$. A positive w'_-, however, means that the $\mathcal{C}_-$ characteristics "fan out" in the $x - t$ plane, while a negative w'_- indicates that they converge and ultimately intersect at one point, cf. Fig. 5.

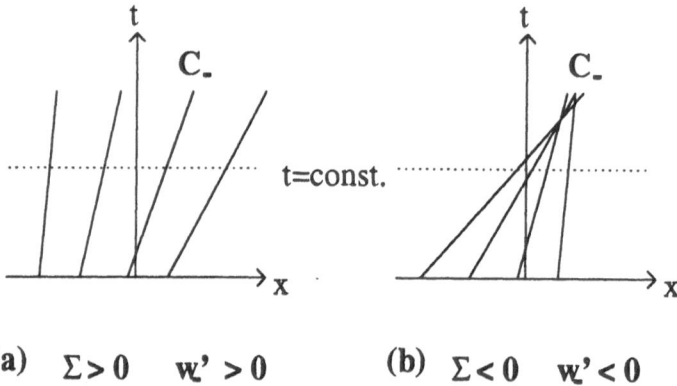

Fig. 5: For a simple wave moving to the right and (a) $\Sigma > 0$ the C_- characteristics fan out, while for (b) $\Sigma < 0$ they converge and intersect.

Intersecting characteristics, however, signal the formation of *shock waves*. Physically speaking, picture a sonic perturbation (travelling along a characteristic) emitted at a point x_1. This perturbation will eventually overtake a perturbation emitted at $x_2 > x_1$ (namely when the corresponding characteristics intersect). Thus, the two small perturbations add up to form a larger one. Imagine this happening for other perturbations (emitted at different points) as well. Eventually, a finite discontinuity (shock wave) is formed from the superposition of a large number of infinitesimal sonic perturbations. Shock waves are discontinuous solutions of ideal fluid dynamics and will be discussed in more detail in the following subsection.

I conclude this subsection by collecting the above arguments in the following classification scheme of one-dimensional wave patterns. *Continuous rarefaction waves* are *stable* in *thermodynamically normal matter* while they are *unstable* in *anomalous matter*. On the other hand, *rarefaction shock waves* are *stable* in *thermodynamically anomalous matter* while they are *unstable* in *thermodynamically normal matter*. If we perform an analogous consideration for a *continuous compression wave* we are led to the conclusion that such waves are *unstable* in *normal* and *stable* in *anomalous* matter, while *compression shock waves* are *stable* in *normal* and *unstable* in *anomalous* matter. These results are summarized in Table 4.1. A "+" sign means "stable" while a "−" sign indicates "unstable".

Most matter is thermodynamically normal. In the presence of phase transitions, however, an equation of state can feature regions where matter is thermodynamically anomalous. As will be seen in Subsections 4.4 and 4.5, this will strongly influence the time evolution of the system in a qualitative and quantitative way.

Table 1: Classification scheme for the stability of one-dimensional wave patterns.

Wave	$\Sigma > 0$	$\Sigma < 0$
Continuous rarefaction	+	−
Rarefaction shock	−	+
Continuous compression	−	+
Compression shock	+	−

4.2   Shock Discontinuities

Shock waves represent discontinuous solutions of ideal fluid dynamics. While the partial derivatives of N_i^μ and $T^{\mu\nu}$ appearing in the conservation equations are ill-defined at the location of such discontinuities, there is still a simple way solve the problem of charge and energy-momentum transport across a shock discontinuity. To this end, let us consider the case of one conserved charge only, and study such a discontinuity in its rest frame. Matter enters the discontinuity with velocity v_0 in a thermodynamic state characterized by the net charge density n_0, the energy density ϵ_0, and the pressure p_0 (which is of course determined by ϵ_0 and n_0 through the equation of state). The task is to determine the velocity v and the thermodynamic state of matter (n, ϵ, and p) emerging from the shock. Imagine a small volume V which encloses the discontinuity, cf. Fig. 6.

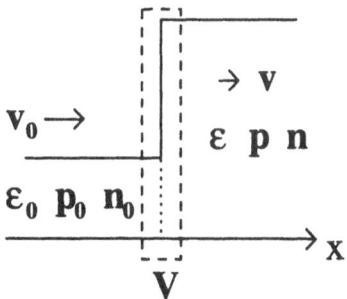

Fig. 6: A shock discontinuity in its rest frame.

Let us now integrate the conservation equations (7) (for a single conserved charge) and (8) for one-dimensional flow over V:

$$\partial_t \int_V \mathrm{d}^3 x \, N^0 + \int_V \mathrm{d}^3 x \, \partial_x \, N^x = 0 \;, \tag{96}$$

$$\partial_t \int_V \mathrm{d}^3x\, T^{00} + \int_V \mathrm{d}^3x\, \partial_x\, T^{x0} = 0 \ , \tag{97}$$

$$\partial_t \int_V \mathrm{d}^3x\, T^{x0} + \int_V \mathrm{d}^3x\, \partial_x\, T^{xx} = 0 \ . \tag{98}$$

In a steady-state situation (a stable, propagating shock discontinuity) the total amount of charge, energy and momentum inside V cannot change with time, therefore, the first terms in these equations vanish. The other terms are integrated by parts to yield the set of equations

$$n_0\gamma_0 v_0 = n\,\gamma\,v \ , \tag{99}$$
$$(\epsilon_0 + p_0)\gamma_0^2 v_0 = (\epsilon + p)\,\gamma^2\,v \ , \tag{100}$$
$$(\epsilon_0 + p_0)\gamma_0^2 v_0^2 + p_0 = (\epsilon + p)\,\gamma^2\,v^2 + p \ . \tag{101}$$

These are the conservation equations for net charge and energy-momentum across a shock discontinuity. They are no longer partial differential equations, but purely algebraic. For a given initial state n_0, ϵ_0, p_0, and velocity v_0, they determine the final state n, ϵ, p, and the velocity v of compressed matter emerging from the shock, if the equation of state $p(\epsilon, n)$ is known.

One can eliminate the velocities from the set of equations (99) – (101) to obtain the so-called *Taub equation* [19]

$$(\epsilon + p)X - (\epsilon_0 + p_0)X_0 = (p - p_0)(X + X_0) \ , \tag{102}$$

where $X \equiv (\epsilon+p)/n^2$ is the so-called *generalized volume*. Once $p(\epsilon, n)$ is fixed, the solution of the Taub equation defines the so-called *Taub adiabat* $p(X)$, cf. Fig. 7. For a given initial state (p_0, X_0) (the so-called *center* of the adiabat) it represents all final states (p, X) for matter emerging from the shock, which are in agreement with net charge and energy-momentum conservation. The actual final state is then selected by specifying v_0. This determines all variables uniquely in the rest frame of the shock. The remaining unknown is, however, the velocity of the shock in an arbitrary calculational frame. For compressional shock waves, such as occur in the initial stage of heavy-ion collisions (cf. [20] for a detailed discussion), this shock velocity can be uniquely determined from the geometry of the collision. For rarefaction shock waves this is not possible, and thus in principle there is a whole region of final states on the Taub adiabat, which are in agreement with energy-momentum and net charge conservation. It turns out, however, that the stationary situation is always given by a rarefaction shock where matter emerges at the so-called *Chapman–Jouguet* point, indicated by "CJ" in Fig. 7 (b) [5]. This point is defined as the point where a chord between the center (p_0, X_0) and a final state on the adiabat is tangential to the adiabat. This then uniquely fixes the state of matter emerging from the shock, as well as the velocity of the shock in the calculational frame. Note that it is also possible to define a Taub adiabat in the case that there is no conserved charge, see [17,21] for details.

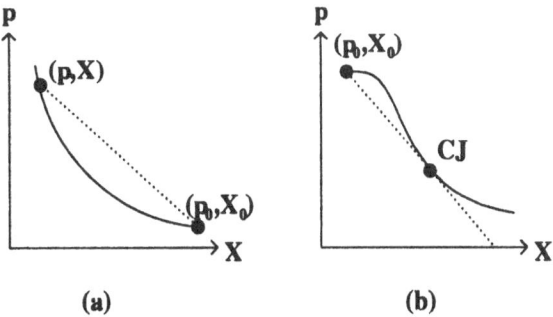

Fig. 7: (a) The Taub adiabat for a compressional shock wave. (p_0, X_0) is the center of the adiabat, (p, X) is one final state on the adiabat which is selected by a choice of v_0. (b) The Taub adiabat for a rarefaction shock wave. "CJ" denotes the Chapman–Jouguet point.

To conclude this subsection, let us consider what happens to the entropy flux across a shock discontinuity. Integrate the conservation equation (35) for the entropy current over the volume V which encloses the shock front in its rest frame,

$$\partial_t \int_V d^3x \, s\, \gamma + \int_V d^3x \, \partial_x \, s\, \gamma v = 0 \ , \qquad (103)$$

and perform an integration by parts in the second term. This yields:

$$s\, \gamma v = s_0 \gamma_0 v_0 + \frac{1}{A_\perp} \partial_t S \ , \qquad (104)$$

where $A_\perp$ is the transverse area of the shock front and $S \equiv \int_V d^3x \, s\, \gamma$ is the total entropy inside the volume V. The second law of thermodynamics tells us that the entropy cannot decrease, $\partial_t S \geq 0$. Consequently,

$$s\, \gamma v \geq s_0 \gamma_0 v_0 \ . \qquad (105)$$

Dividing both sides by (99) one concludes

$$\frac{s}{n} \geq \frac{s_0}{n_0} \ , \qquad (106)$$

i.e., the specific entropy increases across a shock front. This result is remarkable, since we know that the entropy current is conserved in ideal fluid dynamics, Eq. (35). However, this equation holds strictly only for continuous (differentiable) solutions. Shock discontinuities do not belong to this class, and therefore can produce entropy. Physically speaking, microscopic non-equilibrium processes take place inside a shock front which lead to this increase of entropy.

One could object that this conclusion is not stringent in the sense that (106) also allows for the case where $s/n = s_0/n_0$, i.e., where the entropy does not increase across the shock front. However, by explicitly solving the shock equations (99) – (101) with a given equation of state one finds that this case occurs only for infinitesimal shock discontinuities (which then degenerate into sonic perturbations, which in turn preserve entropy). For any finite discontinuity one finds $s/n > s_0/n_0$.

The Chapman–Jouguet point (cf. Fig. 7) is actually special in this respect: it corresponds to that state of matter emerging from the shock, where entropy production is maximized [5]. It is amusing to note that in selecting this state as the final state of matter emerging from a rarefaction shock wave (cf. discussion above), fluid dynamics not only automatically respects the second law of thermodynamics, but even exploits it to the maximum extent.

4.3   Equation of State and Expansion into Vacuum

In this subsection I discuss possible wave patterns for the one-dimensional expansion of semi-infinite matter into the vacuum. To be specific, let us first choose an equation of state which bears relevance to relativistic heavy-ion physics. At zero net baryon number, QCD lattice data [4] suggest the following Ansatz for the entropy density as function of temperature:

$$s(T) = c_{\mathrm{H}}T^3 \frac{1 - \tanh[(T - T_{\mathrm{c}})/\Delta T]}{2} + c_{\mathrm{Q}}T^3 \frac{1 + \tanh[(T - T_{\mathrm{c}})/\Delta T]}{2} \ , \quad (107)$$

where $c_{\mathrm{Q}}/c_{\mathrm{H}}$ is the ratio of degrees of freedom in the quark-gluon phase and the hadronic phase, $T_{\mathrm{c}} \simeq 160$ MeV is the (phase) transition temperature, and ΔT is the width of the transition. Present lattice data are not yet sufficiently precise to decide whether the transition is first (corresponding to $\Delta T = 0$) or higher order, or just a smooth cross-over transition, but they restrict ΔT to be within the range $0 \leq \Delta T \lesssim 0.1\,T_{\mathrm{c}}$. Note that for $\Delta T = 0$ the equation of state becomes that of the well-known MIT bag model [22] with a bag constant $B = (c_{\mathrm{Q}}/c_{\mathrm{H}} - 1)p_{\mathrm{c}}$, where p_{c} is the pressure at the phase transition temperature T_{c}.

To cover the possible range of ΔT, we shall consider the limiting values $\Delta T = 0$ and $\Delta T = 0.1\,T_{\mathrm{c}}$ in the following. Both cases will be compared to results for an equation of state where there is no transition to the quark-gluon phase, i.e., where

$$s(T) \equiv s_{\mathrm{H}}(T) = c_{\mathrm{H}}T^3 \ . \quad (108)$$

Once $s(T)$ is known one can compute other thermodynamic variables from fundamental thermodynamic relations, for instance:

$$p = \int_0^T \mathrm{d}T'\, s(T') \ , \quad \epsilon = Ts - p \ . \quad (109)$$

The three equations of state considered here are explicitly shown in Fig. 8. The ratio of degrees of freedom c_Q/c_H was chosen to be $37/3$, corresponding to an ultrarelativistic gas of u and d quarks and gluons in the quark-gluon phase and a massless pion gas in the hadronic phase.

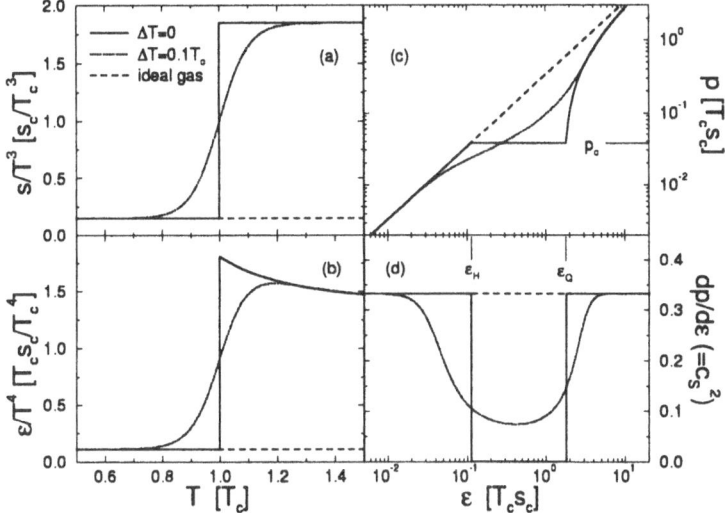

Fig. 8: (a) The entropy density divided by T^3 as a function of T. (b) The energy density divided by T^4 as a function of T. (c) The pressure as a function of energy density. (d) The velocity of sound squared as a function of energy density. $c_Q/c_H = 37/3$. Units of energy are T_c, units of energy density are $T_c s_c$, where $s_c \equiv (c_Q + c_H) T_c^3/2$. Solid line: $\Delta T = 0$, dotted line: $\Delta T = 0.1\,T_c$, dashed line: ideal hadron gas.

Figs. 8 (a,b) show the entropy density divided by T^3 and the energy density divided by T^4 as functions of T. This representation of the equation of state is commonly used by the lattice QCD community. On the other hand, fluid dynamics requires the pressure as a function of energy density, $p(\epsilon)$, which is shown in Fig. 8 (c). The collective evolution of the fluid is, however, controlled by pressure *gradients*. Figure 8 (d) shows the velocity of sound squared $c_s^2 \equiv dp/d\epsilon$ (if there are no conserved charges). This quantity determines the pressure gradient dp for a given gradient in energy density

$d\epsilon$, i.e., it characterizes the capability of the fluid to perform mechanical work, or in other words, it characterizes the *expansion tendency*. Thus, for the equation of state with a first order phase transition, $\Delta T = 0$, in the mixed phase of quark-gluon and hadronic matter, $\epsilon_H \le \epsilon \le \epsilon_Q$, the system does *not* perform mechanical work and therefore has *no* tendency to expand. As will be seen in the following this will have profound influence on the time evolution of the system.

For the equation of state with a smooth cross-over transition, $\Delta T = 0.1\,T_c$, the expansion tendency is not zero, but still greatly reduced in the transition region as compared to the ideal gas equation of state without any transition ($c_s^2 = 1/3 = $ const. for all values of ϵ). The transition region $\epsilon_H \lesssim \epsilon \lesssim \epsilon_Q$ is referred to as the "soft region" of the equation of state [23]. For an equation of state with a first order transition, the point $\epsilon = \epsilon_Q$ is called the "softest point" of the equation of state [24]. (This notion comes from considering the function $p(\epsilon)/\epsilon$ which has a minimum at ϵ_Q.)

Another quantity of interest is Σ, which determines whether matter is thermodynamically normal or anomalous. Figure 9 shows this quantity (times Ts) as computed from (95) for the three equations of state studied here. For $\Delta T = 0$, matter becomes anomalous in the mixed phase, the other two equations of state are thermodynamically normal everywhere. (Strictly speaking, $\Sigma = 0$ only vanishes in the mixed phase, but does not become negative. This is, however, sufficient for the formation of stable rarefaction shock waves.)

Let us now consider the one-dimensional expansion of semi-infinite matter into the vacuum. Figure 10 shows temperature profiles for (a) the expansion of an ideal gas and (b,c) for the expansion with the $\Delta T = 0$ equation of state. In (b) the initial energy density of semi-infinite matter is chosen to be well above ϵ_Q, the phase boundary between the quark-gluon and the mixed phase, in (c) the initial energy density is just below ϵ_Q. The dotted line in (a) indicates the initial temperature profile for all cases. The initial profile indicates a discontinuity at $x = 0$ which separates two regions of constant flow, the semi-infinite slab of matter at rest to the left ($x \le 0$), and the vacuum to the right ($x > 0$). This initial condition is in fact a special case of the Riemann problem discussed in Subsection 3.3. From general arguments (see above) the solution at $t > 0$ can only be a simple wave, connecting these two regions of constant flow. For the ideal hadron gas which is thermodynamically normal matter, we have seen above that this simple wave must be a continuous rarefaction wave, in this case moving to the right. As mentioned above, for such a wave the Riemann invariant $\mathcal{R}_+ = $ const. everywhere, cf. (89), from which we deduce the relationship between the fluid rapidity y and the energy density ϵ on the rarefaction wave:

$$y(\epsilon) = -\frac{c_s}{1 + c_s^2} \ln \frac{\epsilon}{\epsilon_0} \,, \tag{110}$$

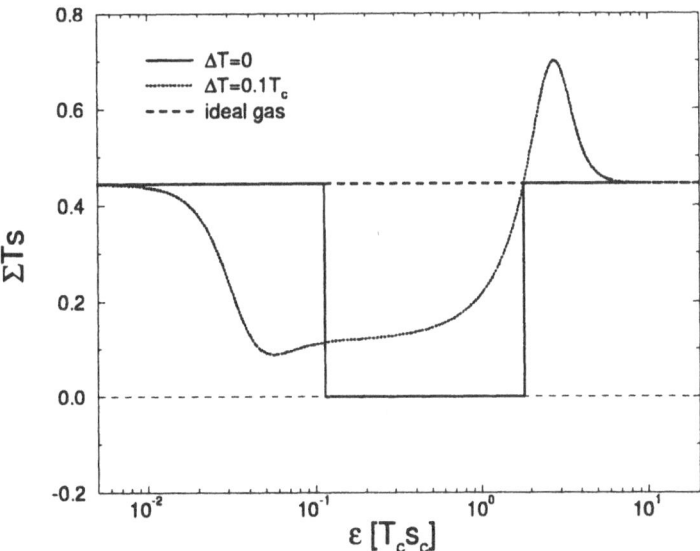

Fig. 9: The quantity Σ (times Ts) as a function of ϵ for $\Delta T = 0$ (solid line), $\Delta T = 0.1\,T_c$ (dotted line), and the ideal hadron gas equation of state (dashed line).

where we have used the fact that for the ideal hadron gas equation of state $p = c_s^2 \epsilon$ and that the initial fluid rapidity of the semi-infinite slab is zero, $y_0 = 0$. The fluid velocity on the rarefaction wave is then given by $v(\epsilon) = \tanh y(\epsilon)$. Finally, the position at which one finds a given energy density ϵ at time t can be deduced by integrating (90) for the non-trivial C_- characteristics:

$$x(\epsilon) = \frac{v(\epsilon) - c_s}{1 - v(\epsilon)\,c_s}\,t \; , \tag{111}$$

where we have used the fact that the initial position of the simple wave is at $x = 0$ and that $c_s = $ const. for the ideal hadron gas equation of state (we have assumed that the hadron gas consists of massless, i.e., ultrarelativistic pions, for which $c_s^2 = 1/3$). Equation (111) tells us that the rarefaction wave moves with sound velocity into the semi-infinite slab of matter (to the left), $x_A = -c_s t$, and with the velocity of light into the vacuum (to the right), $x_B = t$.

The expansion in the case of a first order phase transition, $\Delta T = 0$, proceeds similarly, with the exception that in the region of energy densities corresponding to the mixed phase, matter is thermodynamically anomalous, cf. Fig. 9,

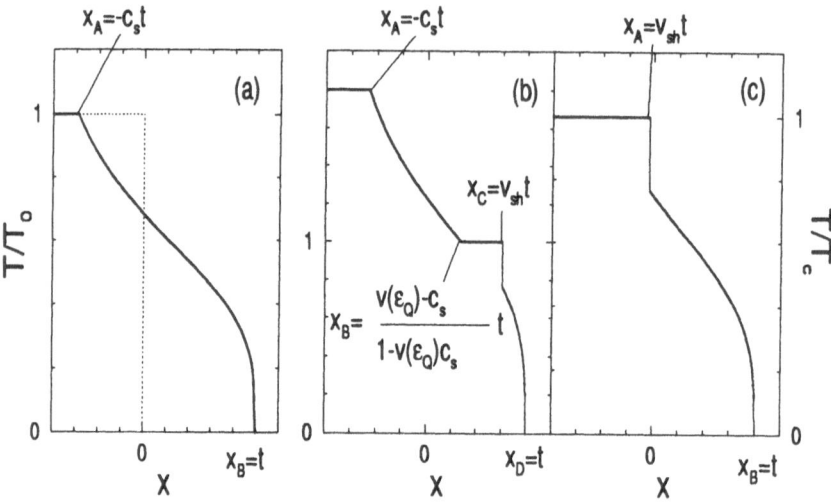

Fig. 10: Temperature profiles for the expansion of semi-infinite matter into vacuum. (a) Ideal hadron gas equation of state, the dotted line indicates the initial state, the temperature is normalized to the initial temperature T_0. (b,c) Equation of state with $\Delta T = 0$, in (b) the initial energy density is well above ϵ_Q, in (c) it is just below ϵ_Q. The temperature in (b,c) is normalized to the critical temperature T_c.

such that from Table 4.1 we conclude that the stable wave pattern is not a continuous rarefaction wave, but a rarefaction shock wave. Thus, as long as matter is in the (thermodynamically normal) quark-gluon phase, the expansion will proceed as a continuous rarefaction wave as in Fig. 10 (a), but upon entering the mixed phase (energy density ϵ_Q, temperature T_c) a rarefaction shock wave will form. The state of matter emerging from this shock wave is determined from the shock equations as described in the previous subsection, i.e., it corresponds to the Chapman–Jouguet point on the Taub adiabat with center located at the phase boundary between quark-gluon and mixed phase (for more details, see [17]). Then, also the velocity of the shock v_{sh} in the calculational frame is determined. In general v_{sh} and the velocity of matter at the base of the continuous rarefaction wave are not equal. This leads to the formation of a plateau of constant flow between x_B and x_C. The state of matter at the Chapman–Jouguet point corresponds to thermodynamically normal hadronic matter, so that the further expansion has to proceed as a continuous rarefaction wave. The emerging wave pattern is shown in Fig. 10 (b).

The only difference between Fig. 10 (b) and (c) is that the initial energy density in (c) is already below ϵ_Q, i.e., in the region corresponding to mixed phase. Therefore, the expansion starts out with a rarefaction shock wave, from which matter emerges at the Chapman–Jouguet point of the respective Taub adiabat with center corresponding to the initial state of matter. (Note that this Taub adiabat differs from the one in (b), since their centers are different.) Further expansion proceeds as a continuous rarefaction wave in hadronic matter.

This completes the discussion of the expansion of semi-infinite matter into vacuum and prepares us to understand the Landau model which is subject of the next subsection.

4.4 The Landau Model

The Landau model is historically the first case where fluid dynamics was applied to describe – at that time – hadron-hadron collisions [25]. Its main focus of application nowadays is, of course, nucleus-nucleus collisions. The main ideas are summarized in Fig. 11. Imagine two nuclei colliding at ultrarelativistic velocities in their center of mass. The nuclei are Lorentz-contracted to a "pancake-like" shape. In the moment of impact, nuclear matter becomes highly excited (the detailed microscopic processes which happen during this stage are of no concern for the following). In the limit that the velocities of the nuclei $v \to 1$, there will be no baryon stopping (due to the limited stopping power of nuclear matter), i.e., the baryon charges will pass through each other unscathed, leaving highly excited, net baryon-free matter in their wake. Due to Lorentz contraction, the initial extension $2L$ in z direction of this slab of highly excited matter is much smaller than the transverse size of the slab, such that the expansion will proceed mainly in the longitudinal direction and is thus essentially one-dimensional. The Landau model assumes that the slab has no initial collective velocity and that rapid thermalization takes place which is completed at $t = 0$. It is also assumed that the equation of state has the simple ultrarelativistic form $p = c_s^2 \epsilon$, $c_s^2 = \text{const.}$, i.e., that matter is thermodynamically normal for all ϵ. (The original idea of Landau actually was that the baryons are immediately stopped in the collision through compressional shock waves. Data from heavy-ion experiments at BNL-AGS and CERN-SPS prove that this picture is unrealistic, due to the aforementioned finite stopping power of nuclear matter. However, since the collision is ultrarelativistic, the thermal energy in the highly excited slab is much larger than the chemical energy associated with the conservation of baryon charge. Therefore, to good approximation, $\mu_B = n_B = 0$, and the further evolution of the slab will be identical to what is discussed here.)

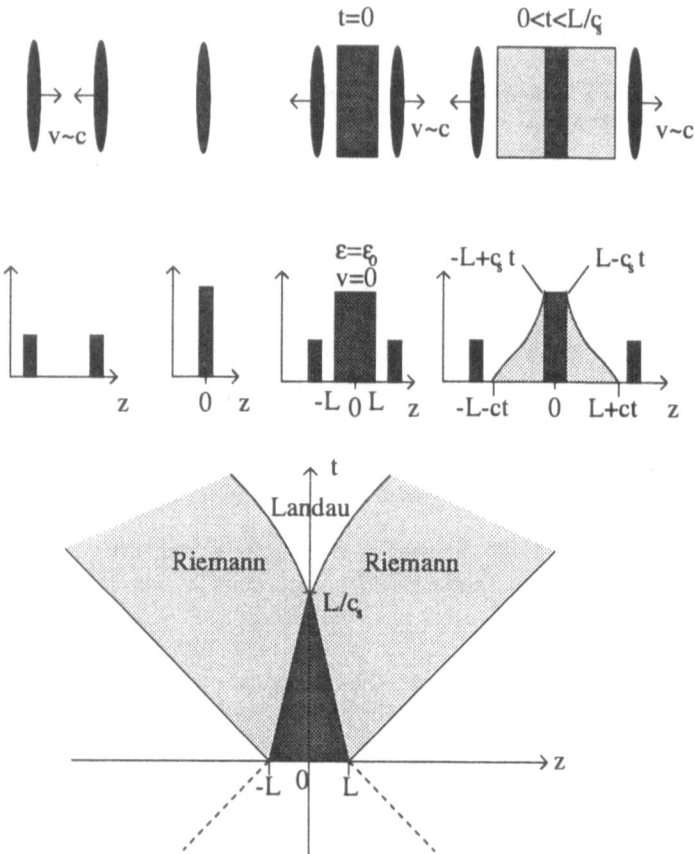

Fig. 11: The Landau model for nuclear collisions. See text for details.

For $t > 0$, the slab starts to expand. As in the expansion of semi-infinite matter discussed in the previous subsection, rarefaction waves will form. For thermodynamically normal matter, these are continuous (Riemann) rarefaction waves which travel into the slab with sound velocity. Therefore, they will meet at the center of the slab (here chosen to be the origin $z = 0$) at a time L/c_s. For times $t > L/c_s$, these waves overlap and the solution becomes more complicated. In a region near the light cone, the solution will remain a Riemann rarefaction wave, therefore we term this region the *Riemann region*. In the center where the Riemann rarefaction waves overlap, however, the solution is no longer a simple wave (indeed, only two regions of constant flow *have* to be connected by a simple wave [18], for two simple waves no such theorem exists). For $c_s^2 = $ const. the solution can still be given in closed analytic form [25], although the derivation is rather complicated. However, since two of our equations of state do not have constant velocities of sound, we have to resort to numerical solution methods, such as the relativistic HLLE discussed above. In principle, numerical algorithms can deal with arbitrary (physically reasonable) equations of state, and are therefore well able to handle this problem (although one should test them thoroughly for test cases where analytical solutions are known [17,20]).

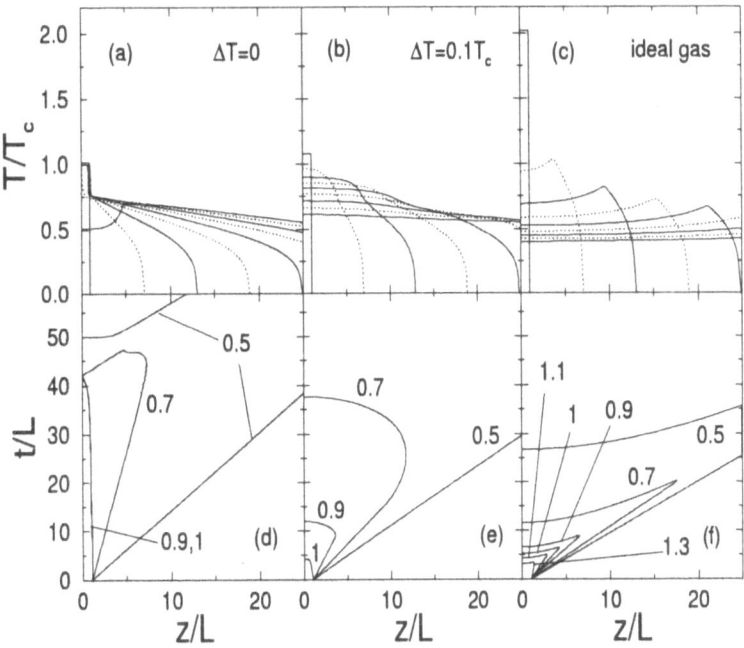

Fig. 12: Expansion in the Landau model for $\Delta T = 0$ (a,d), $\Delta T = 0.1\,T_c$ (b,e), and the ideal gas equation of state (c,f). (a–c) show temperature profiles for different times, (d–f) show the corresponding isotherms in the $t - z$ plane (numbers are temperatures in units of T_c). The initial energy density is $\epsilon_0 = 1.875\,T_c s_c$ in all cases.

In Fig. 12 numerical solutions for the Landau model are presented for the three different equations of state of Fig. 8. The initial energy density is $\epsilon_0 = 1.875\,T_c s_c$ which is slightly larger than ϵ_Q. In Figs. 12 (a–c) temperature profiles are shown for different times t and for the $z > 0$ half plane (the solution in the other half plane is the respective mirror image). For $\Delta T = 0$, Fig. 12 (a), one clearly observes the rarefaction shock wave which, for this initial energy density is almost stationary. Hadronic matter is expelled from the shock until the energy in the interior of the slab decreases below ϵ_H and the shock vanishes. For $\Delta T = 0.1 \, T_c$, Fig. 12 (b), no shock is formed, although the variation of the velocity of sound in the mixed phase, Fig. 8 (d), leads to shapes for the continuous rarefaction waves which differ strongly from those for a constant velocity of sound, Fig. 12 (c). Note the kink in the temperature profiles in the latter case which indicate the position where the Landau solution matches to the Riemann rarefaction wave. Note also the difference in the initial temperatures for the three cases although the initial energy density is the same. This is a consequence of the different number of degrees of freedom for the three equations of state at high energy densities.

In Figs. 12 (d–f) corresponding isotherms are shown in the $t - z$ plane. The most pronounced feature is that due to the small propagation velocity of the rarefaction wave, the system stays hot for a much longer time span for the $\Delta T = 0$ equation of state, Fig. 12 (d), than for the ideal gas, Fig. 12 (f). This is in agreement with the general argument presented earlier that the softening of the equation of state in the mixed phase region leads to a reduced expansion tendency and thus to a "stalled" expansion of the system. The softening of the equation of state is also the reason why the expansion for the $\Delta T = 0.1 \, T_c$ equation of state, Fig. 12 (e), is delayed in comparison to the ideal gas case, although no rarefaction waves are formed. For a quantitative analysis of the delayed expansion in the Landau model see [23].

4.5   The Bjorken Model

One of the main assumptions of Landau's model is that the initial collective velocity of the slab of excited matter vanishes. However, this cannot be quite true on account of the following argument. In the limit $v \to 1$, the size of the nuclei in longitudinal direction goes to zero, and there is no scale in the problem at all. In this case, the collective velocity of matter in the slab *has* to be of the scaling form $v = z/t$. The consequences of this special form for the longitudinal fluid velocity were first investigated in [26,27], again with respect to possible applications in hadron-hadron collisions. Bjorken [28] was the first to discuss it in the framework of nuclear collisions.

The main ideas of the so-called *Bjorken model* are summarized in Fig. 13. As in the Landau model, two ultrarelativistic, Lorentz-contracted nuclei collide at $z = 0$ and $t = 0$ (the moment of complete overlap) in the center of mass frame of the collision. Due to the limited amount of nuclear stopping power, the baryon charges keep on moving along the light cone, while microscopic collision processes (the nature of which is of no concern for the following) lead to the formation of a region of highly excited, net charge free matter in the wake of the nuclei. In contrast to the Landau model, however, the collective velocity in this region is of the scaling form $v = z/t$. The region of highly excited matter is supposed to rapidly equilibrate locally within a time span τ_0 (which is of the order of a fm or less), and the further evolution of the system can be described in terms of ideal fluid dynamics. One important point is that, due to the absence of a scale, physics has to be the same for matter at different longitudinal coordinate z if compared at the same *proper time* $\tau = t\sqrt{1-v^2} = \sqrt{t^2 - z^2}$. (Such curves of constant τ describe hyperbola in space-time.) Thus, the initial thermodynamic state of all fluid elements is the same at the same *proper time* τ_0.

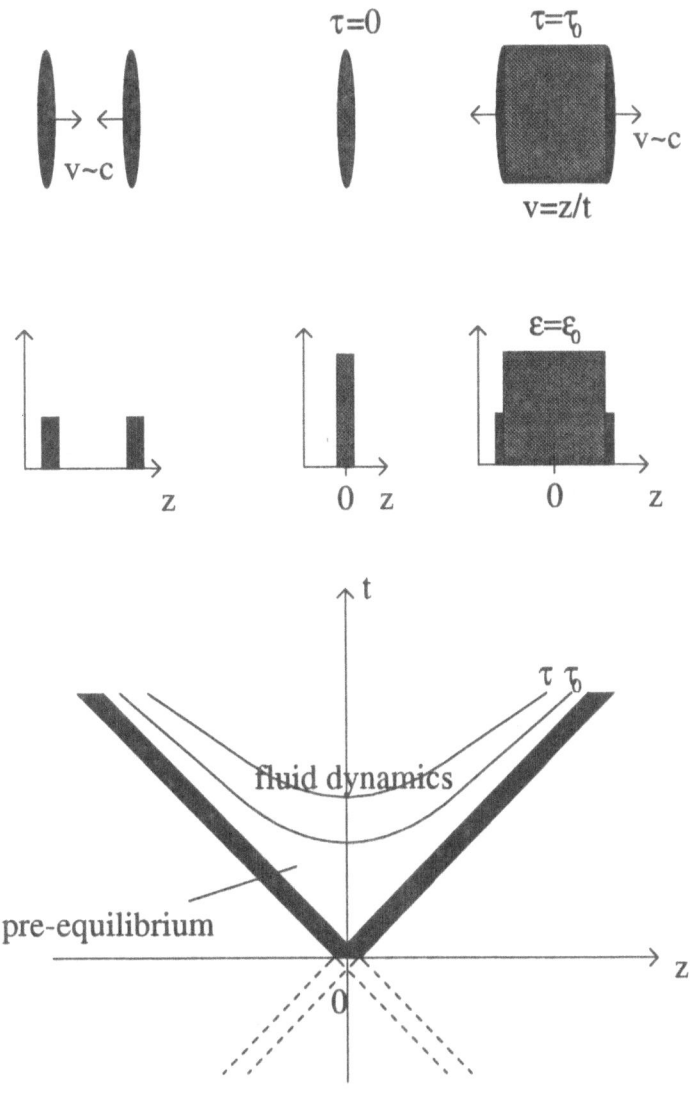

Fig. 13: The Bjorken model for nuclear collisions. See text for details.

If the longitudinal velocity profile is enforced by the scaling argument, the fluid-dynamical solution simplifies in fact considerably. To see this, change the variables t, z in the conservation laws for one-dimensional longitudinal motion in the absence of conserved charges,

$$\partial_t T^{00} + \partial_z T^{z0} = 0 \ , \quad \partial_t T^{0z} + \partial_z T^{zz} = 0 \ , \qquad (112)$$

to the variables $\tau = \sqrt{t^2 - z^2}$, which is the proper time of a fluid element, and $\eta = \operatorname{Artanh} v = \operatorname{Artanh}[z/t]$, which is the rapidity of a fluid element. Then, the coupled system of partial differential equations (112) decouples:

$$\left.\frac{\partial\epsilon}{\partial\tau}\right|_{\eta} + \frac{\epsilon+p}{\tau} = 0 \ , \qquad (113)$$

$$\left.\frac{\partial p}{\partial\eta}\right|_{\tau} = 0 \ . \qquad (114)$$

The second equation (114) has the interesting consequence that there is no pressure gradient between adjacent fluid elements (the one at η and the one at $\eta + d\eta$). At first glance this would seem to indicate that there is no expansion of the fluid at all. This, however, is not true, since the fluid velocity is certainly finite, $v = z/t$. The answer is that the new coordinates (τ, η) already take the scaling expansion into account: a fluid element at η with a width $\Delta\eta$ in fact "grows" in longitudinal direction by an amount $dz = dt \, \Delta\eta$ during the time span dt .

Another consequence of (114) is derived from the Gibbs–Duhem relation:

$$\left.\frac{\partial p}{\partial\eta}\right|_{\tau} = s \left.\frac{\partial T}{\partial\eta}\right|_{\tau} + \sum_{i=1}^{n} n_i \left.\frac{\partial\mu_i}{\partial\eta}\right|_{\tau} = 0 \ . \qquad (115)$$

This equation means that for vanishing conserved charges $n_i = 0$, $i = 1,\ldots,n$, the temperature has to be constant along curves of constant τ, i.e., along the space-time hyperbola shown in Fig. 13 (η varies along these curves). In the general case of non-zero net charges, however, only the particular combination of charge densities, entropy density, and derivatives of T and the μ_i appearing in (115) has to vanish along curves of constant τ. Equation (114) represents the principle of "boost invariance" commonly associated with the Bjorken model: at constant τ the pressure is independent of the longitudinal rapidity, i.e., it is the same in fluid elements with different η, or in other words, it does not change if one performs a longitudinal boost to a different reference frame. This is a consequence of the scaling form for the longitudinal velocity.

Equation (113) also has an interesting consequence. With the first law of thermodynamics, one derives as usual the conservation of the entropy current which now takes the form

$$\left.\frac{\partial s}{\partial \tau}\right|_{\eta} + \frac{s}{\tau} = 0 \ , \tag{116}$$

which can be immediately integrated to give

$$s\,\tau = s_0\tau_0 = \text{const.} \tag{117}$$

at constant η. The constant may in principle differ for different η, but since the initial thermodynamic state along τ_0 was the same for all η, that constant will also be the same for all η at other $\tau > \tau_0$. Equation (117) is interesting because it tells us that the entropy density decreases inversely proportional to τ *independent of the equation of state* of the fluid. The time evolution for energy density, pressure, or temperature might depend on the equation of state, but not the one for the entropy density.

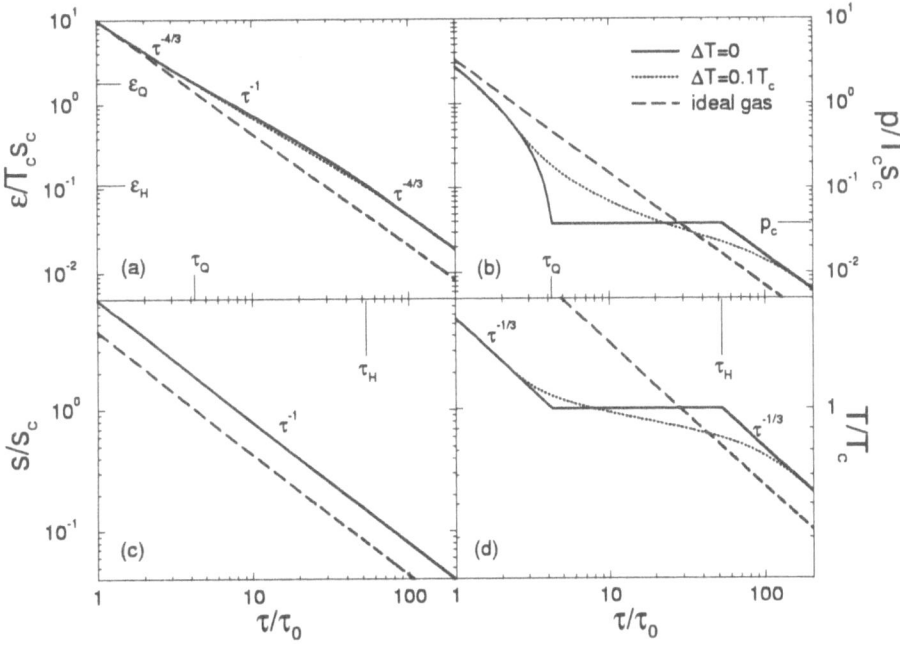

Fig. 14: Proper time evolution for (a) energy density, (b) entropy density, (c) pressure, and (d) temperature in the Bjorken model for nuclear collisions (longitudinal expansion only). Solid line: $\Delta T = 0$, dotted line: $\Delta T = 0.1\,T_c$, dashed line: ideal gas equation of state. The initial energy density is $\epsilon_0 = 10\,T_c s_c$.

This is confirmed in Fig. 14, where the evolution of (a) the energy density, (b) the entropy density, (c) the pressure, and (d) the temperature is shown as a function of proper time τ for the three equations of state ($\Delta T = 0$, $\Delta T = 0.1\,T_c$, and the ideal hadron gas). Note that in the quark-gluon as well as the hadron phase, where $p \sim c_s^2\,\epsilon$ with $c_s^2 = 1/3$, Eq. (113) yields

$$\epsilon \sim \tau^{-4/3} , \tag{118}$$

For the $\Delta T = 0$ equation of state, $p = p_c = \text{const.}$ in the mixed phase, and (113) yields the cooling law

$$\epsilon \sim \tau^{-1} . \tag{119}$$

This is interpreted as follows. The longitudinal scaling expansion dilutes the system $\sim \tau^{-1}$. If no mechanical work is performed, like in the mixed phase where $dp \equiv 0$, only this geometrical dilution determines the (proper) time evolution of the energy density. In the phase where $dp = c_s^2\,d\epsilon$, however, additional mechanical work is performed, and the system cools faster, $\epsilon \sim \tau^{-(1+c_s^2)} = \tau^{-4/3}$. The faster cooling is confirmed studying the temperature evolution, Fig. 14 (d). For $p = c_s^2\,\epsilon$, $c_s^2 = \text{const.}$, and vanishing net charges, one deduces from $dp = c_s^2\,d\epsilon = s\,dT = (\epsilon + p)\,dT/T = (1 + c_s^2)\,\epsilon\,dT/T$, that $\epsilon \sim T^{1+c_s^{-2}}$ and consequently, in the hadron and quark-gluon phase

$$T \sim \tau^{-1/3} , \tag{120}$$

while in the mixed phase one deduces from $dp = s\,dT \equiv 0$ that

$$T = \text{const.} . \tag{121}$$

This expectation is confirmed in Fig. 14 (d).

Of course, in reality the expansion of the system will not only be purely longitudinal. The "Bjorken cylinder" will also expand transversally. The principle of boost invariance allows us to focus on the transverse expansion at $z = \eta = 0$ only, and reconstruct the fluid properties at a different η by performing a longitudinal boost with boost rapidity η. For the sake of simplicity, let us assume that the system is cylindrically symmetric in the transverse direction and that the initial energy density profile is of the form

$$\epsilon(\mathbf{r}, \tau_0, \eta = 0) = \epsilon_0\,\Theta(R - |\mathbf{r}|) , \tag{122}$$

where R is the transverse radius of the Bjorken cylinder. In cylindrical coordinates and at $z = \eta = 0$, the conservation equations read ($T^{00} \equiv E$, $T^{0r} \equiv M$, $v_r \equiv v$):

$$\partial_t E + \partial_r \left[(E + p)v\right] = -\left(\frac{v}{r} + \frac{1}{t}\right)(E + p) , \tag{123}$$

$$\partial_t M + \partial_r \left(Mv + p\right) = -\left(\frac{v}{r} + \frac{1}{t}\right)M . \tag{124}$$

Although these equations have no longer a simple analytical solution, the assumption of cylindrical symmetry has reduced the originally three-dimensional problem to an effectively one-dimensional problem. Indeed, for vanishing right-hand sides the solution of (123), (124) with the initial condition (122) is identical to the one of the Landau model with the substitutions $z \rightarrow r$ and $L \rightarrow R$. The right-hand sides just lead to an additional reduction of E and M from the cylindrical geometry, v/r, and from longitudinal scaling, $1/t$.

This observation, combined with the method of operator splitting discussed previously, suggests the following simple solution scheme (also known as Sod's method [16,29]): equations (123), (124) are of the type

$$\partial_t U + \partial_x F(U) = -G(U) \ . \tag{125}$$

The operator splitting method allows to construct the solution by first solving the one-dimensional *partial* differential equation

$$\partial_t U + \partial_x F(U) = 0 \ , \tag{126}$$

(for instance with the relativistic HLLE scheme discussed above), which yields a *prediction* $\tilde{U}$ for the true solution U. In a second step one *corrects* this prediction by solving the *ordinary* differential equation

$$\frac{dU}{dt} = -G(U) \ , \tag{127}$$

which is numerically realized as [30]

$$U = \tilde{U} - \Delta t\, G(\tilde{U}) \ . \tag{128}$$

The transverse expansion of the Bjorken cylinder at $z = 0$ is shown in Fig. 15 for $r_0 = 0.1\,R$ and $\epsilon_0 = 18.75\,T_c s_c$. One immediately recognizes the qualitative similarities with the Landau expansion, like the delay in the expansion for the two equations of state with a (phase) transition as compared to the expansion with an ideal hadron gas equation of state. The additional geometrical dilution, however, leads in general to a faster cooling overall and quantitatively different shapes for the temperature profiles and the isotherms in the $t - r$ plane.

Let us further quantify the time delay in the expansion induced by the transition in the equation of state. In general, the system will decouple into free-streaming particles once the temperature drops below a certain "freeze-out" temperature T_{fo}, see Section 5 below. From comparison with experimental data, this freeze-out temperature is estimated to be on the order of 100 MeV. Let us therefore define a "lifetime" of the system as the time when the $T = 0.7\,T_c$ isotherm crosses the origin at $r = 0$ in Figs. 15 (d–f). This

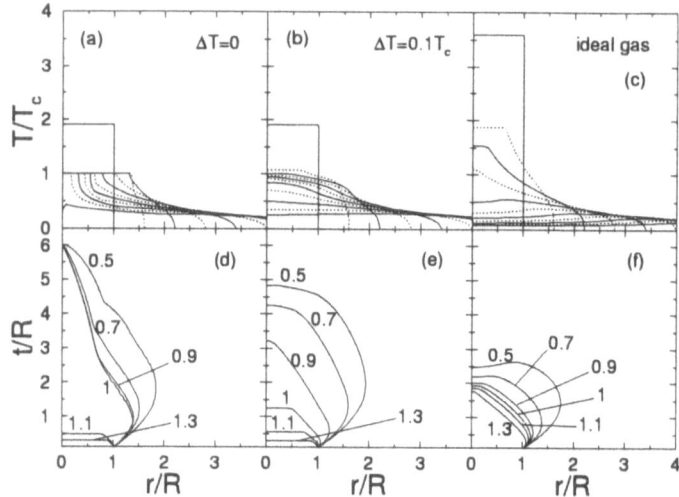

Fig. 15: Transverse expansion of the Bjorken cylinder for $\Delta T = 0$ (a,d), $\Delta T = 0.1\,T_c$ (b,e), and the ideal gas equation of state (c,f). (a–c) show temperature profiles for different times, (d–f) show the corresponding isotherms in the $t-r$ plane (numbers are temperatures in units of T_c). The initial energy density is $\epsilon_0 = 18.75\,T_c s_c$ in all cases.

lifetime is shown in Figs. 16 (a,b) as function of the initial energy density ϵ_0 of the cylinder. One observes a maximum of the such defined lifetime at initial energy densities around $40\,T_c s_c \sim 30\,\mathrm{GeVfm}^{-3}$. At these initial energy densities, the prolongation of the lifetime over the respective ideal hadron gas value is about a factor of 2 (for $\Delta T = 0.1\,T_c$) to 3 (for $\Delta T = 0$).

The prolongation of the lifetime is due to the softening of the equation of state in the phase transition region. It is, however, interesting that the maximum in the lifetime does not occur around initial energy densities corresponding to ϵ_Q (as is the case in the Landau model [23]), but at much larger initial energy densities. The reason for this is the strong longitudinal dilution of the system on account of the scaling profile $v_z = z/t$. In order to see a large effect of the softening of the equation of state in the phase transition region on the expansion dynamics, the transverse (Landau-like) expansion has to be the dominant cooling mechanism for the system. The Bjorken scaling expansion does not account for the reduced expansion tendency of the system in the transition region, it *enforces* an expansion velocity $v_z = z/t$ irrespective of the equation of state. In order to have the transverse expansion dominate the cooling of the system, one has to start the expansion at higher initial energy densities such that the system spends enough time in the mixed phase for the (slow) rarefaction shock to reach the origin. The initial energy density in Fig. 15 was intentionally selected to maximize this effect.

Initial energy densities on the order of $10 - 30\,\mathrm{GeVfm}^{-3}$ are expected to be reached at the RHIC collider. In order to experimentally observe the prolongation of the lifetime as seen in Figs. 16, one has to find a corresponding experimental observable. An obvious candidate is the ratio of the "out" to the "side" radius of two-particle correlation functions. The "out" radius is proportional to the duration of particle emission from a source, while the "side" radius is proportional to the transverse dimension of the source (cf. [31] for a very detailed, pedagogical discussion). Since the transverse radius of the source is approximately the same in all cases, cf. Fig. 15 (a–c), the ratio $R_{\mathrm{out}}/R_{\mathrm{side}}$ seems to be a good generic measure for the lifetime. Moreover, in forming the ratio the dependence on the overall (unknown) spatial size of the source as well as effects from the collective expansion are expected to cancel. The ratio $R_{\mathrm{out}}/R_{\mathrm{side}}$ is plotted in Figs. 16 (c,d) for pions with mean transverse momenta $K_{\perp} = 300$ MeV. (Details on how to compute this quantity can be found in [30,32].) As one observes, $R_{\mathrm{out}}/R_{\mathrm{side}}$ nicely reflects the excitation function of the lifetime of the system.

5   Freeze-Out

In this section I discuss an up to date unsolved problem in the application of relativistic fluid dynamics to describe nuclear collisions, namely the so-called "freeze-out" process. Given an initial condition, fluid dynamics describes the evolution of the system in the whole forward lightcone, Fig. 17 (a). However, as we have seen above, at all times near the boundary to the vacuum, as well as everywhere in the late stage of the evolution, the energy density becomes arbitrarily small, i.e., the system is rather cold and dilute. In this space-time region the assumption of local thermodynamical equilibrium is no longer justified, because the particle scattering cross section σ is finite, such that for small particle densities n the particle scattering rate, $\Gamma \sim \sigma n$, becomes on the order of the inverse system size, $\Gamma \sim R^{-1}$. At this point, the scattering rate is too small to maintain local thermodynamical equilibrium and the particles decouple from the fluid evolution. In this space-time region, a kinetic description for the particle motion would be more appropriate. One should therefore not solve fluid-dynamical equations in the whole forward lightcone, but only inside a space-time region of sufficiently large energy and particle densities, while outside this region, the particle motion should be described by kinetic theory, Fig. 17 (b).

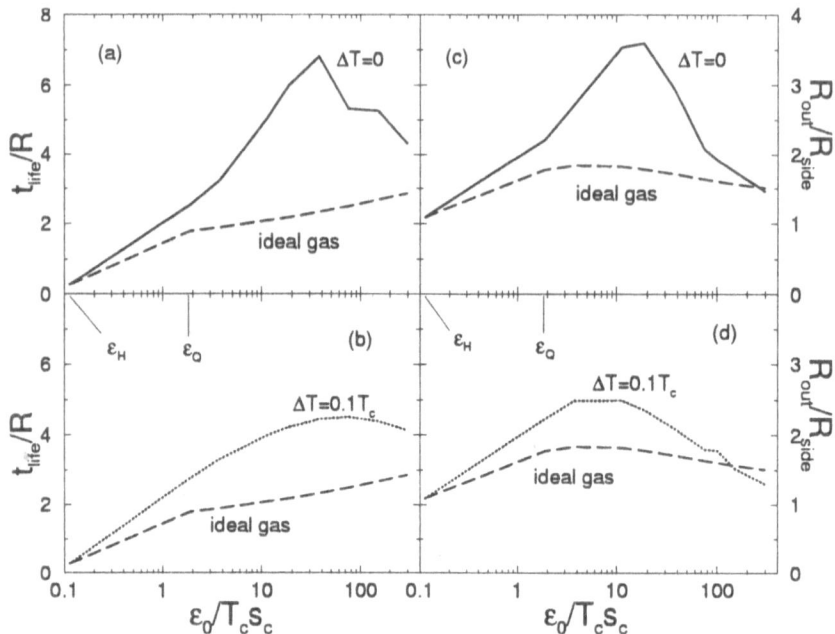

Fig. 16: Lifetime of the system as a function of ϵ_0 for the Bjorken cylinder expansion, $\tau_0 = 0.1\,T_c$. (a) $\Delta T = 0$ (solid) vs. ideal hadron gas (dashed), (b) $\Delta T = 0.1\,T_c$ (dotted) vs. ideal hadron gas (dashed). (c,d) the corresponding ratio R_{out}/R_{side}.

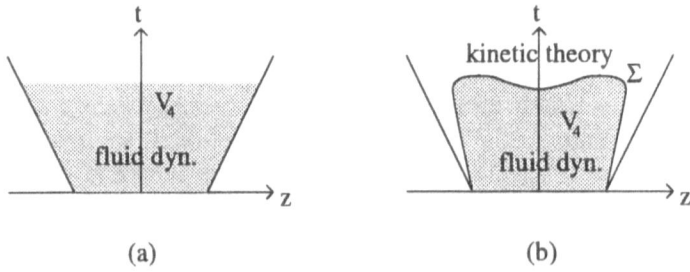

Fig. 17: (a) Conventional fluid-dynamical description in the whole forward lightcone. (b) Fluid dynamics describes the evolution of the system inside V_4, while kinetic theory describes the motion of the frozen-out particles outside V_4.

The boundary Σ between the two regions is determined by a criterion which compares local scattering rates with the system size, as discussed above. The obvious difficulty with this more realistic description of the system's evolution is that this boundary has to be determined *dynamically*, i.e., not only has one to allow for particles decoupling from the fluid, but also for the reverse process of particles entering the fluid from the kinetic region. (This can happen since the particles still, albeit rarely, collide in the kinetic region.) A consistent treatment of this problem is rather complicated, since one has to solve kinetic in addition to the fluid-dynamical equations. No serious attempt has been made so far.

Instead, the following approximate solution has been extensively employed:

1. One assumes that fluid dynamics gives a reasonable description for the evolution of the system in the *whole* forward lightcone.
2. One determines the "decoupling" surface Σ *a posteriori*, once the evolution of the fluid is known.
3. The "thickness" of Σ is assumed to be infinitesimal.
4. One assumes that particles crossing Σ have *completely* decoupled from the system, they stream freely towards the detectors without any further collisional interaction ("freeze-out"). This means that they do not change their momentum and energy once they have crossed Σ.

A very popular argument in order to determine Σ is the following. Since $n \sim T^3$, the scattering rate $\Gamma \sim T^3$ (for constant cross section σ), i.e., if the temperature falls below a certain so-called "freeze-out" temperature T_{fo}, the criterion $\Gamma \lesssim R$ is fulfilled, and particles decouple from the system. In this case, Σ is just given by the isotherm $T = T_{\text{fo}}$ (use of this argument was already made above in the discussion of the "lifetime" of the system).

Note that assumption 3. is a strong idealization and actually rather questionable, because in reality Σ is a space-time region of finite thickness, inside which non-equilibrium, dissipative effects become gradually more and more important (the more dilute the fluid becomes), until ultimately all interactions between particles cease and, when leaving Σ, they stream freely towards the detectors.

Nevertheless, with the above assumptions, one can readily compute the single inclusive spectra of particles reaching the detector. Immediately before the particles decouple from the fluid evolution, i.e., before they cross Σ, they are still in local thermodynamical equilibrium such that their phase space distribution is given by $f_0(k, x)$, Eq. (20). It is reasonable to assume that this phase space distribution is not changed much when they move a small distance along their worldlines, which carries them across Σ into the region of free-streaming. In that region, however, there are no collisions which could further change f_0. Therefore, the phase space distribution of "frozen-out" particles is (approximately) the same as in local equilibrium. The total number of particles crossing a small surface element $\mathrm{d}\Sigma$ of Σ is then given by

$$N_\Sigma \equiv \mathrm{d}\Sigma_\mu N^\mu = \int \frac{\mathrm{d}^3 k}{E}\, \mathrm{d}\Sigma \cdot k\, f_0(k, x)\ , \tag{129}$$

N^μ being the (kinetic) particle number 4-current. The *invariant momentum spectrum* of particles crossing that surface element is consequently

$$E\frac{\mathrm{d}N_\Sigma}{\mathrm{d}^3 k} = \mathrm{d}\Sigma \cdot k\, f_0(k, x)\ . \tag{130}$$

Finally, the invariant momentum spectrum (the *single inclusive* spectrum) of particles crossing the *complete* "freeze-out" surface Σ is

$$E\frac{\mathrm{d}N}{\mathrm{d}^3 k} = \int_\Sigma E\frac{\mathrm{d}N_\Sigma}{\mathrm{d}^3 k} = \int_\Sigma \mathrm{d}\Sigma \cdot k\, f_0(k, x)\ . \tag{131}$$

This equation is known as the *Cooper–Frye formula* [26], and is used in almost all fluid-dynamical applications to heavy-ion collisions to compute the single inclusive spectra of particles.

There is, however, a problem with this formula [33]. For *time-like* surfaces, i.e., where the normal vector $\mathrm{d}\Sigma_\mu$ is *space-like*, $\mathrm{d}\Sigma \cdot k$ may either be positive or negative, depending on the value and direction of k^μ. In other words, the number of particles "freezing out" from a certain time-like surface element $\mathrm{d}\Sigma$ can become negative. This is clearly unphysical, since the number of particles decoupling from the system must be positive definite. For *space-like* surfaces (with a *time-like* normal vector) as well as for time-like surface elements where $\mathrm{d}\Sigma \cdot k > 0$, the Cooper-Frye formula gives a physically reasonable, positive definite result for the number of frozen-out particles. This is illustrated in Fig. 18 which shows the rapidity distribution of particles (i.e., the invariant momentum spectrum integrated over all transverse momenta) for massless particles decoupling from a freeze-out isotherm $T_{\mathrm{fo}} = 0.4\,T_0$ in the Landau model with a $p = \epsilon/3$ equation of state. One clearly notices the negative particle numbers at midrapidity coming from the time-like parts of the isotherm.

This contradiction is readily resolved noting that the Cooper-Frye formula does not really determine the number of particles decoupling from the system, but merely the number of particle *worldlines crossing a surface element* $\mathrm{d}\Sigma$ (and then integrated over the whole surface Σ). For time-like surface elements, there is of course the possibility that for certain k^μ the respective worldlines cross $\mathrm{d}\Sigma$ in the "wrong" direction, i.e., the momenta of the particles point back into the region of fluid, cf. Fig. 19. In particular, for the $T_{\mathrm{fo}} = 0.4\,T_0$ isotherm, which moves *away* from the t-axis in the $t - z$ plane, those are particles with vanishing momentum component in z direction, because their worldlines are parallel to the t-axis. Particles with $p^z = 0$, however, also have vanishing longitudinal rapidity $y = 0$, and that is the reason

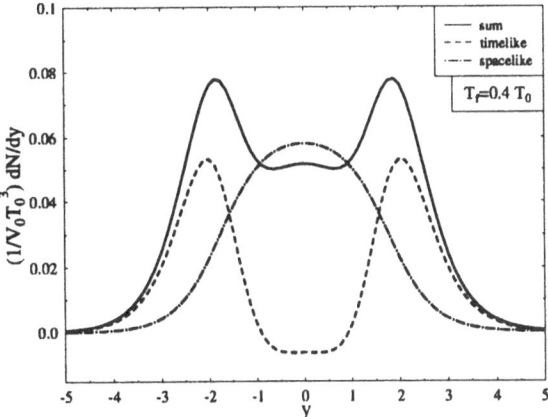

Fig. 18: The rapidity distribution for freeze-out along the $T_{fo} = 0.4\,T_0$ isotherm in the Landau model. Solid: full distribution, dotted: particles from time-like parts of the isotherm, dash-dotted: particles from space-like parts of the isotherm.

why these negative particle numbers appear at midrapidity in Fig. 18. While this explains the negative contributions in the Cooper-Frye formula, it also invalidates this formula as the correct prescription to calculate the spectra of frozen-out particles, if parts of the decoupling surface are time-like.

One suggestion to circumvent this problem was to compute the final spectra only from contribution of particles which cross the space-like parts of Σ. Of course, as can be seen by comparing the dash-dotted with the solid line in Fig. 18, the final spectra are dramatically different. Moreover, by neglecting particles crossing the time-like parts, the absolute number of frozen-out particles will also differ in the two cases. Note that the dN/dy distribution for particles from the space-like parts of the decoupling isotherm has a Gaussian shape in the Landau model. This was already pointed out in Landau's original paper [25] and has since survived as the generic (but wrong) statement that Landau's model gives rise to Gaussian rapidity distributions. In fact, there is *no* decoupling temperature where the full rapidity distribution including particles from the time-like parts resembles a Gaussian, cf. Fig. 20.

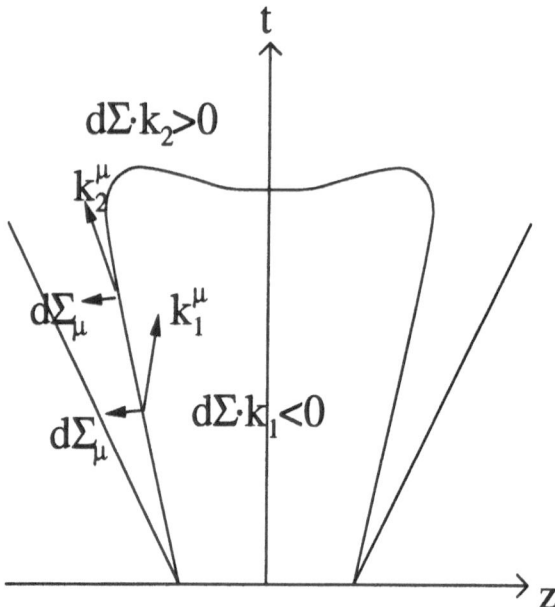

Fig. 19: Explanation for the negative number of frozen-out particles in the Cooper-Frye formula.

Another suggestion to circumvent the problem of negative particle numbers is, instead of freezing out along an isotherm which has time-like parts, to freeze out along a surface which is space-like everywhere, for instance, a curve of constant time in the center-of-mass frame, cf. Fig. 21. In this case, all particles are accounted for, since the decoupling surface is bounded by the lightcone, and no particle can escape through the lightcone. The problem is, that also in this case, the spectra differ considerably from a freeze-out at constant temperature, cf. Fig. 22. This uncertainty is clearly unwanted when one wants to quantitatively compare fluid-dynamical model predictions with experimental data.

The correct formula to compute the number of particles which physically decouple from the system was given in [33]:

$$E\frac{dN}{d^3k} = \int_{\Sigma} d\Sigma \cdot k \, f_0(k, x) \, \Theta(d\Sigma \cdot k) \ . \tag{132}$$

The additional Θ-function ensures that negative contributions to the Cooper–Frye formula are cut off. The problem with this formula is that these negative

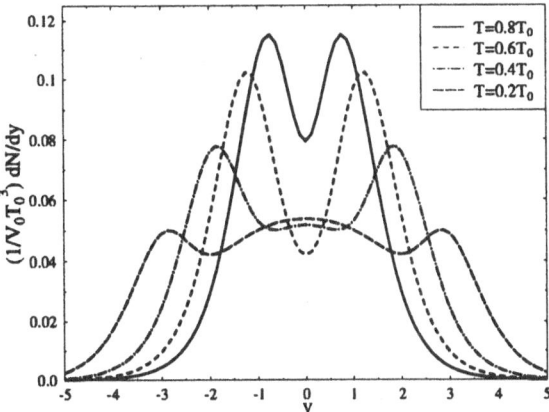

Fig. 20: Rapidity distributions for freeze-out along isotherms with $T_{fo} = 0.8\,T_0$ (solid), $0.6\,T_0$ (dashed), $0.4\,T_0$ (dash-dotted), and $0.2\,T_0$ (long dashed) in the Landau model with a $p = \epsilon/3$ equation of state.

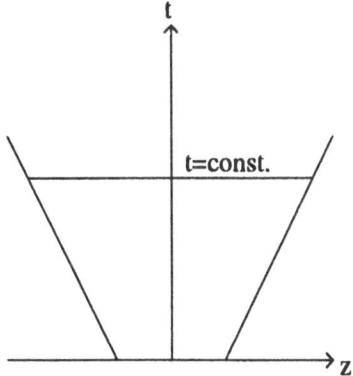

Fig. 21: A curve of constant time in the center-of-mass frame as freeze-out isotherm.

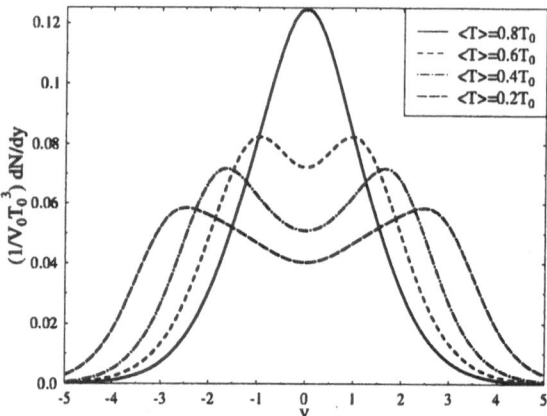

Fig. 22: The rapidity distribution for freeze-out along curves of constant time in the center-of-mass frame defined by requiring the average temperature to be $\langle T \rangle = 0.8\,T_0$ (solid), $0.6\,T_0$ (dashed), $0.4\,T_0$ (dash-dotted), and $0.2\,T_0$ (long dashed) in the Landau model with a $p = \epsilon/3$ equation of state.

contributions were necessary to globally conserve energy, momentum and net charge number, cf. the derivation of the conservation equations in Section 2. The violation of the conservation equations introduced by the freeze-out prescription (132) can, however, be circumvented by adjusting temperature, chemical potential, and the average particle 4-velocity in the single-particle distribution function $f_0(k, x)$ in (132) in such a way as to preserve the conservation laws. In other words, one must not use temperature, chemical potential, and fluid 4-velocity on the fluid side of the freeze-out surface in (132), but modified values which ensure that energy, momentum, and net charge is conserved. One way to achieve this is to assume that the freeze-out surface actually is a conventional fluid-dynamical discontinuity across which energy, momentum, and net charge number are conserved. Solving the corresponding algebraic conservation equations (with energy-momentum tensor and net charge current on the post freeze-out side of the discontinuity constructed from (21,22) with $f_0(k, x)$ replaced by $f_0(k, x)\,\Theta(\mathrm{d}\Sigma \cdot k)$) yields the required modified values for temperature, chemical potential, and average particle 4-velocity on the post freeze-out side. For more details, see [33,34]. However, it still remains to be shown with an explicit calculation whether this suggestion to solve the freeze-out problem is viable in the general case.

References

1. H. Stöcker and W. Greiner, Phys. Rep. **137**, 277 (1986)
2. R.B. Clare and D.D. Strottman, Phys. Rep. **141**, 177 (1986)
3. L.P. Csernai, *Introduction to relativistic heavy ion collisions* (Wiley, New York, 1994)
4. see, for instance: E. Laermann, Nucl. Phys. **A 610**, 1 (1996)
5. L.D. Landau and E.M. Lifshitz, *Fluid mechanics* (Pergamon, New York, 1959)
6. S.R. deGroot, W.A. van Leeuwen, and Ch.G. van Weert, *Relativistic Kinetic Theory* (North-Holland, Amsterdam, 1980)
7. W.A. Hiscock and L. Lindblom, Ann. Phys. **151**, 466 (1983); Phys. Rev. D **31**, 725 (1985); Phys. Rev. D **35**, 3723 (1987)
8. I. Müller, Z. Phys. **198**, 329 (1967)
9. W. Israel, Ann. Phys. **100**, 310 (1976); J.M. Stewart, Proc. Roy. Soc. **A 357**, 59 (1977); W. Israel and J.M. Stewart, Ann. Phys. **118**, 341 (1979)
10. L. Mornas and U. Ornik, Nucl. Phys. **A 587**, 828 (1995)
11. M. Prakash, M. Prakash, R. Venugopalan, and G. Welke, Phys. Rep. **227**, 321 (1993)
12. A.A. Amsden, A.S. Goldhaber, F.H. Harlow, and J.R. Nix, Phys. Rev. C **17**, 2080 (1978); I.N. Mishustin, V.N. Russkikh, and L.M. Satarov, Nucl. Phys. **A 494**, 595 (1989); J. Brachmann, A. Dumitru, J.A. Maruhn, H. Stöcker, W. Greiner, and D.H. Rischke, Nucl. Phys. **A 619**, 391 (1997)
13. To my knowledge, this treatment of the transformation problem is due to B. Kämpfer.
14. V. Schneider *et al.*, J. Comput. Phys. **105**, 92 (1993)
15. A. Harten, P.D. Lax, and B. van Leer, SIAM Rev. **25**, 35 (1983); B. Einfeldt, SIAM J. Numer. Anal. **25**, 294 (1988)
16. M. Holt, *Numerical methods in fluid dynamics*, Springer Series in Comput. Physics (Springer, Berlin, 1977)
17. D.H. Rischke, S. Bernard, and J.A. Maruhn, Nucl. Phys. **A 595**, 346 (1995)
18. R. Courant and K.O. Friedrichs, *Supersonic Flow and Shock Waves* (Springer, New York, 1985)
19. A.M. Taub, Phys. Rev. **74**, 328 (1948)
20. D.H. Rischke, Y. Pürsün, and J.A. Maruhn, Nucl. Phys. **A 595**, 383 (1995)
21. P. Danielewicz and P.V. Ruuskanen, Phys. Rev. D **35**, 344 (1987)
22. A. Chodos, R.L. Jaffe, K. Johnson, C.B. Thorn, and V.F. Weisskopf, Phys. Rev. D **9**, 3471 (1974)
23. D.H. Rischke and M. Gyulassy, Nucl. Phys. **A 597**, 701 (1996)
24. C.M. Hung and E.V. Shuryak, Phys. Rev. Lett. **75**, 4003 (1995)
25. L.D. Landau and S.Z. Belenkii, Uspekhi Fiz. Nauk **56**, 309 (1955)
26. F. Cooper, G. Frye, and E. Schonberg, Phys. Rev. D **11**, 192 (1975)
27. C. Chiu and K.H. Wang, Phys. Rev. D **12**, 272 (1975)
28. J.D. Bjorken, Phys. Rev. D **27**, 140 (1983)
29. G.A. Sod, J. Fluid Mech. **83**, 785 (1977)
30. D.H. Rischke and M. Gyulassy, Nucl. Phys. **A 608**, 479 (1996)
31. U. Heinz, proceedings of the *International Summer School on Correlations and Clustering Phenomena in Subatomic Physics*, Dronten, Netherlands, August 5–16, 1996, e-print archive nucl-th/9609029

32. S. Bernard, D.H. Rischke, J.A. Maruhn, and W. Greiner, Nucl. Phys. **A 625**, 473 (1997)
33. K.A. Bugaev, Nucl. Phys. **A 606**, 559 (1996)
34. C. Anderlik, Z.I. Lazar, V.K. Magas, L.P. Csernai, H. Stöcker, and W. Greiner, e-print archive nucl-th/9808024

The Use of Statistical Mechanics to Describe Hadron Production in High Energy Collisions

Francesco Becattini[1]

Università di Firenze and INFN Sezione di Firenze
Largo E. Fermi 2, I-50125, Florence (Italy)

Abstract. In these lecture notes the application of statistical mechanics to describe hadron production in high energy collisions is reviewed. Special emphasis is given to the necessary assumptions and to point out what can be and what cannot be predicted within this framework. The present status of data analysis is summarized and future tests of the model are outlined; some critical points are addressed.

1 Introduction

The use of statistical mechanics (and thermodynamics) to describe hadron production in pp collisions dates back to the '50s and '60s [1]. Since then much work has been done in this field and various models have been proposed for different kinds of collisions, especially heavy ion collisions as they have been considered for a long time the most natural place where hadron thermalisation could actually take place. In this paper I will show that there is evidence that a statistical description of hadronisation works for a large set of collisions, from e^+e^- to heavy ions, and that definite universal features emerge in the analysis of *elementary collisions* with respect to the more complex heavy ion collisions. A special feature of this approach is the very small number of free parameters needed to reproduce key observables in hadronisation such as the overall particle multiplicities. Moreover, well known facts like baryon to meson suppression naturally arise, with no extra assumption, owing to the interplay between baryon number conservation and the generally heavier baryon masses. This peculiarity of the statistical-mechanical approach is to be compared, for instance, with the popular Monte-Carlo codes used in high energy physics based on implementations of string models, e.g. in e^+e^- collisions [2], which require many more free parameters to reproduce the yields of hadron species. It should be emphasized in advance that the statistical-thermodynamical model alone, in principle, is unable to describe either hadron momentum spectra in high energy elementary collisions, for they are mainly determined by the early perturbative QCD dynamical evolution, or special correlations (e.g. rapidity vs baryon number) unless it is supplemented with further dynamical input. Its predictive capabilities are concerned with global observables, such as particle species multiplicities and correlations, for which the hadronisation process is almost entirely responsible. It is in fact one of the aims of this work to show that those major observables can be made independent of involved dynamical effects and analysed in terms of a pure hadronisation model without using complicated schemes requiring the description of all aspects of hadron production (including spectra etc.) at one time.

2   Basics of the Model

The basic assumption of the model is the existence of a set of hadron gas clusters or fireballs [1] as the final result of a high energy collision. Every cluster has a momentum $\mathbf{P}_i$ and a spacial volume V_i defined in its rest frame as well a mass M_i and a set of quantum numbers $\mathbf{Q}_i^0 = (Q_i, N_i, S_i, C_i, B_i)$ where Q_i is the electric charge, N_i is the baryon number, S_i is the strangeness, C_i is the charm and B_i is the beauty. It should be pointed out that the assumption of a hadron gas in each cluster does not entail that hadrons thermalised via inelastic collisions within the volume V_i, rather that pre-hadronic matter converted into hadrons according to the equiprobability of any multi-hadronic phase space state (Gibbs postulate) where phase space is *locally* defined by the mass and the rest frame volume of the cluster. This is a necessary statement as the emitting source in elementary collisions is believed to be rapidly expanding, thus not allowing thermalisation of a non-equilibrated hadron system; in other words hadrons must be created already at chemical equilibrium [3,4] (except strange hadrons, see Sect. 4).

In order to derive physical observables we start by calculating the canonical partition function of each cluster:

$$Z(T_i, V_i, \mathbf{Q}_i^0) = \sum_{\text{states}} \exp(-E/T_i)\, \delta_{\mathbf{Q},\mathbf{Q}_i^0} \,, \tag{1}$$

where the sum runs over all multihadronic states $\{\mathbf{n}\}$ defined by a set of occupation numbers n_j^k for each species j and for each phase space cell k. The partition function is a Lorentz-invariant one, hence it can be computed in any reference frame; for sake of simplicity we have chosen the cluster rest frame in eq. (1). The $Z(T_i, V_i, \mathbf{Q}_i^0)$ can be worked out by transforming the Kronecker $\delta_{\mathbf{Q},\mathbf{Q}_i^0}$ into an integration:

$$Z(T_i, V_i, \mathbf{Q}_i^0) = \frac{1}{(2\pi)^5} \int_0^{2\pi} \mathrm{d}^5\phi \; \mathrm{e}^{\mathrm{i}\mathbf{Q}_i^0 \cdot \boldsymbol{\phi}} \sum_{\{\mathbf{n}\}} \prod_{j,k} \exp[-n_j^k(\epsilon_j^k/T_i + \mathrm{i}\mathbf{q}_j \cdot \boldsymbol{\phi})] \,. \tag{2}$$

It can be proved that the average multiplicity of the jth hadron species in the ith cluster can now be obtained by multiplying all the exponential factors $\exp[-\epsilon_j^k/T_i - \mathrm{i}\mathbf{q}_j \cdot \boldsymbol{\phi}]$ associated to the jth hadron in eq. (2) by a factor λ_j, then taking the derivative of Z with respect to λ_j for $\lambda_j = 1$ and dividing by Z itself:

---

[1] The words cluster and fireball will be used as synonyms throughout

$$< n_j >_i = \frac{1}{Z(T_i, V_i, \mathbf{Q}_i^0)} \frac{\partial}{\partial \lambda_j} Z(T_i, V_i, \mathbf{Q}_i^0, \lambda_j)\Big|_{\lambda_j=1} . \tag{3}$$

Similarly, the number of particle k-uples can be obtained by taking the higher-order derivatives of Z with respect to similarly defined λ parameters and dividing by Z [3].

In order to derive the overall average multiplicities, the numbers obtained from eq. (3) for each cluster must be summed up. However, the temperatures, the quantum number vectors, the volumes of the clusters, as well as their number N, may fluctuate on an event by event basis, so that the hadron average multiplicities for a given configuration $\{(T_1, V_1, \mathbf{Q}_1^0), \ldots, (T_N, V_N, \mathbf{Q}_N^0)\}$ must be folded with their probabilities of occurrence in order to get the actual observable multiplicities.

As a first step, let us assume that all clusters in an event have the same temperature T and let us group the events having the same set of values (T, V, N) and $V = \sum_{i=1}^{N} V_i$ is the sum of all cluster rest frame volumes; the quantum vector $\mathbf{Q}^0 = \sum_{i=1}^{N} \mathbf{Q}_i^0$ is fixed because it is determined by the initial colliding system. Then, let us define as $w(\mathbf{Q}_1^0, \ldots, \mathbf{Q}_N^0)$ the conditional probability of occurrence of a given configuration $(\mathbf{Q}_1^0, \ldots, \mathbf{Q}_N^0)$ with fixed set of volumes $(V_1, \ldots, V_N)$ and fixed (V, N). Let $f(V_1, \ldots, V_N)$ be the conditional probability of having a set of volumes $(V_1, \ldots, V_N)$ with fixed $V = \sum_{i=1}^{N} V_i$ and let $\rho(V, N)$ be the probability of occurrence of a pair (V, N). Hence:

$$<< n_j >> = \sum_{N=1}^{\infty} \int dT \int dV \, \rho(T, V, N) (\prod_{i=1}^{N} \int dV_i) \, f(V_1, \ldots, V_N)$$

$$\times \sum_{(\mathbf{Q}_1^0, \ldots, \mathbf{Q}_N^0)} w(\mathbf{Q}_1^0, \ldots, \mathbf{Q}_N^0) \sum_{i=1}^{N} < n_j >_i . \tag{4}$$

The functions f, w and ρ are in general unknown. A statistical *ansatz* is taken for w, namely $w(\mathbf{Q}_1^0, \ldots, \mathbf{Q}_N^0)$ is assumed to be the probability of subdividing a global cluster of temperature T, volume V and quantum number vector $\mathbf{Q}^0$ into N clusters with a configuration $(\mathbf{Q}_1^0, \ldots, \mathbf{Q}_N^0)$. Hence, this probability is simply proportional to the number of states contained in the configuration $(\mathbf{Q}_1^0, \ldots, \mathbf{Q}_N^0)$, namely:

$$w(\mathbf{Q}_1^0, \ldots, \mathbf{Q}_N^0) = \frac{\prod_{i=1}^{N} Z(T, V_i, \mathbf{Q}_i^0)}{\sum_{(\mathbf{Q}_1^0, \ldots, \mathbf{Q}_N^0)} \prod_{i=1}^{N} Z(T, V_i, \mathbf{Q}_i^0)} . \tag{5}$$

With this *ansatz*, the sum over all configurations $(\mathbf{Q}_1^0, \ldots, \mathbf{Q}_N^0)$ in the integrand of eq. (4) becomes the average multiplicity of the jth hadron in one single fireball of volume V, temperature T and quantum number vector $\mathbf{Q}^0$. As this quantity depends no longer on the particular set of volumes $(V_1, \ldots, V_N)$

but only on their sum $\sum_{i=1}^{N} V_i \equiv V$ [3] neither on the number of clusters N, while $\mathbf{Q}^0$ is fixed by the initial colliding system, the final expression of the average multiplicity turns out to be:

$$<< n_j >> = \int dT \int dV \, \sigma(T,V) < n_j > (T,V,\mathbf{Q}^0) \,, \qquad (6)$$

where $\sigma(T,V) = \sum_{N=1}^{\infty} \rho(T,V,N)$ and $< n_j > (T,V,\mathbf{Q}^0)$ can be obtained by taking the derivative of the canonical partition function of the whole fireball:

$$< n_j > (T,V,\mathbf{Q}^0) = \frac{1}{Z(T,V,\mathbf{Q}^0)} \frac{\partial}{\partial \lambda_j} Z(T,V,\mathbf{Q}^0,\lambda_j) \Big|_{\lambda_j=1} \,. \qquad (7)$$

It should be stressed that the single fireball with volume V does not actually exist in any of the physical collision events. What is assumed to exist in the physical reality is a set of clusters having different momenta. The single large cluster is only a useful mathematical object whose mathematical existence is owed to the particular choice of the w's probabilities.

If the cluster masses are not large, then their canonical partition function must be replaced with microcanonical ones. This makes calculations more involved but a useful generalisation could be achieved. The possibility of cluster masses fluctuations should be taken into account and one could hopefully end up, by choosing suitable mass fluctuation functions like w in eq. (5), with the microcanonical partition function of a single large fireball as in the previous derivation. This subject is to be studied in more detail.

For large masses M of the single fireball its microcanonical partition function can be tranformed into a canonical one by means of a saddle-point approximation:

$$Z|_{\text{micro}} \to Z|_{\text{can}} = \sum_{\text{states}} e^{-M/T} \delta_{\mathbf{Q},\mathbf{Q}^0} \qquad (8)$$

where the temperature T is related to M and V by the saddle-point equation:

$$M + \frac{\partial}{\partial(1/T)} \log Z|_{\text{can}} = 0 \,. \qquad (9)$$

Anticipating some of the results described in next section, this approximation is indeed possible for all of the elementary collisions examined so far [5,6,3]. In Table 1 a compilation of mean M estimated by using eq. (9) and the fitted T, V and γ_S (see Sect. 4) are compared with the fitted temperatures [6]; in all cases the mean masses turn out to be larger than the temperatures by at least a factor 45, thus justifying the use of the canonical formalism. However, an accurate quantitative estimation of microcanonical corrections is still lacking and it would be valuable to have it in the future.

Table 1: Values of fitted temperatures and estimated mean fireball masses in various high energy collisions

$\sqrt{s}$ (GeV)	T (MeV)	$< M >$ (GeV)
pp collisions		
19.5	190.8 ± 27.4	8.65
23.8	194.4 ± 17.3	9.82
26.0	159.0 ± 9.5	9.57
27.5	169.0 ± 2.1	9.77
p$\bar{\text{p}}$ collisions		
200	175.4 ± 14.8	22.55
546	181.7 ± 17.7	31.73
900	170.2 ± 11.8	36.83
e^+e^- collisions[1]		
29 ÷ 30	163.6 ± 3.6	10.21
34 ÷ 35	165.2 ± 4.4	10.69
42.6 ÷ 44	169.6 ± 9.5	11.89
91.2	160.6 ± 1.7	17.49

1 - The mean masses quoted for e^+e^- collisions are those estimated for light quark events

3   A Brief Summary of the Analysis of Hadron Multiplicities

Aside from the thermodynamical parameters T and V the remaining degrees of freedom reside into possible fluctuations of T and V. If such fluctuations are small, as it is tacitly assumed in most statistical analyses, they can be neglected and a mean value for T and V is taken. In some cases (e.g. refs. [3,7]) the fluctuations of V have been discussed though the actual fit to the data was performed with mean values.

The canonical treatment in statistical mechanics involves the requirement of exact conservation of internal quantities such as the quantum numbers which have been mentioned at the beginning of the previous section. The calculations of partition function within the most general conservation laws related to symmetry groups is a well-known subject [8]. By imposing our quantum number conservation laws associated to simple U(1) groups, we get expressions of observable quantities, in particular hadron multiplicities, differing from the usual grand-canonical thermodynamics by so-called *chemical factors*, which are ratios $Z(\mathbf{Q}^0 - \mathbf{q}_j)/Z(\mathbf{Q}^0)$ of the single-fireball partition function evaluated for two different quantum vectors [6,3]:

$$<< n_j >> = \sum_{n=1}^{\infty} (\mp 1)^{n+1} \gamma_s^{ns_j} z_{j(n)} \frac{Z(\mathbf{Q}^0 - n\mathbf{q}_j)}{Z(\mathbf{Q}^0)} , \qquad (10)$$

where the sign - is for fermions and + for bosons. The functions $z_{j(n)}$ are defined as:

$$z_{j(n)} \equiv (2J_j + 1) \frac{V}{(2\pi)^3} \int d^3p \, \exp\left(-n\sqrt{p^2 + m_j^2}/T\right) =$$
$$(2J_j + 1) \frac{VT}{2\pi^2 n} m_j^2 \, K_2(\frac{nm_j}{T}) . \qquad (11)$$

where J_j is the spin, m_j the mass, $\mathbf{q}_j$ the quantum number vector of the jth hadron; the γ_S factor will be discussed in the next section. For T of the order of 200 MeV only the first term of the above series, corresponding to Boltzmann statistics, can be kept for all hadrons except pions. The formulae (10),(11) yield the *primary j^{th} hadron* multiplicity. In order to fit the free parameters T, V and γ_S it is necessary to let all unstable (or strongly unstable) hadrons decay according to known branching ratios in order to match the experimental measurements.

All details, discussion and results of the canonical analysis of hadron abundances in high energy ($\sqrt{s} > 19.4$ GeV) pp, p$\bar{\text{p}}$ and e^+e^- collisions can be found in refs. [6,3]. Here we just show two plots to demonstrate the good quality of the fits (figs. 1, 2) and a summary plot of the obtained temperatures (fig. 3) including heavy ion collisions at SPS [7]. The apparent constancy of temperatures extracted in different kinds of collisions is an intriguing result indicating universality behaviour of hadronisation. Furthermore, a constant T suggests that hadronisation occurs at a particular value of local energy density (see also Sect. 7).

For asymptotically large volumes (thus multiplicities) the chemical factors in eq. (10) reduce to the usual grand-canonical fugacities [3]. An example of this is shown in fig. 4 where the neutron chemical factor in a completely neutral hadron gas is plotted as a function of the volume [3]: it goes to 1 only at asymptotically large volumes. The effect of the small volume on statistical particle production (so called canonical suppression or enhancement) has been studied in great detail [9] and it is a very important effect in the analysis of elementary collisions [5,6,3]. Due to the large involved multiplicities, grand-canonical calculations naturally apply to heavy ion collisions [10]. In the actual fits, the introduction of one more free parameter, usually the baryon chemical potential, is required because the number of nucleons participating in the collision is not known *a priori* but has to be measured.

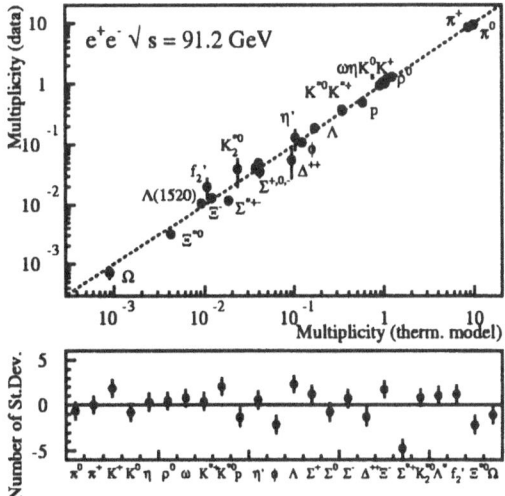

Fig. 1: Comparison between fitted (horizontal axis) and measured (vertical axis) multiplicities in e^+e^- collisions at a centre of mass energy of 91.2 GeV. The dashed line is the quadrant bisector. Below: residual distribution. Taken from ref.[6]

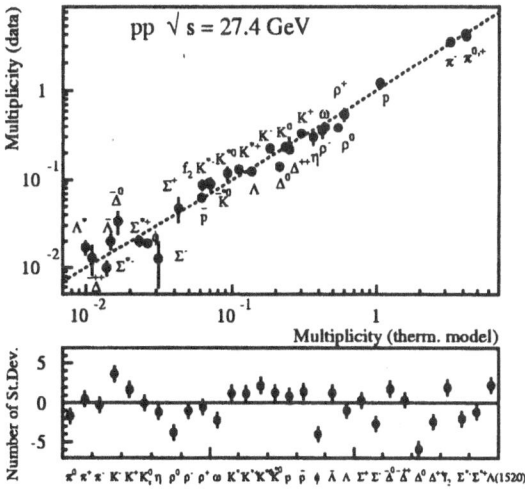

Fig. 2: Comparison between fitted (horizontal axis) and measured (vertical axis) multiplicities in pp collisions at a centre of mass energy of 27.4 GeV. The dashed line is the quadrant bisector. Below: residual distribution. Taken from ref.[3]

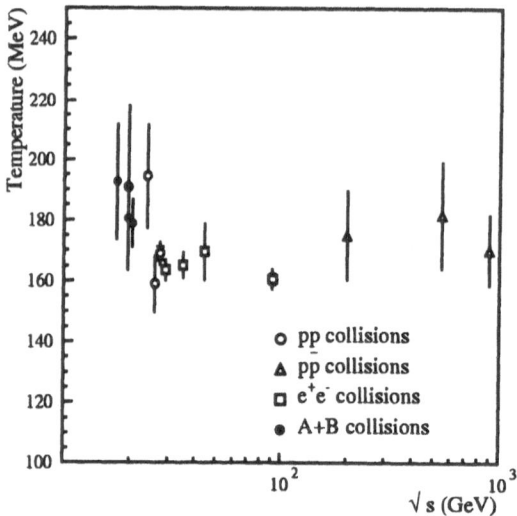

Fig. 3: Fitted temperatures in e^+e^-, pp, p$\bar{\text{p}}$ and heavy ion collisions at SPS. Taken from ref. [11]

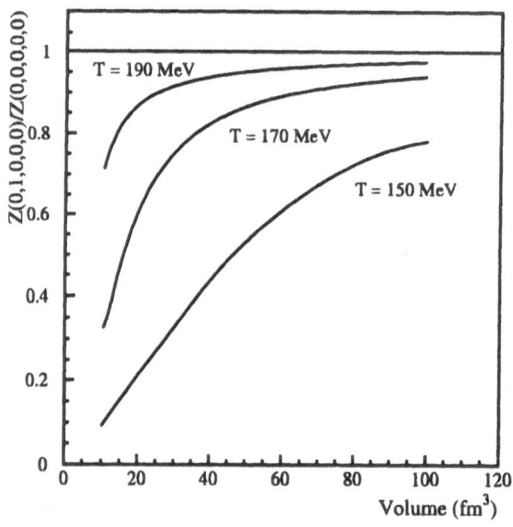

Fig. 4: Neutron chemical factor in a hadron gas as a function of the volume for different temperatures. Taken from ref. [3]

4   Strangeness Suppression

The statistical model described above needs to be supplemented with an extra strangeness suppression factor, which is formally beyond a pure statistical hadronic phase space model, in order to reproduce experimental data. This suppression is implemented by a factor γ_S which is introduced in the partition function as a fugacity related to the valence strange quark content of the hadron [12]. However, this way of regarding γ_S apparently confines its validity to a grand-canonical framework. On the other hand, it is possible to define γ_S in a more general way which is independent of the adopted statistical formalism, i.e. microcanonical, canonical or grand-canonical.

Indeed, for a certain multihadronic state $n_1, \ldots, n_K$ (n_1 is the number of hadrons belonging to species 1, ..., n_K is the number of hadrons belonging to species K) originating from the collision, its phase-space probability is multiplied by a factor γ_S powered to the number of strange+antistrange quarks whose creation out of the vacuum is needed in order to set up that state; in other words, those which have to be newly produced and do not come from the colliding particles. Following this definition, in a canonical framework, the probability P of realizing a multihadronic state $\{n_1, \ldots, n_K\}$ in a collision whose initial state does not have any strange quarks is:

$$P \propto \exp(-E/T)\, \gamma_S^{\sum_{j=1}^{K} n_j |s_j|}\, \delta_{\mathbf{Q},\mathbf{Q}^0}\,, \tag{12}$$

where $|s_j|$ is the number of valence strange+antistrange quarks contained in the jth hadron. This probability can be worked out to calculate the canonical partition function, eventually yielding:

$$Z(\mathbf{Q}^0) = \frac{1}{(2\pi)^5} \int d^5\phi\, e^{i\,\mathbf{Q}^0 \cdot \phi}$$
$$\times \exp\left[V \sum_j \frac{(2J_j+1)}{(2\pi)^3} \int d^3p\, \log\left(1 \pm \gamma_s^{s_j} e^{-\sqrt{p^2+m_j^2}/T_i - i\mathbf{q}_j \cdot \phi}\right)^{\pm 1}\right]. \tag{13}$$

which is exactly the same partition function [6,3] obtained by introducing γ_S as a fugacity. The definition in eq. (12) can be easily generalized within the microcanonical or grand-canonical framework. Furthermore, it should be emphasized that this definition is more general and more appropriate for collisions with initial strange quarks (for instance K p) in which the use of the same parition function (13) obtained for colliding systems devoid of valence strange quarks would lead to odd results. In this special case, the probability (12) becomes:

$$P \propto \exp(-E/T)\, \gamma_S^{\sum_{j=1}^{K} n_j |s_j| - |S|}\, \delta_{\mathbf{Q},\mathbf{Q}^0}\,, \tag{14}$$

where $|S|$ is the initial absolute strangeness, so that the canonical partition function will differ by a factor $1/\gamma_S^{|S|}$ from the (13). As already mentioned,

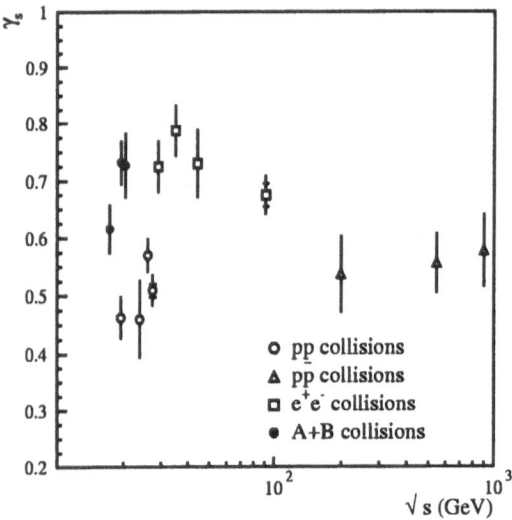

Fig. 5: Fitted γ_S in e^+e^-, pp, p$\bar{\text{p}}$ and heavy ion collisions at SPS. Taken from ref. [11]

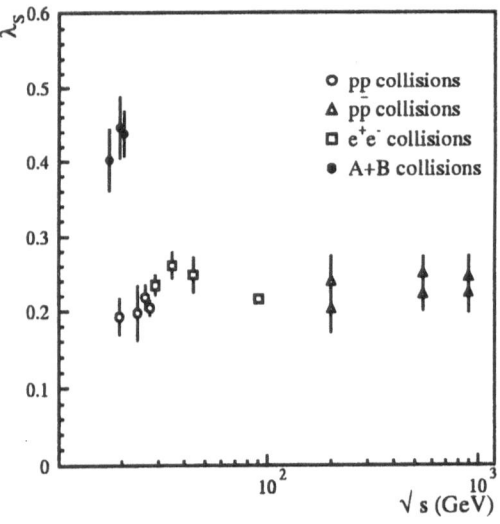

Fig. 6: Estimated $\lambda_S = 2\langle s\bar{s}\rangle/(\langle u\bar{u}\rangle + \langle d\bar{d}\rangle)$ in e^+e^-, pp, p$\bar{\text{p}}$ and heavy ion collisions at SPS. Taken from ref. [11]

in actual analyses of hadron multiplicities, γ_S is a free parameter besides V and T. However, there are considerable clues [7,13] that γ_S is not the most appropriate quantity to parametrise strangeness suppression because it turns out to be different in e^+e^-, $p\bar{p}$ and pp collisions (see fig. 5). On the contrary, the universality of hadronisation in those collisions in the strangeness sector shows up when plotting the best estimate of the ratio λ_S between newly produced $s\bar{s}$ pairs and half the sum of newly produced $u\bar{u}$ and $d\bar{d}$ pairs (see fig. 6) which has a pretty constant value of about 0.2. The model predictions about primary production rate of hadrons allow to make such a valence quark counting before strong decays take place. This result suggests that hadronisation in all kinds of elementary collisions works in a such a way that the average number of strange quarks picked from the vacuum is in a constant ratio with u or d quarks. Even more interesting is the fact that λ_S is definitely higher in heavy ion collisions at SPS energies by a almost factor 2 whereas γ_S has nearly the same value as in e^+e^- collisions. This gap is a clear indication that something different happens in heavy ion collisions as far as strangeness production is concerned.

The constancy of λ_S in elementary collisions indicates that the best way of parametrizing strangeness suppression would amount to fix the *mean* absolute value of strangeness with possible superimposed fluctuations. The mathematical translation of this idea requires the extension of the demanded conservation laws from 5 (Q, N, S, C, B) to 6 $(Q, N, S, C, B, |S|)$, and, consequently, the extension of the integration from 5-dimensional to 6-dimensional in eq. (13) in order to calculate the partition functions needed to evaluate chemical factors in eq. (10). The actual calculation would be even more involved because the absolute value of strangeness certainly undergoes fluctuations event by event, so that the number of integrals to be computed would turn out to be very large. The implementation of this new kind of calculation is a point deserving detailed investigation, especially because it might lead to a further qualitative improvement of thermal canonical fits. Furthermore, it would be instructive to understand to what extent γ_S acts anyhow as a good parametrisation of strangeness suppression though less fundamental than λ_S.

5   Multiplicity Distributions

Not only hadron species average multiplicities but also the probability of having k-uples of different hadrons and, as a consequence, multiplicity distributions, can be reduced to a folding integral like in eq. (6) where the set of clusters is replaced by a single fireball. To prove that, an argument can be used which is similar to that used for the average multiplicities; for a fixed configuration $(\mathbf{Q}_1^0, \ldots, \mathbf{Q}_N^0)$, volumes $(V_1, \ldots, V_N)$, number of clusters N, the conditional probability P_c of getting a k-uple $\{n_1, \ldots, n_K\}$ is given by :

$$P_c(\{n_1,\ldots,n_K\}) = (\prod_{j=1}^{K} \frac{1}{2\pi i} \oint \frac{d\lambda_j}{\lambda_j^{n_j+1}}) \prod_i \frac{Z_i(V_i,\mathbf{Q}_i^0,\lambda_1,\ldots,\lambda_K)}{Z_i(V_i,\mathbf{Q}_i^0)}, \quad (15)$$

where the λ_j are the factors defined in Sect. 2. This expression follows from a very general statement about generating functions and from the chain of equations:

$$\sum_{\{n_1,\ldots,n_K\}} P(\{n_1,\ldots,n_K\})\lambda_1^{n_1}\ldots\lambda_K^{n_K} =$$

$$\sum_{\{n_1,\ldots,n_K\}} (\sum_{\text{states}} P(\text{state})|_{\text{fixed}\{n_1,\ldots,n_K\}})\lambda_1^{n_1}\ldots\lambda_K^{n_K} =$$

$$\sum_{\text{states}} P(\text{state})\lambda_1^{n_1}\ldots\lambda_K^{n_K} =$$

$$\prod_{i=1}^{N} Z_i(V_i,\mathbf{Q}_i^0,\lambda_1,\ldots,\lambda_K)/Z_i(V_i,\mathbf{Q}_i^0) \quad (16)$$

implying that $\prod_{i=1}^{N} Z_i(V_i,\mathbf{Q}_i^0,\lambda_1,\ldots,\lambda_K)/Z_i(V_i,\mathbf{Q}_i^0)$ is indeed the generating function of the distribution P_c. Hence, the observable multihadronic probability distribution $P(\{n_1,\ldots,n_K\})$ will be the convolution:

$$P(\{n_1,\ldots,n_K\}) = \sum_{N=1}^{\infty} \int dT \int dV\, \rho(T,V,N)(\prod_{i=1}^{N}\int dV_i)\, f(V_1,\ldots,V_N)$$

$$\times \sum_{(\mathbf{Q}_1^0,\ldots,\mathbf{Q}_N^0)} w(\mathbf{Q}_1^0,\ldots,\mathbf{Q}_N^0)P_c(\mathbf{Q}_1^0,\ldots,\mathbf{Q}_N^0). \quad (17)$$

By using eq. (15) and the probabilities (5) the sum over all configurations becomes the multiple integral of the single fireball generating function:

$$P(\{n_1,\ldots,n_K\}) = \int dT \int dV\, \sigma(T,V)$$

$$\times(\prod_{j=1}^{K}\frac{1}{2\pi i}\oint\frac{d\lambda_j}{\lambda_j^{n_j+1}})\frac{Z(V,\mathbf{Q}^0,\lambda_1,\ldots,\lambda_K)}{Z(V,\mathbf{Q}^0)}. \quad (18)$$

This proof can be repeated in the canonical framework as well.
To summarize, in order to study not only average multiplicities but also all *global correlations* (i.e. independent of momenta) among particles at any order, the use of a single cluster having mass, volume and quantum numbers equal to the sum of masses, volumes and quantum numbers of the produced

clusters is still valid, provided that the probabilities w's are assumed to be those in eq. (5). Even with this simplifying assumption, multiplicity distributions are affected by possible fluctuations of T and V. For fixed T and V the fluctuations of single hadron species are not independent of other species due to conservation laws. Only in the high multiplicity regime, where conservation laws are weakened, hadron multiplicity distribution become poissonian. In ref. [14] a study of charged particle multiplicity distributions in e^+e^- collisions has been performed with the statistical-thermodynamical model by neglecting V or T fluctuations. A fairly good agreement has been found with the data though some discrepancy emerged suggesting that the fixed V approximation is not fully satisfactory.

6   Heavy Flavours and Spectra

The statistical-thermal model gives rise to some other predictions concerning various physical observables. One prediction which turned out to be in striking agreement with the data is the relative abundance of heavy flavoured hadrons [15] in e^+e^- collisions (see Table 2). In order to produce statistical-thermal predictions of heavy flavoured hadron production rates it is necessary to assume that primarily created heavy quark pairs (like in Z boson decay) do not reannihilate and appear as open flavours in the final hadrons, according to experimental observations. This amounts to a modification of the partition function in a fraction of events whose value is used as input. Specifically, in $e^+e^- \rightarrow b\bar{b}$ events, the canonical partition function is taken to be [3]:

$$Z = Z_1(\mathbf{Q}^0) - Z_2(\mathbf{Q}^0, |B| = 0) , \qquad (19)$$

where $|B|$ is the absolute value of beauty and:

$$Z_1(\mathbf{Q}^0) = \sum_{states} \exp(-E/T)\, \delta_{\mathbf{Q},\mathbf{Q}^0}$$
$$Z_2(\mathbf{Q}^0, |B| = 0) = \sum_{states} \exp(-E/T)\, \delta_{\mathbf{Q},\mathbf{Q}^0}\, \delta_{|B|,0} \qquad (20)$$

implying that states devoid of bottom quarks are excluded from the sum. The second term in eq. (19) involves a 6-dimensional integration as the absolute value of beauty is treated as a new independent quantum number to be conserved in the multihadronic state. It is worth deriving eq. (19) on the basis of the same argument about the reduction of a set of clusters to one single fireball which has been used in Sect. 2. The statistical *ansatz* of the clusters configuration probabilities, like in eq. (5) leads to the following partition function:

$$\sum_{\{\mathbf{Q}_i^0, |B_i|\}; \sum_i |B_i| > 0} \delta_{\mathbf{Q}^0, \sum_i \mathbf{Q}_i^0} \prod_{i=1}^{N} Z(\mathbf{Q}_i^0, |B_i|) \qquad (21)$$

where $\{\mathbf{Q}_i^0, |B_i|\} \equiv (\mathbf{Q}_1^0, |B_1|, \dots, \mathbf{Q}_N^0, |B_N|)$. The above sum can be equivalently written as a difference:

$$\sum_{|B|=0}^{\infty} \sum_{\{\mathbf{Q}_i^0, |B_i|\}} \delta_{\mathbf{Q}^0, \sum_i \mathbf{Q}_i^0} \prod_{i=1}^{N} Z(\mathbf{Q}_i^0, |B_i|) \, \delta_{|B|, \sum_i |B_i|} \, -$$

$$\sum_{\{\mathbf{Q}_i^0, |B_i|\}} \delta_{\mathbf{Q}^0, \sum_i \mathbf{Q}_i^0} \prod_{i=1}^{N} Z(\mathbf{Q}_i^0, |B_i|) \, \delta_{0, \sum_i |B_i|} \, . \qquad (22)$$

The first term in eq. (22) becomes the partition function of a single fireball:

$$\sum_{|B|=0}^{\infty} Z(\mathbf{Q}^0, |B|) = Z_1(\mathbf{Q}^0) \qquad (23)$$

and so does the second term which becomes the second term in eq. (20). The actual values of T and V extracted from the fits to the examined high energy elementary collisions predict an overwhelmingly low pure statistical production of heavy flavoured hadrons owing to the high value of their mass compared to the temperature [5]. The predicted rates are around 10^{-8}/event for charmed hadrons and 10^{-20} for bottomed hadrons in e^+e^- collisions at $\sqrt{s} = 91.2$ GeV. In fact, advantage is taken of the smallness of the thermal functions z_j in eq. (11) to perform analytical integrations in the variables ϕ linked to conservation of charm, beauty and their absolute values, by means of a first-order power expansion [5,3]. This procedure leads to much simpler expressions for the abundances of primary heavy flavoured hadrons normalized to the corresponding heavy flavoured quark pair production rate [3]:

$$<< n_j >> = \gamma_s^{s_j} z_j \frac{\sum_i \gamma_S^{s_i} z_i \zeta(\mathbf{Q}^0 - q_j - q_i)}{\sum_{i,k} \gamma_S^{s_i} \gamma_S^{s_k} z_i z_k \zeta(\mathbf{Q}^0 - q_i - q_k)} \, , \qquad (24)$$

where the ζ are now reduced partitions functions computed without heavy flavours implying only a 3-dimensional integration.

Unlike global observables such as multiplicities and multiplicity distributions, momentum spectra cannot be predicted without supplementing this model with further dynamical assumptions. Indeed, at high centre of mass energies, momentum spectra are mainly determined by cluster collective momentum reflecting early QCD dynamics, as it has been observed in e^+e^- collisions. This results from the fact that, in the cluster rest frame, particles emerge at a momentum scale of the order of T, namely some hundreds MeV, whereas clusters have much higher momenta. As a consequence, hadronisation brings about only a little smearing of the original dynamical distributions. However, this hadronisation 'noise' could be observable if all clusters had momenta along one direction, as it seems to be in pp collisions along the beam

Table 2: Predictions of heavy flavoured hadron abundances at $\sqrt{s} = 91.2$ GeV obtained by the T, V, γ_S parameters fitted from light flavoured hadron abundances. The B_s^{**} prediction is affected by the interpretation of the observed peaks as four different states or two different states (within brackets). Taken from ref. [15]

Hadron	Prediction	Measured	Residual
D^+	0.0926	0.087 ± 0.008	-0.67
D^0	0.233	0.227 ± 0.012	-0.50
D_s	0.0579	0.066 ± 0.010	+0.81
D^{*+}	0.108	0.0880 ± 0.0054	-3.7
D_s^+/c-jet	0.103	0.128 ± 0.027	+0.92
D_1/c-jet	0.0347	0.038 ± 0.009	+0.37
D_2^*/c-jet	0.0471	0.135 ± 0.052	+1.7
D_{s1}/c-jet	0.00536	0.016 ± 0.0058	+1.8
B^0/b-jet	0.412	0.384 ± 0.026	-1.1
B^*/B	0.692	0.747 ± 0.067	+0.82
B^*/b-jet	0.642	0.65 ± 0.06	+0.13
B_s/b-jet	0.106	0.122 ± 0.031	+0.52
$B_{u,d}^{**}$/b-jet	0.206	0.26 ± 0.05	+1.0
B^{**}/B	0.251	0.27 ± 0.06	+0.32
B_s^{**}/b-jet	0.021(0.011)	0.048 ± 0.017	+1.6
B_s^{**0}/B^+	0.026(0.013)	0.052 ± 0.016	+1.6
Λ_c^+	0.0248	0.0395 ± 0.0084	+1.7
b-baryon/b-jet	0.0717	0.115 ± 0.040	+1.1
$(\Sigma_b + \Sigma_b^*)$/b-jet	0.0404	0.048 ± 0.016	+0.48
$\Sigma_b/(\Sigma_b^* + \Sigma_b)$	0.411	0.24 ± 0.12	-1.4

line or in e^+e^- collisions along the jet axis. In this case the transverse momentum p_T spectrum of any hadron except pions is predicted to be proportional to:

$$\sqrt{p_T^2 + m^2} \; p_T \, K_1(\frac{\sqrt{p_T^2 + m^2}}{T}) \tag{25}$$

at temperatures of the order of 200 MeV; for pions the formula is more complicated due to non-negligible Bose-Einstein statistics effects. Indeed, spectra like (25) have been observed for a long time in pp collisions but a conclusive data analysis is still lacking because most final detectable hadrons actually emerge from strong decays, making it very difficult to disentangle the primary component, whose spectrum should behave like (25), from the secondary one. Furthermore, even primary hadrons p_T spectra are broadened by possible transverse momenta of the clusters related to hard gluon radiation or minijets etc. To summarize, this prediction is very difficult to test, yet it would be certainly worthwile as a definite connection between spectra and abundances in hadronisation would be established.

An interesting issue related to the spectra concerns the consistency between the assumed probabilities w's and the fact that rapidity distributions in pp collisions, for instance, have very different shapes for baryons and mesons [16]. As it has been assumed that the probabilities of quantum number configurations w's are those arising by the splitting of a whole fireball, this maximal random choice may seem apparently inconsistent with the experimental observation of different spectra for differently 'charged' particles. Indeed this is not the case; let $p(\mathbf{Q}_1^0, \ldots, \mathbf{Q}_N^0)$ be the actual probabilities for a set of N clusters having the same volume and let us order the clusters in rapidity $y_1 > \ldots > y_N$. In this case, in a canonical model where all clusters have the same temperature, the probabilities $w(\mathbf{Q}_1^0, \ldots, \mathbf{Q}_N^0)$ chosen according to an equation like eq. (5) are symmetric:

$$w(\mathbf{Q}_{\sigma(1)}^0, \ldots, \mathbf{Q}_{\sigma(N)}^0) = w(\mathbf{Q}_1^0, \ldots, \mathbf{Q}_N^0) \qquad (26)$$

for any permutation σ of the integers $1, \ldots, N$. Therefore, if $p(\mathbf{Q}_1^0, \ldots, \mathbf{Q}_N^0)$ are the actual weights, in order that our previous derivations hold, the condition to be fulfilled is:

$$w[\mathbf{Q}_1^0, \ldots, \mathbf{Q}_N^0] = \frac{1}{N!} \sum_\sigma p(\mathbf{Q}_{\sigma(1)}^0, \ldots, \mathbf{Q}_{\sigma(N)}^0) , \qquad (27)$$

where the square brackets mean that the set $[\mathbf{Q}_1^0, \ldots, \mathbf{Q}_N^0]$ is a not–ordered one. This condition is weaker than a strict equality between $w(\mathbf{Q}_1^0, \ldots, \mathbf{Q}_N^0)$ and $p(\mathbf{Q}_1^0, \ldots, \mathbf{Q}_N^0)$. Thus, the consistency between previously obtained expressions of hadron multiplicities (10) and rapidity distributions can be achieved in a canonical model by generating clusters with the same volume and temperature with random quantum numbers (provided that their sum fulfills the initial state constraint), then giving them a suitably chosen rapidity that is allowed, on the basis of permutational symmetry (26), to be strongly correlated with their quantum numbers. An example of this procedure is a toy Monte-Carlo calculation shown in figure 6 where outcoming rapidity distributions of protons and pions in pp collisions at $\sqrt{s} = 27$ GeV are shown. Although they look very different, overall particle multiplicities keep the same value determined by the simple statistical *ansatz* for the w's.

7 Outlook

The application of a statistical-thermodynamical model of hadronisation to the analysis of hadron multiplicities allows to reproduce accurately hadron

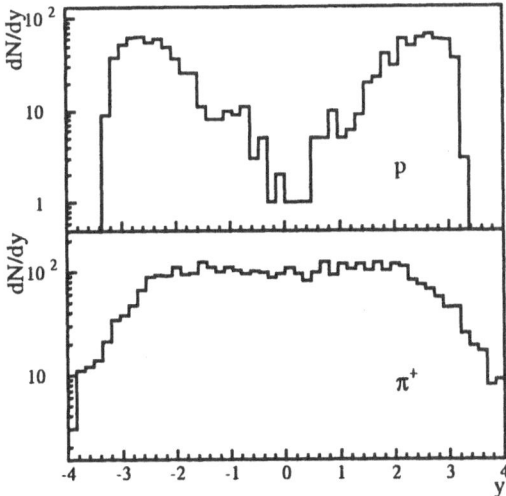

Fig. 7: Example of a Monte-Carlo calculation of rapidity distributions in pp collisions at $\sqrt{s} = 27.4$ GeV for protons (above) and pions (below)

multiplicities with only three free parameters. The most interesting results emerged from this study are some universality features such as the constancy of T and λ_S in elementary collisions (figs. 3, 6). Particularly the constancy of temperature suggests that the local energy density (or some closely related physical quantity) has a constant critical value at the hadronisation [3]. In other words, the local comoving volume available for hadron production, hence the local phase space, is determined by the mass of hadronising clusters; the temperature of the resulting hadron gas is thereby constant. These results are also very useful in the perspective of the study of hadron production in heavy ion collisions.

More tests of this model are needed and are on the way; for the average multiplicities and multiplicity distributions at lower energies, where a microcanonical treatment will be presumably necessary, to check the onset of such statistical behaviour; for the transverse momenta spectra in high energy collisions, to put in evidence the contribution of hadronisation and, hopefully, establish a link between kinematics and chemical composition.

Acknowledgements

I warmly thank all the organizers of the XI Chris Engelbrecht summer school, especially J. Cleymans, for their friendliness and their kind hospitality. Special thanks to J. Baldry for the manuscript revision.

References

1. E. Fermi: Progr. Theor. Phys. **5** (1950) 570
   L. D. Landau: Izv. Akad. Nauk SSSR, Ser. Fiz. **17** (1953) 51
   R. Hagedorn: Nuov. Cim. **15** (1960) 434
2. T. Sjöstrand: Comp. Phys. Comm. **28** (1983) 229
   T. Sjöstrand: Pythia 5.7 and. Jetset 7.4, CERN/TH 7112/93 (1993)
3. F. Becattini, U. Heinz: Z. Phys. C **76** (1997) 269.
4. U. Heinz: talk given at Quark Matter 97, December 1997, Tsukuba (Japan), proc. in press (nucl/th 9801107)
5. F. Becattini: Z. Phys. C **69** (1996) 485.
6. F. Becattini: Proc. of XXXIII Eloisatron Workshop on *Universality feature in multihadron production and the leading effect* (1996), p. 74 (hep/ph 9701275)
7. F. Becattini, M. Gaździcki, J. Sollfrank: hep/ph 9710529, Eur. J. Phys. C in press
8. L. Turko: Phys. Lett. B **104** (1981) 153
   M. I. Gorenshtein et al.: Phys. Lett. B **123** (1983) 437
   R. Hagedorn, K. Redlich: Z. Phys. C **27** (1985) 541
9. see for instance:
   J. Cleymans et al.: Phys. Rev. C **57** (1998) 3319 and references therein
   A. Muronga, J. Cleymans: nucl/th 9709045 and references therein
10. These are some of the grand-canonical analyses of hadron multiplicities in heavy ion collisions:
    J. Sollfrank et al.: Z. Phys. C **61** (1994) 659
    J. Cleymans et al.: Z. Phys. C **74** (1997) 319
    P. B. Munzinger et al.: Phys. Lett. B **344** (1995) 43
    P. B. Munzinger et al.: Phys. Lett. B **365** (1996) 1
    J. Letessier, J. Rafelski and A. Tounsi: Phys. Lett. **410** (1997) 315
    G. D. Yen, M. I. Gorenstein: nucl/th 9808012
    A. S. Kapoyannis, C. N. Ktorides and A. Panagiotou: hep/ph 9806241, submitted to Phys. Rev. D
11. F. Becattini: talk given at Quark Matter 97, December 1997, Tsukuba (Japan), proc. in press
12. J. Rafelski: Phys. Lett. B **262** (1991) 333
    C. Slotta, J. Sollfrank and U. Heinz: Proc. of *Strangeness in Hadronic Matter* (1995) p. 462
13. F. Becattini: talk given at *Strangeness in Quark Matter 98*, July 1998, Padova (Italy)
14. F. Becattini, A. Giovannini, S. Lupia: Z. Phys. C **72** (1996) 491
15. F. Becattini: Proc. of *Strangeness in Quark Matter 97*, J. Phys. G **23** (1997) 1933
16. M. Aguilar-Benitez et al, NA27 coll.: Z. Phys. C **50** (1991) 405

Introduction to Light Cone Field Theory and High Energy Scattering

Raju Venugopalan[1]

Niels Bohr Institute,
Blegdamsvej 17, Copenhagen, Denmark, DK–2100

Abstract. In this set of four lectures, we provide an elementary introduction to
light cone field theory and some of its applications in high energy scattering.

1 Introduction

In these lectures, we will attempt to provide a "hands on" introduction to
some of the ideas and methods in light cone field theory and its application
to high energy scattering. Light cone quantization as an approach to study
the Hamilton dymanics of fields was first investigated by Dirac, who pointed
out several of its elegant features in a landmark paper [1]. It was first applied
to high energy physics in the 60's in the context of current algebra [2]. Light
cone field theory currently finds applications in most areas of high energy
physics, from perturbative QCD to string theories.

The elegance and simplicity of the light cone approach results from the
analogy of relativistic field theories quantized on the light cone to non–
relativistic quantum mechanics. In fact, this correspondence runs deep and
it was shown by Susskind that there is an exact isomorphism between the
Galilean subgroup of the Poincaré group and the symmetry group of two dimensional quantum mechanics [3]. Furthermore, as was first shown by Weinberg [4], the vacuum structure of field theories simplifies greatly in the infinite
momentum limit. The combination of the non–relativistic kinematics of light
cone field theories as well as their simple vacuum structure, has given rise
to the belief that potential methods of quantum mechanics can be applied
to field theories quantized on the light cone. This observation is at the heart
of recent attempts to understand bound state problems in QCD in the light
cone formalism [5]. Indeed, beginning with the t'Hooft model [6] for mesons in
1+1–dimensional large N_c QCD, which made use of the light cone formalism,
there have been many attempts to study confinement and chiral symmetry
breaking in this approach (see Ref. [7] and references therein).

Light cone field theory also provides much of the intellectual support for
the intuitive quark–parton picture of high energy scattering. Frequently, the
phrases 'the theory of strong interactions, QCD' and the 'quark–parton picture of strong interactions' are used interchangeably. However, it is only in
light cone quantization (and light cone gauge) that the quark–parton structure of QCD is manifest and multi–parton Fock states can be constructed as

eigenstates of the QCD Hamiltonian [8]. One can therefore construct Lorentz invariant light cone wavefunctions– a fact which has been particularly useful in the study of exclusive processes in QCD [9]. Further, in deeply inelastic scattering, the experimentally measured structure functions are simply related (in leading twist) to the light cone quark distribution functions. The partonic picture of light cone quantum field theory was demonstrated very clearly in the papers of Kogut and Soper [10] and of Bjorken, Kogut and Soper [11].

The goal of these lectures is to illustrate both of the above points, the attractive features of light cone field theory and its applications to high energy scattering, in the simplest possible fashion by working out concrete examples. In the first lecture, we begin by introducing the light cone notation and the two component formalism. We then define the light cone Fock states and their equal light cone time commutation relations. We conclude by discussing the structure of the Poincaré group and demonstrate the above mentioned isomorphism to two dimensional quantum mechanics. In the second lecture, we explicitly derive the light cone QCD Hamiltonian in the two component formalism making use of the light cone constraint equations. It is shown that the Hamiltonian can be expressed as the sum of non–interacting and "potential" terms. For simplicity, in lecture three, we specialize to the case of QED and use the form of the Hamiltonian derived in lecture 2 to illustrate the parton picture of high energy scattering. In particular, we study high energy scattering off an external potential in the eikonal approximation in QED. In the fourth and final lecture we show how Bjorken scaling can be derived in QCD using the light cone commutation relations and briefly discuss the relation of light cone distribution functions to structure functions.

There are several reviews that the reader may study to learn more about the subject. An introductory review which also includes a guide to the literature for beginners is that by Harindranath [12]. Another introductory review which stresses recent advances is that by Burkardt [7]. The most recent and comprehensive review of the subject is by Brodsky, Pauli and Pinsky [13]. A part of our lectures relies heavily on the classic papers of Kogut and Soper [10] and Bjorken, Kogut and Soper [11]. The reader should keep in mind that a wide variety of conventions are in use in the literature. Some of these are discussed in the review of Brodsky, Pauli and Pinsky.

The lectures below were delivered at the Cape Town lecture school and for spacetime reasons are the "short" form of lectures delivered previously at the University of Jyväskylä international summer school. The topics that were omitted in the short version include light cone perturbation theory, the renormalization group and the operator product expansion, and small x physics. The longer version of these lectures will be published separately at a later date [14].

2 Light Cone Quantization and the Light Cone Algebra

We begin by defining our convention and notations. Our metric here is the $+2$ metric $\hat{g}^{\mu\nu} = (-,+,+,+)$. Note: for my convenience (and unfortunately, your inconvenience) I may change notations in the latter lectures. But you will have fair warning! The gamma matrices in usual space–time co–ordinates are denoted by carets. In the chiral representation,

$$\hat{\gamma}^0 = \begin{pmatrix} 0 & I \\ I & 0 \end{pmatrix} \; ; \; \hat{\gamma}^i = \begin{pmatrix} 0 & \sigma^i \\ -\sigma^i & 0 \end{pmatrix} \; ; \; \hat{\gamma}^5 = \begin{pmatrix} I & 0 \\ 0 & -I \end{pmatrix} ,$$

and $\{\hat{\gamma}^\mu, \hat{\gamma}^\nu\} = -2\hat{g}^{\mu\nu}$. Above, $\sigma^i, i = 1, 2, 3$ are the usual 2×2 Pauli matrices and I is the 2×2 identity matrix. In light cone co–ordinates, $\gamma^\pm = (\hat{\gamma}^0 \pm \hat{\gamma}^3)/\sqrt{2}$ and $\{\gamma^\mu, \gamma^\nu\} = -2g^{\mu\nu}$, where $g^{++} = g^{--} = 0$, $g^{+-} = g^{-+} = -1$. Also, $g_{t_1,t_2} = 1$ with $t_1, t_2 = 1, 2$ denoting the two transverse co–ordinates. We define $x^\mu \equiv (x^0, x^1, x^2, x^3) = (t, \boldsymbol{x})$ and

$$x^\pm = \frac{(t+z)}{\sqrt{2}} \; ; \; \partial_\pm \equiv \frac{\partial}{\partial x^\pm} = \frac{1}{\sqrt{2}}(\partial_t \pm \partial_z) \; ; \; A^\pm = \frac{(A^0 \pm A^z)}{\sqrt{2}}. \qquad (1)$$

Note for instance that in this convention $A_+ = -A^-$ and $A_t = +A^t$. Also, $q^2 = -2q^-q^+ + q_t^2$. Hence, a "space–like" q^2 implying large space–like components would correspond to $q^2 > 0$.

We now define the projection operators

$$\alpha^\pm = \frac{\hat{\gamma}^0 \gamma^\pm}{\sqrt{2}} \equiv \frac{\gamma^\mp \gamma^\pm}{2}, \qquad (2)$$

which project out the two component spinors $\psi_\pm = \alpha^\pm \psi$, [10]

$$\psi_+ = \begin{pmatrix} 0 \\ \psi_2 \\ \psi_3 \\ 0 \end{pmatrix} \; ; \; \psi_- = \begin{pmatrix} \psi_1 \\ 0 \\ 0 \\ \psi_4 \end{pmatrix}, \qquad (3)$$

where $\psi_1, \cdots \psi_4$ are the four components of ψ. It follows from the above that $\psi_+ + \psi_- = \psi$.

Some relevant properties of the projection operators $\alpha^\pm$ are

$$(\alpha^\pm)^2 = \alpha^\pm \; ; \; \alpha^\pm \alpha^\mp = 0 \; ; \; \alpha^+ + \alpha^- = 1 \; ; \; (\alpha^\pm)^\dagger = \alpha^\pm. \qquad (4)$$

We can use these to show that

$$\alpha^\pm \psi_\mp = 0 \; ; \; \alpha^\pm \psi_\pm = \psi_\pm \; ; \; \alpha^\pm \hat{\gamma}^0 = \frac{1}{2}\hat{\gamma}^0 \alpha^\pm \; ; \; \alpha^\pm \gamma_\perp = \gamma_\perp \alpha^\pm. \qquad (5)$$

We will make liberal use of these identities in deriving the light cone Hamiltonian in lecture 2.

A particular property of light cone quantization is that it is the two component spinor ψ_+ above that is the dynamical spinor in the light cone QCD Hamiltonian P_{QCD}^-. Interestingly, the same feature is observed for fermion fields which obey equal time commutation relations when they are boosted to the infinite momentum frame. The dynamical spinors ψ_+ are defined in terms of creation and annihilation operators as

$$\psi_+ = \int_{k^+>0} \frac{d^3 k}{2^{1/4}(2\pi)^3} \sum_{s=\pm\frac{1}{2}} \left[e^{ik\cdot x} b_s(k;x^+) + e^{-ik\cdot x} d_s^\dagger(k;x^+) \right] , \qquad (6)$$

where $b_s(k)$ is a quark destruction operator and destroys a quark with momentum k while $d_s^\dagger(k)$ is an anti–quark creation operator and creates an anti–quark with momentum k. They obey the equal light cone time (x^+) anti–commutation relations

$$\{b_s(k, x^+), b_{s'}^\dagger(k', x^+)\} = \{d_s(k, x^+), d_{s'}^\dagger(k', x^+)\}$$
$$= (2\pi)^3 \delta^{(3)}(k - k')\delta_{ss'} . \qquad (7)$$

The above definitions ensure that the fermionic contribution to the light cone QCD Hamiltonian can be written as the sum of kinetic and potential pieces, $P_{f,QCD}^- = P_{f,0}^- + V_{QCD}$, where the kinetic piece of the Hamiltonian is defined as

$$P_{f,0}^- = \int \frac{d^3 k}{(2\pi)^3} \sum_{s=\pm\frac{1}{2}} \frac{(k_t^2 + M^2)}{2k^+} \left(b_s^\dagger(k) b_s(k) + d_s^\dagger(k) d_s(k) \right) . \qquad (8)$$

These points will become clearer when we explicitly derive the QCD light cone Hamiltonian in lecture 2.

The gauge field A^μ has two dynamical components $A_i^a(x)$ with $i = 1, 2$ in light cone gauge $A^+ = 0$. These are defined in terms of creation–annihilation operators as

$$A_i^a(x) = \int_{k^+>0} \frac{d^3 k}{\sqrt{2|k^+|}(2\pi)^3}$$
$$\sum_{\lambda=1,2} \delta_{\lambda i} \left[e^{ik\cdot x} a_\lambda^a(k;x^+) + e^{-ik\cdot x} a_\lambda^{a\dagger}(k;x^+) \right] , \qquad (9)$$

where the λ's here correspond to the two independent polarizations and $a_\lambda^a \dagger$ (a_λ^a) creates (destroys) a gluon with momentum k. They obey the commutation relations

$$[a_\lambda^a(k), a_{\lambda'}^{b\,\dagger}(k')] = (2\pi)^3 \delta^{(3)}(k - k')\delta_{ab}\delta_{\lambda\lambda'} . \qquad (10)$$

In an analogous fashion to Eq. 8, the bosonic kinetic energy can be written (after normal ordering) as

$$\int \frac{d^3 k}{(2\pi)^3} \sum_{\lambda=1,2} \frac{(k_t^2 + M^2)}{2k^+} a_\lambda^\dagger(k) a_\lambda(k) . \qquad (11)$$

We will now discuss the structure of the Poincaré group on the light cone. For the field $\hat{\phi}_r$, which here denotes vector or scalar bosons, we can define the stress–energy tensor

$$\hat{T}^{\lambda\nu} = \hat{\Pi}_r^{\lambda}\partial^{\nu}\hat{\phi}_r - \hat{g}^{\lambda\nu}\mathcal{L}\,, \tag{12}$$

where $\mathcal{L}$ is the Lagrangean density, and $\hat{\Pi}_r^{\lambda}$ is the generalized momentum

$$\hat{\Pi}_r^{\lambda} = \frac{\delta\mathcal{L}}{\delta(\partial_{\lambda}\hat{\phi}_r)}\,. \tag{13}$$

Keep in mind that the carets denote quantities in the usual spacetime co-ordinates. Define now the following generalized quantity

$$\hat{\Sigma}_{\alpha\beta}^{\mu\nu} = \begin{cases} \frac{1}{4}[\hat{\gamma}^{\mu},\hat{\gamma}^{\nu}]^{\alpha\beta} & \text{for spinors} \\ (\hat{g}_{\alpha}^{\mu}\hat{g}_{\beta}^{\nu} - \hat{g}_{\beta}^{\mu}\hat{g}_{\alpha}^{\nu}) & \text{for vectors} \end{cases} \tag{14}$$

One can then define the boost–angular momentum stress tensor

$$\hat{J}^{\lambda\mu\nu} = \hat{x}^{\mu}\hat{T}^{\lambda\nu} - \hat{x}^{\nu}\hat{T}^{\lambda\mu} + \hat{\Pi}_r^{\lambda}\hat{\Sigma}_{rs}^{\mu\nu}\hat{\phi}_s\,. \tag{15}$$

There are ten conserved currents

$$\partial_{\lambda}\hat{T}^{\lambda\mu} = 0\,,$$
$$\partial_{\lambda}\hat{J}^{\lambda\mu\nu} = 0\,, \tag{16}$$

and correspondingly, ten conserved charges,

$$\hat{P}^{\mu} = \int d\hat{x}^1\, d\hat{x}^2\, d\hat{x}^3\, \hat{T}^{0\mu}\,,$$
$$\hat{M}^{\mu\nu} = \int d\hat{x}^1\, d\hat{x}^2\, d\hat{x}^3\, \left(\hat{x}^{\mu}\hat{T}^{0\nu} - \hat{x}^{\nu}\hat{T}^{0\mu} + \hat{\Pi}_r^0\hat{\Sigma}_{rs}^{\mu\nu}\hat{\phi}_s\right)\,. \tag{17}$$

The four components of the energy–momentum vector $\hat{P}^{\mu}$ and the six components of the boost–angular momentum $\hat{M}^{\mu\nu}$ comprise the ten generators of the Poincaré group [1]. These generators satisfy the Poincaré algebra

$$[\hat{P}^{\mu},\hat{P}^{\nu}] = 0\;;\; [\hat{M}^{\mu\nu},\hat{P}^{\rho}] = i\left(\hat{g}^{\nu\rho}\hat{P}^{\mu} - \hat{g}^{\mu\rho}\hat{P}^{\nu}\right)\,,$$
$$\left[\hat{M}^{\mu\nu},\hat{M}^{\rho\sigma}\right] = i\left(\hat{g}^{\mu\sigma}\hat{M}^{\nu\rho} + \hat{g}^{\nu\rho}\hat{M}^{\mu\sigma} - \hat{g}^{\mu\rho}\hat{M}^{\nu\sigma} - \hat{g}^{\nu\sigma}\hat{M}^{\mu\rho}\right)\,. \tag{18}$$

The six components of the boost–angular momentum can be further split into the three generators of rotations $\hat{M}^{ij} = \epsilon^{ijk}\hat{J}^k$ (where $\hat{J}^k$ is the angular momentum operator) and three generators of boosts $\hat{K}^i = \hat{M}^{i0}$. In the

---

[1] The Poincaré group is a sub–group of the conformal group which contains 15 generators, the additional generators being 4 conformal transformations and 1 dilatation.

language of Ref. [1], these are referred to as kinematic and dynamic operators, respectively, since the former is independent of the interaction while the
latter isn't.

Transforming the above to light cone co-ordinates, we obtain $P^\mu \equiv (P^+, P^1, P^2, P^-)$, where $P^\pm = (\hat{P}^0 \pm \hat{P}^3)/\sqrt{2}$, and

$$M^{\mu\nu} = \begin{pmatrix} 0 & -S_1 & -S_2 & K_3 \\ S_1 & 0 & J_3 & B_1 \\ S_2 & -J_3 & 0 & B_2 \\ -K_3 & -B_1 & -B_2 & 0 \end{pmatrix}. \tag{19}$$

Above we used the following definitions

$$B_1 = \frac{(K_1 + J_2)}{\sqrt{2}} \; ; \; B_2 = \frac{(K_2 - J_1)}{\sqrt{2}},$$

$$S_1 = \frac{(K_1 - J_2)}{\sqrt{2}} \; ; \; S_2 = \frac{(K_2 + J_1)}{\sqrt{2}}.$$

The commutation relations among the $M^{\mu\nu}$'s and the P^μ's are of course the
same as in Eq. 18. The operators B_1 and B_2 are kinematic and boost the
system in the x and y directions respectively. In addition, the operators J_3
and interestingly, K_3 are kinematic and rotate the system in the x–y plane
and boost it in the longitudinal direction respectively.

An interesting observation by Susskind [3] related to the above is that the
commutation relations among the seven generators $P^\pm, P_t, J_3, B_1$ and B_2
are the same as the commutation relations among the symmetry operators
of non–relativistic quantum mechanics in two dimensions. Indeed, one can
formally make the correspondence,

- $P^- \longrightarrow$ Hamiltonian.
- $P_t \longrightarrow$ Momenta.
- $P^+ \longrightarrow$ Mass.
- $J_3 \longrightarrow$ Angular Momentum.
- $B_t \longrightarrow$ generators of Galilean boosts in x–y plane.

These seven generators obey the commutation relations

$$[P^-, P_t] = [P^-, P^+] = [P_t, P^+]$$
$$= [J_3, P^-] = [J_3, P^+] = [B_t, P^+] = 0. \tag{20}$$

and

$$[J_3, P^t] = i\epsilon_{tl} P^l,$$
$$[J_3, B^t] = i\epsilon_{tl} B^l,$$
$$[B^t, P^-] = -iP^t, [B^t, P^l] = -i\delta_{tl} P^+.$$

Above ϵ_{ij} is the Levi–Civita tensor in two dimensions. Since they are kinematic operators, they leave the planes of $x^+ = constant$ invariant under their
operations.

Susskind, Bardacki & Halpern [15] and Kogut & Soper have shown that the above mentioned isomorphism is responsible for the non–relativistic quantum–mechanical structure of quantum field theories on the light cone. The simplest illustration of this isomorphism is the fact that the free particle Hamiltonian takes the form

$$H \equiv P^- = \frac{P_t^2 + M^2}{2P+}.$$

Recalling the form of the energy in two dimensional quantum mechanics, we obtain the isomorphisms above. For QED and QCD, the above form is modified by the addition of a potential term which we will discuss in detail in lecture 2. Finally, we should mention that the other kinematic operator, K_3, the boost operator in the longitudinal direction, serves to rescale the other operators

$$\exp(i\omega K_3)P^- \exp(-i\omega K_3) = \exp(\omega)P^-.$$
$$\exp(i\omega K_3)J_3 \exp(-i\omega K_3) = J_3.$$
$$\exp(i\omega K_3)S \exp(-i\omega K_3) = \exp(-\omega)S. \tag{21}$$

This property of K_3 will come in handy in lecture 3.

3   The Light Cone QCD Hamiltonian

In this lecture, we will derive an explicit form for the light cone QCD Hamiltonian making use of the light cone constraint relations. Consider first the fermionic part of the QCD action

$$S_F = \int d^4x\, \bar{\psi}\,(\not{P}+ M)\,\psi.$$

Above, $P^\mu = -iD^\mu \equiv -i(\partial^\mu - igA^\mu)$. For convenience, we will not write the integral $\int d^4x$, in the following but it must be understood to be there. Then writing out the above action explicitly,

$$S_F = \bar{\psi}\gamma^- P_-\psi + \bar{\psi}\gamma^+ P_+\psi + \bar{\psi}\gamma^t P_t\psi + \bar{\psi}M\psi.$$

Consider now the first term in the above:

$$\bar{\psi}\gamma^- P\psi = \psi^\dagger \hat{\gamma}^0 \gamma^- P_-\psi \rightarrow \sqrt{2}\psi^\dagger_- P_-\psi_-. \tag{22}$$

To dissect the above, we first decomposed $\hat{\gamma}^0 = (\gamma^+ + \gamma^-)/\sqrt{2}$, made use of $(\gamma^-)^2 = 0$, and the properties of the projector α^+ in Eq. 4 to obtain the RHS. Similarly, it is recommended to the serious student that he or she show that

$$\bar{\psi}\gamma^+ P_+\psi = \sqrt{2}\psi^\dagger_+ P_+\psi_+,$$
$$\bar{\psi}\gamma^t P_t\psi = \frac{1}{\sqrt{2}}\left(\psi^\dagger_-\gamma^+ P_t + \psi^\dagger_+\gamma^- P_t\psi_-\right),$$
$$\bar{\psi}M\psi = M\left(\psi^\dagger_+ \hat{\gamma}^0\psi_- + \psi^\dagger_- \hat{\gamma}^0\psi_+\right). \tag{23}$$

One then obtains

$$S_F = \sqrt{2}\psi_-^\dagger P_-\psi_- + \sqrt{2}\psi_+^\dagger P_+\psi_+ + \frac{1}{\sqrt{2}}\left(\psi_-^\dagger\gamma^+\slashed{A}_t + \psi_+^\dagger\gamma^-\slashed{A}\psi_-\right)$$

$$+ M\left(\psi_+^\dagger\hat{\gamma}^0\psi_- + \psi_-^\dagger\hat{\gamma}^0\psi_+\right),\tag{24}$$

where we have written the fermionic piece of the action in terms of the two–spinors ψ_- and ψ_+ and their hermitean conjugates.

Following Eq. 13, the momenta conjugate to these two–spinor fields are

$$\Pi_+ = \frac{\delta\mathcal{L}}{\delta(\partial_+\psi_+)} = \left(\frac{\sqrt{2}}{i}\right)\psi_+^\dagger,$$

$$\Pi_- = \frac{\delta\mathcal{L}}{\delta(\partial_+\psi_-)} = 0.\tag{25}$$

Since trivially $[\Pi_-,\psi_-] = 0$, the two–spinor ψ_- is, unlike ψ_+, not an independent quantum field. We will now show that one may derive a constraint equation (i.e., independent of the light cone time x^+) for ψ_- in terms of the dynamical field ψ_+. The light cone constraint relations can be obtained from the operator equations of motion. In this case, it is the Dirac equation $(\slashed{P}+ M) = 0$, or

$$(-i\partial_- - gA_-)\gamma^-\psi + (-i\partial_+ - gA_+)\gamma^+\psi$$

$$+ \sum_{j=1}^{2}(-i\partial_j - gA_j)\gamma^j\psi + M\psi = 0.\tag{26}$$

Multiply the above through by γ^+. Since $(\gamma^+)^2 = 0$, this projects out the x^+–light cone time–dependence in the above and we obtain (after liberally using our projection operator tricks from Eq. 4) the equation

$$\sqrt{2}P_-\psi_- = -\hat{\gamma}^0(\slashed{A}_t + M)\psi_+.\tag{27}$$

In light cone gauge, $A_- = -A^+ = 0$, hence $P_- = (-i\partial_- - gA_-) \to -i\partial_-$. With this gauge condition therefore, one can easily invert the $P_- = -P^+$ operator and one obtains the light cone constraint equation

$$\psi_- = \frac{\hat{\gamma}^0}{\sqrt{2}P^+}\left(\slashed{A}_t + M\right)\psi_+.\tag{28}$$

Thus for light cone time x^+, ψ_- is determined completely by ψ_+ at that time. Only the two components of the spinor ψ corresponding to ψ_+ are independent dynamical fields on the light cone.

We can now use the above obtained constraint equation to replace ψ_- in Eq. 24 for S_F. For instance,

$$\sqrt{2}\psi_-^\dagger P_-\psi_- = \sqrt{2}\left(\frac{\hat{\gamma}^0}{\sqrt{2}P^+}(\slashed{A}_t + M)\psi_+\right)^\dagger P_-\left(\frac{\hat{\gamma}^0}{\sqrt{2}P^+}(\slashed{A}_t + M)\psi_+\right)$$

$$\longrightarrow \frac{(-)}{\sqrt{2}} \psi_+^\dagger (M - \slashed{P}_t) \frac{1}{P^+} (\slashed{P}_t + M) \psi_+ \,.$$

Now rescale the fields $\psi \to 2^{-1/4} \psi$ [2]. As an exercise, the reader should use the light cone constraint equation above, the properties of the projection operators $\alpha^\pm$ in Eq. 4 and those of the light cone gamma matrices to first substitute for ψ_- everywhere and then demonstrate the following identities,

$$\frac{M}{\sqrt{2}} \left(\psi_+^\dagger \hat{\gamma}^0 \psi_- + \psi_-^\dagger \hat{\gamma}^0 \psi_+ \right) = \frac{M}{2} \left(\psi_+^\dagger \gamma^- \psi_- + \psi_-^\dagger \gamma^+ \psi_+ \right)$$

$$\frac{1}{2} \left(\psi_-^\dagger \gamma^+ (\slashed{P}_t + M) \psi_+ + \psi_+^\dagger \gamma^- (\slashed{P}_t + M) \psi_- \right)$$
$$= \psi_+^\dagger (M - \slashed{P}_t) \frac{1}{P^+} (\slashed{P}_t + M) \psi_+ \,.$$

Putting these together with the other term above, our result for the fermionic action expressed solely in terms of the dynamical two–spinor ψ_+ is

$$S_F = -\psi_+^\dagger P^- \psi_+ + \frac{1}{2} \psi_+^\dagger (M - \slashed{P}_t) \frac{1}{P^+} (\slashed{P}_t + M) \psi_+ \,. \qquad (29)$$

We now turn to the bosonic contribution to the action,

$$S_B = \frac{1}{4} F_{\mu\nu}^a F^{\mu\nu,a} \,, \qquad (30)$$

and following a procedure analogous to the fermionic case, shall write it in terms of A_t, the two transverse, dynamical components of the gauge field A^μ. We have seen earlier that the choice of light cone gauge $A_- = -A^+ = 0$ greatly simplifies the light cone constraint relation for the fermions. In this gauge, the various components of the field strength tensor also simplify to

$$F_{+-}^a = -\partial_- A_+^a \,,$$
$$F_{t+}^a = \partial_t A_+^a - \partial_+ A_t^a + g f^{abc} A_t^b A_+^c \,,$$
$$F_{t-}^a = -\partial_- A_t^a \equiv -E_t^a \,.$$

In addition, there are of course the purely transverse pieces F_{ij}; with $i, j = 1, 2$. ¿From the above it is evident that there is no (light cone) time derivative $\partial_+ A_+$ in the action. The field A_+ therefore has no momentum conjugate and we may use the operator equations of motion to eliminate the field $A_+ = -A^-$.

---

[2] This rescaling gets rid of the $\sqrt{2}$ factors in the action. This also explains the peculiar $2^{-1/4}$ normalization factor in Eq. 6 for the properly normalized ψ_+ field.

The equations of motion are the Yang–Mills equations of course. The light cone constraint equation is just Gauss' law on the light cone since it must be valid at all times. This condition is then

$$(D_t F^{t+})^a + (D_- F^{-+})^a = -J^{+,a} \implies -\partial_-^2 A^{-,a} = J^{+,a} + (D_t E_t)^a \,,$$

where E_t^a are the two transverse components of the electric field and D_t is the covariant derivative $\partial_t - ig A_t$. We can write our light cone constraint equation for A^- compactly below as

$$A^{-,a} = \frac{1}{(P^+)^2} \left(J^{+,a} + (D_t E_t)^a \right) . \tag{31}$$

Returning to the action

$$S_B = \frac{1}{4} F^2 = \frac{1}{4} F_t^2 - F_{t+}^a F_{t-}^a - \frac{1}{2} F_{+-}^a F_{+-}^a \,,$$

substituting the expressions for the field strength components in terms of the gauge fields and performing an integration by parts, we obtain

$$S_B = \frac{1}{4} F_t^2 - A_+^a (D_t E_t)^a - \frac{1}{2} (\partial_- A_+)^2 - (\partial_- A_t^a)(\partial_+ A_t^a) . \tag{32}$$

Before we substitute for A_+ above, we will first write out the full action $S_{QCD} = S_F + S_B - J_{ext} \cdot A$:

$$S_{QCD} = -\psi_+^\dagger(-i\partial^- - gA^-)\psi_+ + \frac{1}{2}\psi_+^\dagger(M - F_t)\frac{1}{P^+}(F_t + M)\psi_+ + \frac{1}{4}F_t^2$$
$$- A_+^a (D_t E_t)^a - \frac{1}{2}(\partial_- A_+)^2 - (\partial_- A_t^a)(\partial_+ A_t^a) - J_{ext}^+ A_+ .$$

Consider the first term above. We can write this as

$$-\psi_+^\dagger \left(-i\partial^- - gA^-\right)\psi_+ = -i\psi_+^\dagger \partial_+ \psi_+ - J_{dyn}^+ A_+ \,,$$

where $J_{dyn}^{+,a} = \psi_+^{a\dagger}\lambda^a\psi_+$. (The λ^a are the Gell–Mann SU(3) matrices.) We now substitute the above result in our expression for the action and after
a) defining $J^+ = J_{dyn}^+ + J_{ext}^+$,
b) performing an integration by parts,
c) making use of the constraint relation Eq. 31 to eliminate A_+,
we obtain finally,

$$S_{QCD} = -i\psi_+^\dagger \partial_+ \psi_+$$
$$-(\partial_- A_t)(\partial_+ A_t) + \frac{1}{4}F_t^2 + \frac{1}{2}\psi_+^\dagger(M - F_t)\frac{1}{P^+}(F_t + M)\psi_+$$
$$+ \frac{1}{2}(J^+ + D_t E_t)\frac{1}{(P^+)^2}(J^+ + D_t E_t) . \tag{33}$$

The final step before we obtain the Hamiltonian is to identify the momenta conjugate to the dynamical fields (now with the proper normalization!),

$$\Pi_+^{fermi} = \frac{\delta S_{QCD}}{\delta(\partial_+\psi_+)} = -i\psi_+^\dagger\,.$$

$$\Pi_t^{bose} = \frac{\delta S_{QCD}}{\delta(\partial_+ A_t)} = -\partial_- A_t \equiv -E_t\,. \tag{34}$$

Writing out the fields and their momentum conjugates in terms of the creation and annihilation operators introduced in Eqs. 6 and 9, and making use of their commutation relations, the reader may confirm that

$$[\Pi_t^{bose}(x), A_t(x')] = \{\Pi_+^{fermi}(x), \psi_+(x')\} = i\delta^{(3)}(x - x')\,. \tag{35}$$

The Hamiltonian density in our convention is defined as

$$H \equiv P_{QCD}^- = S_{QCD} - \Pi_+^{fermi}\partial_+\psi_+ - \Pi_t^{bose}\partial_+ A_t\,,$$

We can therefore write our final expression for the Hamiltonian density as

$$P_{QCD}^- = \frac{1}{4}F_t^2 + \frac{1}{2}(J^+ + D_t E_t)\frac{1}{(P^+)^2}(J^+ + D_t E_t)$$

$$+ \frac{1}{2}\psi_+^\dagger(M - \not{F}_t)\frac{1}{P^+}(\not{F}_t + M)\psi_+\,. \tag{36}$$

We have therefore succeeded in obtaining the light cone Hamiltonian in QCD, expressed solely in terms of the two–spinor ψ_+ and A_t, the two transverse components of the gauge field. The following observations can be made regarding the above expression. Firstly, one can show straightforwardly that the light cone Hamiltonian can be written as

$$P_{QCD}^- = P_0^- + V_{QCD}\,, \tag{37}$$

where P_0 (the sum of the RHS of Eqs. 8 and 11) is the piece of the Hamiltonian not containing any factors of the coupling g and V_{QCD} is the rest, which can also be written out in terms of creation–annihilation operators. Furthermore, the ground state of the non–interacting Hamiltonian P_0^- is also, remarkably, the ground state of the full Hamiltonian. This is the meaning behind statements one may have heard that the light cone vacuum is 'trivial'. Because the vacuum is trivial, one may simply construct any eigenstate of the full Hamiltonian in terms of a complete Fock eigen–basis corresponding to eigenstates of the non–interacting Hamiltonian. As we shall demonstrate in the next lecture with a specific example, this point forms the basis for the quark–parton model in quantum field theory.

Just as in non–relativistic quantum mechanics then, one can use light cone time ordered perturbation theory to construct these states. Unfortunately, there is no room to discuss time ordered perturbation theory here but it will be discussed in the "long" version of these lectures [14].

There is one point we have not mentioned thus far but it threatens the entire pretty picture above. This has to do with the terms $1/P^+$ and $1/(P^+)^2$ above. Recall that they were obtained by inverting the light cone constraint equations in light cone gauge. Clearly, that operation and these terms are not well defined for $P^+ = 0$. The simple vacuum is thus only deceptively so and all the complications are now hidden in the zero–mode. That this would be the case should have been clearer in retrospect. Defining the operator $1/P^+$ requires knowing the boundary conditions of the fields at large distances and therefore, should be sensitive to confining and chiral symmetry breaking effects.. Attempts to regulate the zero mode, a well know example of which is discretized light cone quantization [13], also result in a non–trivial vacuum. On the other hand, perturbative physics should not be terribly sensitive to how fields are regulated at large distances. Different 'epsilon' prescriptions corresponding to different boundary conditions at infinity give the same short distance physics [16]. The justification of the above approach is therefore the success of the parton model in describing physics at large transverse momenta in QCD. The program to describe non–perturbative physics in the same framework is very advanced and we refer the reader to Ref. [13] to read of the latest developments.

4   High-Energy Eikonal Scattering and the Parton Model in QED

In the previous lectures we developed some of the basic formalism of light cone field theory. We will now apply this formalism to a specific example; high energy scattering of an electron from an external potential in QED. We will show how one recovers the standard Eikonal picture in this formalism. More importantly, our results clearly can be interpreted in terms of a parton model picture of high energy scattering. This lecture closely follows the excellent paper of Bjorken, Kogut and Soper [11] where this example and others are discussed. For convenience, we will also use their "-2" convention (for eg., $A_- = A^+$ and $A_t = -A^t$).

The light cone Hamiltonian in QED is similar to the QCD Hamiltonian derived above in Eq. 36 and of course much simpler. To treat the problem of scattering off an external potential, we introduce an external potential $\mathbf{a}_\mu$ using the gauge invariant minimal substitution $p_\mu \to p_\mu - g\mathbf{a}_\mu$. The QED Hamiltonian including the external potential $\mathbf{a}_\mu$ is then

$$
P_{scatt}^-(x^+) = \int d^2 x_t\, dx^- \left\{ e\mathbf{a}_+ \psi_+^\dagger \psi_+ + \frac{1}{2} e^2 \psi_+^\dagger \psi_+ \frac{1}{(p^+ - e\mathbf{a}^+)^2} \psi_+^\dagger \psi_+ \right.
$$

$$
\left. + \psi_+^\dagger (M - i\sigma \cdot (\mathbf{p} - e\mathbf{A} - e\mathbf{a})) \frac{1}{2(p^+ - \mathbf{a}^+)} (M + i\sigma \cdot (\mathbf{p} - e\mathbf{A} - e\mathbf{a})) \psi_+ \right.
$$

$$+ e\psi_+^\dagger \psi_+ \frac{1}{p^+ - ea^+} \boldsymbol{p} \cdot \boldsymbol{A} + \frac{1}{2} \sum_{t=1,2} A^t p^2 A^t \bigg\} . \tag{38}$$

Note that one can define

$$\left[\frac{1}{p^+ - a^+} \psi_+ \right] (x) = \int d\xi \, \frac{1}{2i} \epsilon(x^- - \xi)$$

$$\exp\left(-ie \int_\xi^{x^-} d\xi' a^+ (x^+, x_t, \xi') \right) \psi_+ (x^+, x_t, \xi) , \tag{39}$$

where ϵ is the sign function. This can be checked by multiplying through by $p^+ - ea^+$ [3].

Now write $P_{scatt}^- = P_{QED}^- + V$, where P_{QED}^- is the usual time independent QED Hamiltonian with $a^\mu = 0$ and $V(x^+) = P_{scatt}^- - P_{QED}^-$. We wish to construct the scattering matrix S_{fi} in the interaction picture. In the usual quantum mechanical treatment,

$$\psi_I(x^+, x_t, x^-) = e^{iP_{QED}^- x^+} \psi_+(0, x_t, x^-) e^{-iP_{QED}^- x^+} .$$

$$A_I(x^+, x_t, x^-) = e^{iP_{QED}^- x^+} A(0, x_t, x^-) e^{-iP_{QED}^- x^+} . \tag{40}$$

Then, the scattering matrix for the scattering of an electron off an external potential is given by

$$S_{fi} = < f | T \left\{ \exp\left(-i \int dx^+ V(x^+) \right) \right\} | i > , \tag{41}$$

where 'T' denotes light cone time ordering and $|i>$ and $|f>$ are asymptotic states which are eigenstates of the QED Hamiltonian P_{QED}^-. They can thus be evaluated in Rayleigh–Schrödinger perturbation theory (see the discussion at the end of lecture 2).

We want to compute the scattering matrix in the high energy scattering limit $P_i, P_f \to \infty$. Consider the states $|I>$ and $|F>$, which may be states in the rest frame of the electron. They are related by boosts to the states $|i>$ and $|f>$ above. Then $|i> = e^{-i\omega K_3}|I>$ and $|f> = e^{-i\omega K_3}|F>$, where K_3 (defined previously in lecture 1) is the generator of boosts in the longitudinal direction. In QED, K_3 is the operator

$$K_3 = \int d^2x_t \, dx^- \, x^- \left[\frac{i}{2} \psi_+^\dagger \overleftrightarrow{\partial}_- \psi_+ + (\partial_- A_t)(\partial_- A_t) \right]_{x^+ = 0} , \tag{42}$$

and ω is the rapidity corresponding to the boost. The scattering matrix element between the scattering states in the rest frame is then

$$< F | e^{i\omega K_3} T \left\{ \exp\left(-i \int dx^+ V(x^+) \right) \right\} e^{-i\omega K_3} | I > .$$

---

[3] In QCD, the sole change is to replace the exponential on the RHS by a path ordered exponential.

Using the definition of path ordered exponentials, this relation can be written as

$$< F|T\left\{\exp\left(-i\int dx^+ e^{i\omega K_3}V(x^+)e^{-i\omega K_3}\right)\right\}|I>. \tag{43}$$

A great advantage of the light cone formalism is that the fields transform simply under boosts. We have

$$e^{i\omega K_3}\psi_I(x^+,x_t,x^-)e^{-i\omega K_3} = e^{\omega/2}\psi_I(e^{-\omega}x^+,x_t,e^\omega x^-).$$
$$e^{i\omega K_3}A_I(x^+,x_t,x^-)e^{-i\omega K_3} = A_I(e^{-\omega}x^+,x_t,e^\omega x^-).$$

The above can be shown explicitly by computing the commutators $[K_3,\psi_I]$ and $[K_3,A_I]$ using the definition of K_3 in Eq. 42. The field a however commutes with K_3 and therefore does not transform under boosts.

Consider now the argument of the exponential in Eq. 43. We can show that all but one of the terms in $V(x^+)$ are invariant under the boost operation. For example,

$$e^{i\omega K_3}\,\psi_I^\dagger\psi_I\frac{1}{(P^+)^2}\psi_I^\dagger\psi_I\,e^{-i\omega K_3}\longrightarrow \psi_I^\dagger\psi_I\frac{1}{(P^+)^2}\psi_I^\dagger\psi_I.$$

Above, we have used the fact that elements of the Lorentz group are simply rescaled by boosts, $e^{i\omega K_3}P^+e^{-i\omega K_3} = e^\omega P^+$, as well as Eq. 44. The only term that does not remain invariant is

$$ea_+\psi_I^\dagger\psi_I \xrightarrow{K_3} e^\omega ea\psi_I^\dagger\psi_I.$$

Hence,

$$e^{i\omega K_3}V(x^+)e^{-i\omega K_3}$$
$$=\int d^2x_t\,dx^-\,e^\omega e\,a_+(x^+,x_t,x^-)\psi_I^\dagger(e^{-\omega}x^+,x_t,e^\omega x^-)\psi(e^{-\omega}x^+,x_t,e^\omega x^-)$$
$$+O(e^{-\omega}). \tag{44}$$

Now let $x^- \to e^\omega x^-$ above. Then

$$e^{i\omega K_3}V(x^+)e^{-i\omega K_3}$$
$$=\int d^2x_t\,dx^-\,ea_+(x^+,x_t,e^{-\omega}x^-)\psi_I^\dagger(e^{-\omega}x^+,x_t,x^-)\psi(e^{-\omega}x^+,x_t,x^-)$$
$$+O(e^{-\omega}). \tag{45}$$

Going to the infinite rapidity limit $\omega \to \infty$ corresponding to very high energy scattering, we note from the above that the operators are all evaluated at $x^+ = 0$ so the time ordering in x^+ is irrelevant in that limit. Then one can show in that limit (and this is a subtle point) that

$$S_{fi} =< F|\mathcal{P}|I > +O(e^{-\omega}) \equiv< f|\mathcal{P}|i > +O(e^{-\omega}). \tag{46}$$

Thus we have expressed S_{fi} again in terms of the states $|i>, |f>$, thereby demonstrating the Lorentz invariance of these states in the infinite momentum limit. Also, above

$$\mathcal{P} = \exp\left(-i \int d^2 x_t \chi(x_t) \rho(x_t)\right), \tag{47}$$

where

$$\chi(x_t) = e \int dx^+ \mathbf{a}_+(x^+, x_t, 0), \tag{48}$$

and

$$\rho(x_t) = \int dx^- \psi_I^\dagger(0, x_t, x^-)\psi_I(0, x_t, x^-). \tag{49}$$

We have therefore recovered the well known eikonal scattering limit in QED.

We shall now show that the above derivation has a deep connection with the parton model. The asymptotic 'in' state of the electron, $|i>$, is an eigenstate of the QED Hamiltonian P^-_{QED}. We can expand $|i>$ in terms of the "bare" quanta [4] associated with the fields $\psi_+(0, x_t, x^-)$ and $A_t(0, x_t, x^-)$ at $x^+ = 0$:

$$|i> = \int d^2 k_t \int_{k^+>0} \frac{dk^+}{k^+} \sum_\lambda \left\{ g(k_t, k^+, \lambda)a^\dagger(k_t, k^+, \lambda)|0>\right.$$
$$+ \int d^2 k_{t1} \frac{dk_1^+}{k_1^+} \int d^2 k_{t2} \frac{dk_2^+}{k_2^+}$$
$$\left. \sum_{s_1, s_2} h(k_1, k_2; s_1, s_2)b^\dagger(k_1; s_1)d^\dagger(k_2; s_2)|0> + \cdots \right\}. \tag{50}$$

The creation and annihilation operators introduced here are the same as those in lecture 1. The coefficient h above can be interpreted simply as the amplitude for $|i>$ to contain a *bare electron* with momenta k_1 and spin s_1, and a *bare positron* with momenta k_2 and spin s_2. It was shown first by Drell, Levy and Yan that the amplitude squared for an arbitrary number of parton eigenstates, integrated over phase space could be simply related to the structure functions W_1, W_2 [17].

We can also see this here if we similarly expand $|f>$ in terms of the bare quanta. The scattering matrix S_{fi} in Eq. 46 can then be evaluated if we move $\mathcal{P}$ past the creation–annihilation operators till it acts on $|0>$:

$$\mathcal{P}b^\dagger d^\dagger a^\dagger \cdots a^\dagger|0> = \mathcal{P}b^\dagger \mathcal{P}^{-1} \cdots \mathcal{P}a^\dagger \mathcal{P}^{-1}\mathcal{P}|0>. \tag{51}$$

Since it is evident that $\mathcal{P}$ is invariant under translations in the x^- direction, it commutes with the generator of x^- translations–P^+. One can check that

---

[4] These are eigenstates (in QED!) of P_0^- in Eq. 37.

$\mathcal{P}|0>= |0>$. This follows formally by expanding $\mathcal{P}$ and requiring that the operators in $\rho(x_t)$ are normal ordered. How do the creation–annihilation operators transform with $\mathcal{P}$? Using the light cone commutation relations, $\{\psi_+(x), \psi_+^\dagger(x')\} = \delta^{(3)}(x - x')$, we find

$$\mathcal{P}\psi_+^\dagger(0, x_t, x^-)\mathcal{P}^{-1} = \exp{(-i\chi)}\,\psi_+^\dagger(0, x_t, x^-). \tag{52}$$

Fourier transforming the above, and using Eq. 6, we obtain for the electron creation operator

$$\mathcal{P}b^\dagger(k_t, k^+, s)\mathcal{P}^{-1} = \int \frac{d^2 k_t'}{(2\pi)^2} b^\dagger(k_t', k^+, s)\tilde{\mathcal{P}}(k_t' - k_t), \tag{53}$$

where (with $q_t = k_t' - k_t$)

$$\tilde{\mathcal{P}}(q_t) = \int d^2 x_t e^{-iq_t \cdot x_t} e^{-i\chi(x_t)}. \tag{54}$$

Similarly for the positron creation operator

$$\mathcal{P}d^\dagger(k_t, k^+, s)\mathcal{P}^{-1} = \int \frac{d^2 k_t'}{(2\pi)^2} d^\dagger(k_t', k^+, s)\tilde{\mathcal{P}}_c(k_t' - k_t), \tag{55}$$

with

$$\tilde{\mathcal{P}}_c(q_t) = \int d^2 x_t e^{-iq_t \cdot x_t} e^{+i\chi(x_t)}. \tag{56}$$

Finally, $\mathcal{P}a^\dagger\mathcal{P}^{-1}$, since all the operators in $\mathcal{P}$ commute with $a^\dagger$.

What we have learnt from the above is that when a high energy *bare electron* or *bare positron* interacts with a potential at x_t, the net effect is to multiply its wavefunction by the eikonal phase $e^{-i\chi}$ or $e^{+i\chi}$ respectively. The following physical picture then emerges from our manipulations above.

- The scattering of high energy particles (denoted here by '$|i>$', which is an eigenstate of the Hamiltonian P_{QED}^-) is not simple–i.e., it cannot be described by a simple overall phase.
- However, due to the "potential" structure of QED on the light cone, the *physical* particle states ($|i>$) can be expanded in a complete basis of multi–parton eigenstates (eigenstates of $P_{0,QED}^-$).
- The scattering of these partons is simple–they acquire an eikonal phase in the scattering.
- The mutual interactions of partons in the physical state $|i>$ is complex, but as the rapidity $\omega \to \infty$, these interactions are slowed down by time dilation. Recall that in Eq. 44, the only term that survives is the one that contains the coupling to the external field **a** and all the other terms which contain the interactions of the partons with each other are suppressed.

Chronologically, one can view the scattering as follows. Partons in the initial state interact strongly for $-\infty < x^+ < 0$ with the potential V_{QED}. At $x^+ = 0$, each individual parton scatters simply off the external potential, acquiring an eikonal phase. For $0 < x^+ < \infty$, the partons then again interact among each other with the potential V_{QED}. This picture of scattering is also known as the impulse approximation. It explains the striking phenomenon of Bjorken scaling observed in deep inelastic scattering at very large momentum transfers.

Finally, for completeness, we will mention that the cross–section for electron scattering off an external potential is given by

$$d\sigma = \int_{k_1^+, \cdots, k_n^+ > 0} \frac{d^2 k_{t1} dk_1^+}{(2\pi)^3 k_1^+} \cdots \frac{d^2 k_{tn} dk_n^+}{(2\pi)^3 k_n^+}$$

$$(2\pi)\delta(k^+ - \sum_{i=1}^n k_i^+) \times | < f|\mathcal{T}|i > |^2 , \qquad (57)$$

where the transition amplitude is defined as

$$< f|\mathcal{T}|i > = < f|U(\infty, 0)[\mathcal{P} - 1]U(0, -\infty)|i > . \qquad (58)$$

Above, U is the light cone analog of the usual unitary evolution operator in quantum mechanics.

5   Bjorken Scaling and Light Cone Fock Distributions

In this last lecture, we will discuss the "light cone" limit $x^2 \to 0$ of deep inelastic scattering, in QCD. For very large momentum transfers, in this limit, one observes the phenomenon known as Bjorken scaling. Unfortunately, we will not have room for a discussion of the renormalization group ideas which predict, in QCD, the experimentally observed logarithmic violations of Bjorken scaling. These will be presented in the "longer" version of these lectures at a later date [14].

In deep inelastic scattering of an incident lepton off a hadron or nucleus, the kinematic invariants are the square of the momentum carried by the "space–like" virtual photon $q^2 = -Q^2 < 0$, (note: we use the ' -2 ' convention here) and $x_{Bj} = \frac{Q^2}{2P \cdot q}$, where P^μ is the four–momentum of the target. The cross–section expressed in terms of these invariants is a product of the point particle Rutherford cross section times a form factor, the electromagnetic form factor of the hadron F_2. In general, $F_2 \equiv F_2(x_{Bj}, Q^2)$, but in QCD, as $Q^2 \to \infty$, $F_2(x_{Bj}, Q^2) \to F_2(x_{Bj})$. The scaling of the structure function as a function of x_{Bj} is what is known as Bjorken scaling. In this lecture, we will derive Bjorken scaling using the free field commutation relations.

The cross section for the inclusive deep inelastic scattering process $l(k) + (h, A)(P) \to l(k') + X$, where X denotes undetected final states, is a tensor product of the leptonic tensor $l^{\mu\nu}$ and the hadronic tensor $W_{\mu\nu}$. The hadronic tensor is defined as [19]

$$
W_{\mu\nu}(q^2, P\cdot q) = \sum_n (2\pi)^4 \delta^{(4)}(q + P - p_n) < P|J_\mu(0)|n >< n|J_\nu(0)|P >
$$
$$
\to \int d^4x\, e^{iq\cdot x} < P|J_\mu(x)J_\nu(0)|P > . \tag{59}
$$

The sum above is over all hadronic final states with momenta p_n. Since $q^0 + P^0$ and p_n^0 are +ve, we can write the above as

$$
W_{\mu\nu} = \int d^4x\, e^{iq\cdot x} < P|[J_\mu(x), J_\nu(0)]|P > . \tag{60}
$$

Since the commutator vanishes outside the forward light cone, we will write the above as

$$
W_{\mu\nu} = \int dx^- e^{iq^+ x^-} \int dx^+ e^{iq^- x^+}
$$
$$
\int_{x_t^2 < 2x^+ x^-} d^2x_t < P|[J_\mu(x), J_\nu(0)]|P > . \tag{61}
$$

Above, $J_\mu^a = \bar{\psi}\gamma_\mu \lambda^a \psi(x)$.

In the high energy limit $q^+ \to \infty$, $q^- =$ fixed, the largest contribution to $W_{\mu\nu}$ comes from the region of the integral with the smallest oscillations, or x^+ finite, $x^- \to 0$. Since causality demands that $x^2 = 2x^+ x^- - x_t^2 < 2x^+ x^-$, the largest contribution to $W_{\mu\nu}$ is from the region of the light cone $x^2 \to 0$. In other words, the structure function is dominated by the light cone singularities of the commutator of currents. The limit $q^+ \to \infty$ and $q^- =$ fixed, corresponds to the limit $\nu = P\cdot q/M \to \infty$, $Q^2 \to \infty$ and $x_{Bj} = \frac{Q^2}{2P\cdot q}$ fixed.

Let us examine the commutator in Eq. 61 in the limit $x^2 \to 0$. Here, using the "free field" current commutation relation which is reasonable in the weak coupling limit,

$$
\{\bar{\psi}(x), \psi(-x)\} = \frac{1}{8\pi}\gamma^\mu \partial_\mu \epsilon(x^0)\delta(x^2) + O(M^2 x^2), \tag{62}
$$

we obtain

$$
[J_\mu(x), J_\nu(-x)] \approx [\bar{\psi}(x)\gamma_\mu\gamma_\alpha\gamma_\nu\psi(-x) - \bar{\psi}(-x)\gamma_\nu\gamma_\alpha\gamma_\mu\psi(x)]
$$
$$
\frac{1}{8\pi}\partial^\alpha \epsilon(x^0)\delta(x^2). \tag{63}
$$

We now use the identity

$$
\gamma_\mu\gamma_\alpha\gamma_\nu = S_{\mu\nu\alpha\beta}\gamma^\beta + i\epsilon_{\mu\nu\alpha\beta}\gamma^\beta\gamma^5, \tag{64}
$$

where

$$S_{\mu\nu\alpha\beta} = (g_{\mu\alpha}g_{\nu\beta} + g_{\nu\alpha}g_{\mu\beta} - g_{\mu\nu}g_{\alpha\beta}) , \qquad (65)$$

and $\epsilon_{\mu\nu\alpha\beta}$ is the anti–symmetric Levi–Civita tensor in four dimensions. Substituting this identity in the current commutator, we obtain

$$[J_\mu(x), J_\nu(-x)] \overset{x^2\to 0}{\Longrightarrow} \left[\bar\psi(x)S_{\mu\nu\alpha\beta}\gamma^\beta\psi(-x) + i\epsilon_{\mu\nu\alpha\beta}\bar\psi(x)\gamma^\beta\gamma^5\psi(-x) \right.$$

$$\left. - \bar\psi(-x)S_{\nu\mu\alpha\beta}\gamma^\beta\psi(x) - i\epsilon_{\nu\mu\alpha\beta}\bar\psi(-x)\gamma^\beta\gamma^5\psi(x) \right] \frac{1}{8\pi}\partial^\alpha\epsilon(x^0)\delta(x^2) . \qquad (66)$$

We now perform a Taylor expansion on ψ and $\bar\psi$,

$$\bar\psi(x)\psi(-x) = \sum_n \frac{1}{n!}x^{\mu_1}\cdots x^{\mu_n}\bar\psi(0)\overleftrightarrow{\partial_{\mu_1}}\cdots\overleftrightarrow{\partial_{\mu_n}}\psi(0) . \qquad (67)$$

Putting this back into our expression for the commutator, we obtain

$$[J_\mu(x), J_\nu(-x)] = \sum_{n=1,3}^\infty \frac{1}{n!}x^{\mu_1}\cdots x^{\mu_n}O^{(n+1)}_{\beta\mu_1,\cdots,\mu_n}(0)S_{\mu\nu\alpha\beta}$$

$$\frac{1}{4\pi}\partial^\alpha\epsilon(x^0)\delta(x^2) , \qquad (68)$$

where

$$O^{(n+1)}_{\beta\mu_1,\cdots,\mu_n}(0) = \bar\psi(0)\gamma^\beta\overleftrightarrow{\partial_{\mu_1}}\cdots\overleftrightarrow{\partial_{\mu_n}}\psi(0) . \qquad (69)$$

We may note the following points regarding the above result.

- Only the odd terms in the sum survive. The even terms cancel out.
- We have expanded the operators in the vicinity of the light cone in a series of local operators–each of which multiplies the same singular function.
- Only a particular combination of Lorentz indices appears. We are interested only in the parity conserving terms, which is why the terms multiplying the anti–symmetric tensor $\epsilon_{\mu\nu\alpha\beta}$ do not appear. In general however, there will be an additional piece proportional to $\epsilon_{\mu\nu\alpha\beta}$ which contributes to $W_{\mu\nu}$. The corresponding structure function often referred to as F_3 is measured by parity violating currents, as for example is the case in deep inelastic neutrino scattering.
- O is a twist two operator. Twist is a term which refers to the 'dimension' - 'spin' of an operator. Our operator above has dimension $= 3/2 \times 2 + n$ and spin $= n + 1$. In general, the expansion of the operators on the light cone can be organized into an expansion over successively higher twists, called the operator product expansion (often known by its acronym OPE) the coefficients of higher twist operators being suppressed by powers of x^2. The dominant operators at short distances are those with the smallest twist. There are a finite number of twist two operators.

In general, the naive dimensions of the operators are modified by interactions and they acquire 'anomalous dimensions', which may be determined by a renormalization group analysis. We will not discuss the OPE any further, but refer the reader to some of the textbooks with excellent discussions of the topic [18–20].

We return from this digression to topic of immediate interest: the derivation of Bjorken scaling. Recall that we had

$$W_{\mu\nu} = \int d^4y\, e^{iq\cdot y} < P|[J_\mu(y), J_\nu(-y)]|P > .$$

We now substitute Eq. 68 in the RHS of the above. The matrix element of the symmetric, traceless operator $O^{(n+1)}$ between the hadronic states, has the tensorial structure,

$$< P|O^{(n+1)}_{\beta\mu_1,\cdots,\mu_n}(0)|P > = A_{n+1}\, p_\beta p_{\mu_1}\cdots p_{\mu_n} + B_{n+1}\, \delta_{\mu_1\mu_2}p_\beta p_{\mu_3}\cdots p_{\mu_n}$$
$$+ \text{ less singular terms}. \tag{70}$$

The second term above gives an additional power of x^2 when contracted with the coefficients and is therefore suppressed. The leading contribution then is

$$W_{\mu\nu} = \int d^4y\, e^{iq\cdot y} \sum_{n=1,3}^{\infty} \frac{(p\cdot y)^n}{n!} A_{n+1} S_{\mu\nu\alpha\beta}\frac{1}{4\pi}p^\beta \partial^\alpha \epsilon(x^0)\delta(x^2). \tag{71}$$

Define a function and its Fourier transform

$$\tilde{f}(p\cdot y) = \sum_{n=1,3}^{\infty} \frac{(p\cdot y)^n}{n!} A_{n+1} = \int \frac{dx}{2\pi} e^{ixy\cdot p}\frac{f(x)}{x}. \tag{72}$$

Substituting the above into $W_{\mu\nu}$ and using the identity

$$\int d^4y\, e^{iky}\delta(y^2)\epsilon(y^0) = (2\pi)^2\epsilon(k^0)\delta(k^2), \tag{73}$$

we obtain

$$W_{\mu\nu} = \int \frac{dx}{2\pi}\frac{f(x)}{x}p^\beta(q + xP)^\alpha S_{\mu\nu\alpha\beta}(2\pi)^2\epsilon(xP^0 + q^0)$$
$$\frac{1}{4\pi}\delta((xP + q)^2). \tag{74}$$

Using the definition of $S_{\mu\nu\alpha\beta}$ in Eq. 65 and performing the delta function integration which sets $x \equiv x_{Bj} = -q^2/2P\cdot q$, we can write the above finally as

$$W_{\mu\nu} = \frac{f(x)}{(P\cdot q)}\left(P_\mu - \frac{(P\cdot q)q_\mu}{q^2}\right)\left(P_\nu - \frac{(P\cdot q)q_\nu}{q^2}\right)$$
$$- \frac{f(x)}{2x}\left(g_{\mu\nu} - \frac{q_\mu q_\nu}{q^2}\right). \tag{75}$$

The electromagnetic tensor $W_{\mu\nu}$ has the most general tensorial decomposition,

$$W_{\mu\nu} = a_1 P_\mu P_\nu + a_2 P_\mu q_\nu + a_3 P_\nu q_\mu + a_4 q_\mu q_\nu + a_5 g_{\mu\nu}.$$

The symmetry properties require however that $a_2 = a_3$ and from current conservation $q^\mu W_{\mu\nu} = 0$, and similarly for $q^\nu W_{\mu\nu} = 0$, we obtain,

$$W_{\mu\nu} = \frac{F_2}{(p \cdot q)} \left(p_\mu - \frac{(p \cdot q)q_\mu}{q^2} \right) \left(p_\nu - \frac{(p \cdot q)q_\nu}{q^2} \right)$$
$$-F_1 \left(g_{\mu\nu} - \frac{q_\mu q_\nu}{q^2} \right), \tag{76}$$

where F_1 and F_2 are the structure functions. Comparing the above to our result Eq. 75, we observe that $F_2 = f(x_{Bj})$, which is the famous scaling phenomenon known as Bjorken scaling. Further, in this limit $F_1 = F_2/2x$– this result is known as the Callan–Gross relation.

We shall now show that the structure functions derived above can be simply related, in leading twist, to the light cone parton distributions and further show that F_2 thereby has the intepretation of being the probability that a quark has a fraction x of the total hadron momentum p^+ on the light front.

Consider the forward Compton scattering amplitude for the virtual photon scattering of the hadron in deep inelastic scattering,

$$T_{\mu\nu}(q^2, p \cdot q) = i \int d^4 z e^{iq \cdot z} < P \mid T(J_\mu(z)J_\nu(0)) \mid P > \equiv 2\mathrm{Im}W_{\mu\nu}. \tag{77}$$

This can be decomposed into longitudinal and transverse pieces

$$T_{\mu\nu} = \frac{p_\mu p_\nu}{M^2} t_\perp(x, q^2) - g_{\mu\nu} t_L(x, q^2), \tag{78}$$

just as for $W_{\mu\nu}$. Now, in the Bjorken limit, the Callan–Gross relation implies that the longitudinal piece above vanishes. To leading twist then, just as for the hadronic tensor, we can decompose the transverse component of the Compton amplitude as

$$t_\perp^{T=2} = \sum_{n=1}^{\infty} \int d^4 z \, e^{iq \cdot z} C_n^\beta(z^2) \, z^{\mu_1, \cdots, \mu_n} < P \mid \mathbf{O}_{\beta\mu_1, \cdots, \mu_n} \mid P >, \tag{79}$$

where, making the analogy to Eq. 68, the coefficient functions $C_n^\beta(z^2)$ are the same for all odd values of n and zero otherwise. Also, $\mathbf{O}$ is the operator defined in Eq. 69 [5]. One can define

$$\frac{2^n q^{\mu_1} \cdots q^{\mu_n}}{(-q^2)^{n+1}} \tilde{C}_n^\beta(q^2) = i \int d^4 z \exp(iq \cdot z) z^{\mu_1} \cdots z^{\mu_n} C_n^\beta(z^2). \tag{80}$$

---

[5] In general, the partial derivatives in Eq. 69 should be replaced by covariant derivatives.

Typically, the functions $\tilde{C}_n^\beta(q^2)$ are different and are the coefficient functions in the operator product expansion. However, in the scaling limit, they are constants. Substituting the above identity into our expression for $t_\perp^{T=2}$, we obtain

$$\frac{p \cdot q}{M^2} t_\perp^{T=2}(x, q^2) = \frac{-2q^2}{p \cdot q} \sum_{n=1,3}^{\infty} \left(\frac{2q_\beta}{q^2}\right) \left(\frac{2q_{\mu_1}}{q^2}\right) \cdots \left(\frac{2q_{\mu_n}}{q^2}\right)$$

$$< P \mid O^{\beta \mu_1 \cdots \mu_n} \mid P > . \qquad (81)$$

Since O is traceless and symmetric, we can again use the tensorial decomposition in Eq. 70. Then, since $x = -q^2/2p \cdot q$, we obtain

$$\frac{p \cdot q}{M^2} t_\perp^{T=2} = 4x \sum_{n=1,3}^{\infty} \left(\frac{-1}{x}\right)^{n+1} A_{n+1} . \qquad (82)$$

We can determine A_{n+1} by setting all the Lorentz indices in Eq. 70 to $+$. Then,

$$A_{n+1} = \left(\frac{1}{p^+}\right)^{n+1} < P \mid O^{++\cdots+} \mid P >_C . \qquad (83)$$

From the definition of the operator O in Eq. 69, the matrix element above is given, in light cone gauge $A^+ = 0$, by all two particle irreducible insertions of the vertex $\bar{\psi}\gamma^+(k^+)^n\psi$ (see Ref. [21] and references therein).

Let us now digress a little to discuss the light cone Fock space distribution. We will relate it subsequently to the structure functions above. Recall the decomposition we had in Eq. 6 of lecture 1 of the dynamical 2–spinor ψ_+. We can then define the light cone parton distribution function as

$$\frac{dN}{d^3k} = \frac{1}{(2\pi)^3} \sum_\lambda \left[b_\lambda^\dagger b_\lambda + d_\lambda^\dagger d_\lambda\right] . \qquad (84)$$

Writing this in terms of ψ_+ and using the light cone identity

$$\text{Tr}\left[\sum_\lambda \gamma^+ \psi_\lambda(x)\bar{\psi}(y)\right] = \sqrt{2}\text{Tr}\left[\sum_\lambda \psi_{+,\lambda}(x)\psi_{+,\lambda}^\dagger(y)\right] , \qquad (85)$$

we obtain

$$\frac{dN}{d^3k} = \frac{2}{(2\pi)^3} \int d^3x d^3y e^{-ik \cdot (x-y)} \text{Tr}\left[\gamma^+ S(x, y)\right] , \qquad (86)$$

where $S(x, y) = -i < T(\psi(x)\bar{\psi}(y)) >$. The light cone distribution function integrated over all momenta is the function

$$H(\alpha) = \int \frac{d^4k}{(2\pi)^4} \delta(\alpha - \frac{k^+}{p^+})\frac{1}{p^+} \text{Tr}\left[\gamma^+ \tilde{S}(p, k)\right] , \qquad (87)$$

where $\tilde{S}(p, k)$ is the fermion Green's function in momentum space. We will now show that the function $H(\alpha)$ is, in leading twist, the structure function F_2.

Returning now to Eq. 83, we find

$$\mathbf{A}_{n+1} = \frac{1}{(p^+)^{n+1}} \int \frac{d^4 k}{(2\pi)^4} (k^+)^n \text{Tr}\left[\gamma^+ \tilde{S}(p, k)\right]. \tag{88}$$

In terms of $H(\alpha)$ then,

$$\mathbf{A}_{n+1} = \int_{-\infty}^{\infty} d\alpha\, \alpha^n H(\alpha). \tag{89}$$

¿From the analytic properties of the function $H(\alpha)$, specifically the anti-commutation properties of the operators $\bar{\psi}\gamma^+$ and ψ on the light cone [21], one may conclude that $H(\alpha) = 0$ for $|\alpha| > 1$. Substituting the expression for $\mathbf{A}_{n+1}$ in the transverse Compton amplitude, we obtain,

$$\frac{p \cdot q}{M^2} t_\perp^{T=2} = 4 \int_{-1}^{1} d\alpha \sum_{n=1,3}^{\infty} \left(\frac{\alpha}{x}\right)^n H(\alpha). \tag{90}$$

Performing the sum over n and analytically continuing $t_\perp$ to the physical region $x \to x - i\epsilon$, with x real and $0 < x \le 1$,

$$\frac{p \cdot q}{M^2} t_\perp^{T=2} = 2x \int_{-1}^{1} d\alpha\, H(\alpha) \left\{ \frac{1}{x - \alpha - i\epsilon} - \frac{1}{x + \alpha - i\epsilon} \right\}. \tag{91}$$

Taking the imaginary part of the amplitude to obtain the structure functions, we get

$$F_2(x) = x(H(x) - H(-x)). \tag{92}$$

¿From the definition of $H(x)$ in Eq. 87, it is the probability to find a quark with momentum $k^+ = xp^+$ in the target. The function $-H(-x)$ has the interpretation of finding an anti–quark with momentum $k^+ = xp^+$ in the target. We have therefore, with Eq. 92, obtained the usual parton model interpretation of structure functions. In general, for a large but finite Q^2, the above result can be slightly modified to read

$$F_2(x, Q^2) = \int_0^{Q^2} d^2 k_t \frac{dN}{d^2 k_t dx}. \tag{93}$$

This follows simply from putting an upper cut–off Q^2 on the k_t integration in Eq. 87. Finally, we should mention that the multi–parton Fock distributions discussed in lecture 3 can be related by a similar analysis to the higher twist contributions to the forward Compton scattering amplitude [21,22].

Acknowledgements

I would like to thank the organizers of the Eleventh Chris Engelbrecht Summer School in Theoretical Physics for inviting me deliver these lectures in Cape Town, SA. In particular, I would like to thank Prof. Jean Cleymans for his gracious hospitality. I would also like to thank the students and other participants at the school who helped create a stimulating environment. This work was supported by the Danish Research Council and the Niels Bohr Institute.

References

1. P. A. M. Dirac, *Rev. Mod. Phys.* **21** (1949) 392.
2. S. Fubini and G. Furlan, *Physics* **229** (1965); see also, V. de Alfaro et al., *Currents in Hadron Physics*, North Holland, (1973).
3. L. Susskind, *Phys. Rev.* **165** (1968) 1535.
4. S. Weinberg, *Phys. Rev.* **150** (1966) 1313.
5. K. G. Wilson et al., *Phys. Rev.* **D49** (1994) 6720.
6. G. 'tHooft, *Nucl. Phys.* **B72** (1974) 461; *ibid.***B75** (1974) 461.
7. M. Burkardt, *Adv. Nucl. Phys.* **23** (1996) 1.
8. S. J. Brodsky and H-C. Pauli, *Invited lectures at 30th Schladming Winter School in Particle Physics*, Schladming, Austria, (1991).
9. S. J. Brodsky and G. P. Lepage, *Phys. Rev.* **D22** (1980) 2157.
10. J. Kogut and D. Soper, *Phys. Rev.* **D1** (1970) 2901.
11. J. Bjorken, J. Kogut and D. Soper, *Phys. Rev.* **D3** (1971) 1382.
12. A. Harindranath, *Lectures at International School on Light-Front Quantization and Non-Perturbative QCD*, Ames, IA, Jun (1996); hep-ph/9612244.
13. S. Brodsky, H-C. Pauli and S. S. Pinsky, *Phys. Rept.* **301** (1998) 299.
14. R. Venugopalan, to be published.
15. K. Bardakci and M. B. Halpern, *Phys. Rev.* **176** (1968) 1686.
16. Paul Hoyer, *Invited talk at APCTP - ICTP Joint International Conference (AIJIC 97) on Recent Developments in Nonperturbative Quantum Field Theory*, Seoul, Korea, May (1997).
17. S. D. Drell, D. J. Levy, and T.M. Yan, *Phys. Rev.* **187** (1969) 2159; *Phys. Rev. Lett* **22**, (1969) 744.
18. D. J. Gross, *Les Houches Lectures, session XXVIII*, Ed. R. Balian and Z. Zinn-Justin, (1975).
19. S. Pokorski, *Gauge field theories*, Cambridge Univ. Press, (1987).
20. G. Sterman, *An Introduction to Quantum Field Theory*, Cambridge Univ. Press, (1993).
21. R. L. Jaffe, *Nucl. Phys.* **B229** (1983) 205.
22. R. K. Ellis, W. Furmanski, and R. Petronzio, *Nucl. Phys.* **B207** (1982) 1; **B212** (1983) 29.

Chiral Symmetry Breaking in Hot Matter

Sandi P. Klevansky[1]

Institut für Theoretische Physik,
Philosophenweg 19, D-69120 Heidelberg,
Germany

Abstract. This series of three lectures covers (a) a basic introduction to symmetry breaking in general and chiral symmetry breaking in QCD, (b) an overview of the present status of lattice data and the knowledge that we have at finite temperature from chiral perturbation theory. (c) Results obtained from the Nambu–Jona-Lasinio model describing static mesonic properties are discussed as well as the bulk thermodynamic quantities. Divergences that are observed in the elastic quark-antiquark scattering cross-section, reminiscent of the phenomenon of critical opalescence in light scattering, is also discussed. (d) Finally, we deal with the realm of systems out of equilibrium, and examine the effects of a medium dependent condensate in a system of interacting quarks.

1 Introduction

Chiral symmetry is the symmetry of Quantum Chromodynamics (QCD) that dictates the static properties of the low lying mesonic sector, in particular those pertaining to the pseudoscalar nonet (π, K, η). This symmetry is responsible for the fact that, in its broken phase, quarks acquire mass (and are termed "constituent" quarks, as they form parts of hadrons, while, in the restored phase, quarks have only their small or current mass values. It is believed that at finite temperature this symmetry is restored, a feature that is strongly motivated by numerical studies of QCD on the lattice. Concomitantly with this picture, it is believed that another phase transition from a deconfined phase of matter (consisting of a hot fireball of quarks and gluons) to a confined phase can occur, in which only the final state of hadrons is observed. Given these two features, a large amount of scientific endeavor has been and will continue to be invested in the study of heavy ion collisions, in which high temperatures can be attained. In particular, the low-lying mesons are copiously produced, and since these provide the testing ground for chiral symmetry at $T = 0$, it is hoped that (with enough theoretical and experimental study), a clear signal of this phase transition will emerge. To be quite precise, one requires unambiguous signals of *both* phase transitions, that of confinement/deconfinement, as well as chiral symmetry breaking/restoration. Thus far, however, there are no unambiguous signals known that are experimentally measurable for either of these transitions. In this paper, we shall confine ourselves to a discussion of chiral symmetry and its associated aspects, leaving the difficulties of confinement to a later stage.

This series of three lectures is intended to introduce the concepts of chiral symmetry starting from basics. There is a short guide for the uninitiated into the ideas of what symmetry breaking is, and then an attempt to summarize the current status of what we know to be fact, taken on the theoretical level, at finite temperature. This involves examining firstly the lattice gauge simulations of QCD at finite temperature and then examining how far we can go with chiral perturbation theory [1,2]. From lattice gauge simulations, the existence of the chiral and deconfinement phase transitions is inferred. Critical exponents for the chiral transition have been obtained, but are as yet not conclusive. Temperature dependence of the mesonic screening masses have also been calculated, and the question of $U_A(1)$ symmetry restoration addressed. Bulk thermodynamic properties have been studied over several years, with larger and larger lattices, and this represents the state of the art of what we know today about these quantities in QCD. By contrast, while chiral perturbation theory gives a superb description of the low energy sector, and also gives the leading behavior expected of the order parameter as a function of temperature [5,6], it cannot *per se* be used to describe the phase transition region, which is non-analytic. The level of accuracy of CHPT at finite temperatures is illustrated in the calculation of the pion masses as a function of temperature in a recent publication [6].

Note that we restrict ourselves mainly to finite temperature and not to finite density. The first lattice simulations at finite density have already been performed [7]. However, there are many technical difficulties that are not yet under control, and as such no results are completely reliable as yet. For this reason, we will also not attempt to make any model discussions at finite density at this stage, although there are of course several.

In the second lecture, a simple chiral model, the Nambu–Jona-Lasinio (NJL) [8–10] model is discussed, in which it becomes evident that features relating to static properties of the low-lying mesons are excellently reproduced. This includes charge radii, meson-meson scattering lengths, polarizabilities, *etc*, and one can validate that the expected results of chiral perturbation theory are recovered, here with very few parameters. In addition, the variation of the meson masses with temperature, although calculated in this model as *pole* masses, shows the same qualitative behavior as was observed by the lattice gauge groups. Given these successes with this model, one is encouraged to study the dependence of all static mesonic properties as a function of the temperature in order to investigate whether abrupt behavior occurs at the phase transition point. For two flavors of quarks, one finds that the pseudoscalar sector in particular is typified by an almost constant behavior in all static properties (such as the mean pion radius $\langle r_\pi^2 \rangle^{1/2}$, the polarizabilities α_π and the scattering lengths a_π) for a wide range of the temperature shortly up until the point at which the chiral phase transition occurs, and then these quantities show a sharp divergence. This is true for the case in which the current quark mass of the up and down quarks $m_0^u = m_0^d = m_0 = 0$, and a phase transition can occur. When $m_0 \neq 0$, only a crossover can be observed in the order parameter. A new transition temperature $T_M = T_{\text{Mott}}$ is defined as being the temperature at which the mesonic states become unbound, or resonances. It thus respresents a delocalization of the mesons, rather than their deconfinement. The static properties of the pionic sector then remain constant for most of the temperature range, and diverge at the Mott temperature. One thus still observes a dramatic structure – either directly at the phase transition temperature itself in the case of $m_0 = 0$ or at $T = T_M$ for $m_0 \neq 0$.

It is also extremely interesting to study dynamical quantities such as the elastic cross-section for $q\bar{q} \rightarrow q\bar{q}$ scattering. This particular quantity displays a divergence at the critical or Mott temperature in a similar fashion as occurs in the phenomenon of critical opalescence that is observed in the scattering of light. However, although this feature and those observed for the static properties are exciting, their direct measurement is elusive if not downright impossible.

The scalar mesonic sector within the NJL model is observed to display a completely different behavior. Here the mass drops relatively quickly with temperature. Nevertheless, experimentally, the scalar mesons constitute a multiplet that appears to have the symmetry badly broken, and the lowest meson of which (the σ) has an extremely large width. Consequently only indirect information on this sector is useful.

How then can one hope to observe the chiral phase transition? To attempt to answer this question, we recall that the chiral phase transition appears to be intimately linked with the confinement/deconfinement phase transition, i.e. they appear to take place at the same temperature [11]. A heuristic understanding of this feature is quite satisfactory – it implies that at high temperatures, one should have chiral symmetry restored in a plasma phase, with free (current) quarks and gluons being the ingredients, while at $T < T_c$, the confined phase contains only hadrons that are made up of constituent (massive) quarks. Experimental effort to detect the quark-gluon plasma phase is concentrated on contructing hot and dense matter via heavy-ion collisions such as $Pb + Pb$ at increasingly high energies, and will form a main part of the program of the two accelerators RHIC at Brookhaven and the LHC (Geneva) that are currently under construction. Given the fact now that heavy-ion collisions take place over a small time scale, it is conceivable that the features of divergences occurring in both static and dynamical quantities might enter realistically into a *non-equilibrium* treatment of such collisions, which of course involves many particles, the lightest of which are the pions, and thus to measurable observables.

For this reason, the final lecture is devoted to a discussion of non-equilibrium physics of an interacting fermionic Lagrangian, and which is then applied to the Nambu–Jona-Lasinio model in the lowest possible terms in an appropriate double expansion in both $\hbar$ and the inverse number of colors

N_c [12,13]. Using the simplest approximations that lead to a semi-classical result, one can recover a Boltzmann like equation for the quark distribution function. Here one sees that the problems are simply open ended. The issue of constructing interlinked equations dealing with several species of particle must be confronted and the issue of multiparticle production (hadronization) must be addressed, since the usual Boltzman collision scenario that incorporates only binary collisions is inadequate for a relativistic description.

Obviously it is an impossible task to discuss all aspects of chiral symmetry breaking and restoration within three lectures, and for this reason I have been highly selective in the material presented. There are many, many studies in the literature involving chiral symmetry, and I am in no way attempting in this paper to be comprehensive. The interested reader may also refer to the work of Refs.[14] for treatments of the linear sigma model at finite temperature, for example, and to the work of Ref. [15] for discussions in the baryonic sector, in addition to the other general references that are given in the text.

The structure of this manuscript reflects the three lectures directly: in Section 2, current factual information on the chiral transition, taken from lattice gauge simulations and chiral perturbation theory is presented. In Section 3, the Nambu–Jona-Lasinio (NJL) model is used to present the ramifications of symmetry breaking at the critical temperature. In Section 4, a non-equilibrium formulation of a theory of interacting fermions is described and the equations are investigated for the NJL model. In the concluding section, we discuss where this could possibly lead to observable consequences.

2   Equilibrium Thermodynamics

In this section, we attempt to present those aspects of chiral symmetry at finite temperature that are regarded as being "exact" or factual, that is to say, they are derived from QCD itself, or from considerations thereof. We start by briefly introducing the reader to the general concept of symmetry breaking at $T = 0$. Following this, chiral symmetry breaking in the QCD Lagrangian is analysed. In the following subsection, the simulations of lattice gauge theory are discussed, dealing firstly with the temperature dependence of the order parameter, the critical exponents obtained at the phase transition, meson screening masses and the question of whether $U_A(1)$ symmetry is restored at high temperatures or not. Secondly, we indicate what is known from the lattice about bulk thermodynamic properties. The pressure density, energy density and entropy densities have been calculated on the lattice. These quantities give rather indications of the confinement/deconfinement transition, and as we will show in Section 3, cannot be described well by a model that contains chiral symmetry alone, and which ignores the confinement aspect.

In the final subsection, we briefly introduce the concepts of chiral perturbation theory (CHPT) and we describe the state of the art results at finite temperature. As will be seen, these give an important functional dependence at low temperatures, but cannot be expected to cope with the phase transition region, which is non-analytic.

2.1   Introduction to Chiral Symmetry at $T = 0$

The fact that a Hamiltonian, or equivalently a Lagrangian, is invariant under a symmetry transformation results in a degeneracy within the spectrum that is observed. Mathematically, one expresses the fact that a Hamilton function H is invariant under a specific symmetry via the statement

$$UHU^\dagger = H \tag{1}$$

where U is an element of the group corresponding to this symmetry. Now if one considers the states $|A\rangle$ and $|B\rangle$ that are related by the transformation U,

$$|B\rangle = U|A\rangle, \tag{2}$$

it follows that $|B\rangle$ and $|A\rangle$ are degenerate, since

$$E_A = \langle A|H|A\rangle = \langle B|H|B\rangle = E_B. \tag{3}$$

In order that this degeneracy manifest itself, however, it is necessary that the ground state of the system be invariant under such a transformation. Writing $|A\rangle$ and $|B\rangle$ in terms of creation operators,

$$|A\rangle = \phi_A|0\rangle \qquad \text{and} \qquad |B\rangle = \phi_B|0\rangle \tag{4}$$

with

$$U\phi_A U^\dagger = \phi_B, \tag{5}$$

one sees that Eq.(2) holds only if

$$|0\rangle = U|0\rangle, \tag{6}$$

i.e. the ground state is invariant under the symmetry group. Should this *not* be the case, one speaks of a *spontaneously broken symmetry*.

Denoting U as $U = \exp(i\varepsilon^a Q^a)$ in terms of the (continuous) group parameters ε^a and the generators of the symmetry

$$Q^a = \int d^3x\, J_0^a(x), \tag{7}$$

the statement Eq.(6) is seen to coincide with the equivalent form

$$Q^a|0\rangle \neq 0 \tag{8}$$

although

$$[Q^a, H] = 0. \tag{9}$$

The direct consequence of this statement is that $HQ^a|0\rangle = 0$, or that there must exist a spectrum of massless particles with quantum numbers specified by the generators of the symmetry. This constitutes the *Goldstone theorem*. To be more precise, one can formulate this as follows: given that a Hamiltonian has continuous symmetries described by groups G_1 requiring N_{G_1} generators, while the ground state is invariant under groups G_2 requiring $N_{G_2} < N_{G_1}$ generators, the spontaneous breakdown of chiral symmetry leads to the existence of $N_{G_1} - N_{G_2}$ Goldstone bosons [16].

Let us investigate now how this is applied to QCD.

2.2   Chiral Symmetry in QCD

In this section, we analyse the symmetries of quantum chromodynamics, and compare this with the symmetry of the vacuum, determined purely by viewing the experimental spectrum. Start by examining the QCD Lagrangian itself, which can be written in a compact fashion as

$$\mathcal{L}_{QCD} = \bar{\psi}(i\not{D} - m_0)\psi - \frac{1}{4}tr_c G_{\mu\nu}^a G_a^{\mu\nu}, \tag{10}$$

where $G_{\mu\nu}$ is the field strength tensor of the gluon field,

$$G_{\mu\nu}^a = \partial_\mu G_\nu^a - \partial_\nu G_\mu^a - g f_{abc} G_\mu^b G_\nu^c, \tag{11}$$

D^μ is the covariant derivative,

$$D_\mu = \partial_\mu + ig(\frac{1}{2}\lambda_a)G_\mu^a(x) \tag{12}$$

and f_{abc} are the structure constants of the symmetry group SU(3) [17]. The quark field is a *vector* in flavor space,

$$\psi = \begin{pmatrix} \psi_u(x) \\ \psi_d(x) \\ \psi_s(x) \\ \cdot \\ \cdot \\ \cdot \end{pmatrix} \tag{13}$$

and the (current) quark mass matrix is a diagonal matrix in flavor space,

$$m_0 = diag[m_0^u, m_0^d, m_0^s, ...], \tag{14}$$

so that the second term in Eq.(10) is

$$\bar{\psi}m_0\psi = \sum_f m_0^f \bar{\psi}_f \psi_f. \tag{15}$$

If the quarks are massless, then the Lagrangian Eq.(10) contains no term of the form Eq.(15) which can mix left and right handed components of the quark fields, that are defined as

$$\psi_{R,L} = \frac{1}{2}(1 \pm \gamma^5)\psi, \tag{16}$$

i.e. these two fields are independent, and the Lagrangian remains invariant under transformations that individually transform these fields,

$$\psi_{R,L} \to U_{R,L}\psi, \qquad U_{R,L} \in U(N_f), \tag{17}$$

and these are called chiral symmetries. However, a mass term of the form Eq.(15) spoils this invariance since

$$\text{Terms} \sim \bar{\psi}\psi = \bar{\psi}_L\psi_R + \bar{\psi}_R\psi_L \tag{18}$$

mix left and right handed fields. Thus the term $m_0\bar{\psi}\psi$ constitutes an *explicit* symmetry breaking.

The QCD Lagrangian Eq.(10) is invariant under several transformations, such as

$$\psi \to \psi' = e^{i\alpha}\psi$$
$$\psi \to \psi' = e^{i\alpha\lambda^a}\psi \tag{19}$$

etc. Accordingly, there are conserved Noether currents that correspond to these symmetries. They are

$$V_0^\mu = \bar{\psi}\gamma^\mu\psi \qquad A_0^\mu = \bar{\psi}\gamma^\mu\gamma^5\psi \tag{20}$$

$$V_a^\mu = \bar{\psi}\gamma^\mu\frac{\lambda_a}{2}\psi \qquad A_a^\mu = \bar{\psi}\gamma^\mu\gamma_5\frac{\lambda_a}{2}\psi \tag{21}$$

Among these is the current A_0^μ, which corresponds to the transformation $\psi \to \psi' = \exp(i\gamma_5\alpha)\psi$, where α is a continuous parameter. However, despite its appearance, this current is *not* conserved,

$$\partial^\mu A_\mu^0 = \frac{N_c}{8\pi^2} tr_c G_{\mu\nu}\tilde{G}^{\mu\nu}. \tag{22}$$

This means that it does not reflect an underlying symmetry of the Lagrangian and its breaking was resolved by 't Hooft as being due to the presence of instantons [18].

One may thus identify the (continuous) symmetry groups of QCD as being generated by the charges of the remaining symmetries, and this is

$$G_1 = SU_L(N_f) \otimes SU_R(N_f) \otimes U_V(1). \tag{23}$$

On the other hand, by examining the particle spectrum that is observed experimentally, one finds that the symmetry of the vacuum is

$$G_2 = SU_V(N_f) \otimes U_V(1). \tag{24}$$

Accordingly, there must be $N_f^2 - 1$ massless Goldstone particles and these have the quantum numbers obtained from applying the axial charge operators to the vacuum, i.e. $J^P = O^-$. In the case of two flavors, there are three such states, which are identified as corresponding to the charged and neutral pions. For three flavors, one identifies the eight pseudoscalars as the pions, kaons and eta. One sees that the explicit symmetry breaking in this case is larger: $m_0^s \simeq 150 \text{MeV}$ in comparison with $m_0^u \simeq m_0^d \simeq 5$ MeV.

The phase in which a system finds itself is usually characterized by an *order parameter*. This is an operator that transforms in a non-trivial fashion under the broken symmetry. Generally order parameters have the property of being zero in the symmetric or restored phase and non-zero in the spontaneously broken phase, but this is not necessarily so. There are many possible ways of choosing an order parameter. The major criterion for doing so is that the order parameter should display that same invariances as the ground state. In the case of quantum chromodynamics, the ground state of QCD is invariant under Lorentz transformations and spatial reflections. The order parameter must thus be invariant under these same symmetries, and as such must be a scalar. The operator $\bar{\psi}\psi$ is the simplest choice. One thus makes the choice of $\langle \bar{\psi}\psi \rangle$, which is referred to as the quark condensate.

2.3   Lattice Gauge Simulations

Simulations of QCD on the lattice provide the most exact knowledge that we have of this theory that is derived from the QCD Lagrangian itself. The Lagrangian is discretized in space and time dimensions, and the variation with respect to the temperature of physical quantities is formally controlled by varying the size of the lattice in the temporal direction [3,4], since

$$T = 1/N_\tau a, \tag{25}$$

where a is the lattice size and N_τ the temporal extent. In what follows, we simply list the major results that have been extracted via this methodology over the past few years. We show the temperature dependence of the chiral and deconfinement order parameters, discuss critical exponents, meson screening masses and $U_A(1)$ resotration. Finally, we show plots of the bulk thermodynamic quantities.

Order parameters   The following recent results [3,4] have emerged from the lattice gauge studies:

- The pure gauge sector of QCD displays a well-established first order chiral transition at a rather high critical value of the temperature, $T_c = 270(5)$ MeV. The bulk properties for such a system are also well known [19].
- Full QCD including fermions displays a chiral phase transition at far lower critical temperature than that observed for pure gluonic systems. One finds $T_c \simeq 150$ MeV for two flavors of quark.

- Studies of the Polyakov loop for quenched QCD places the critical temperature determined from the order parameter for deconfinement, T_D at about the same temperature at which the chiral transition T_c occurs [11], i.e.

$$T_D \simeq T_c \qquad (26)$$

This can be directly seen from Fig. 1, in which the order parameter for the chiral and deconfinement transitions are shown, together with their susceptibilities, as a function of $\beta = 6/g^2$, over the transition region. Large (small) values of β represent the high (low) temperature regime.

Based on these points, our physical (but heuristic!) understanding of the situation is that, at low energies, one has only hadronic states. These can be thought of as consisting of quarks carrying a dynamically generated quark mass $m = m_u = m_d$ for two flavors, and constructed into baryonic states or mesonic states according to the Goldstone theorem. At the temperature at which where chiral symmetry is restored, T_c, and the constituent quarks take on their current mass value, deconfinement occurs simultaneously. The hadronic states dissolve, and one moves to a plasma containing only quark and gluonic degrees of freedom.

Critical Exponents An obvious question that one may pose, when faced with a phase transition, is what are the critical exponents that govern the transition? Pisarski and Wilczek suggested that the dynamics of QCD is controlled by an effective scalar Lagrangian, constructed along the lines of the linear σ model [21], and, which for two flavors of quarks, has $SU(2) \otimes SU(2) = O(4)$ symmetry. Now, according to arguments of universality [22], only the symmetry structure and dimensionality determine the values of the critical exponents, i.e. one expects that one should obtain the critical exponents of a 3D $O(4)$ symmetric spin model. The task of studying the critical exponents has been undertaken by a lattice group [4]. Noting that the masses, which are responsible for *explicit* chiral symmetry breaking, play an analogous role to that of a magnetic field in the superconducting transition, these authors [3,4] have adopted the convention of defining a scaled quark mass as $h = m_q/T$ and the reduced temperature $t = (T - T_c)/T_c$. With this convention, the free energy density scales as

$$f(t, h) = -\frac{T}{V} \ln Z = b^{-1} f(b^{y_t} t, b^{y_h} h), \qquad (27)$$

introducing the thermal (y_t) and magnetic (y_h) critical exponents. Here b is an arbitrary scaling factor. There are various scaling relations that can be derived using Eq.(27). In particular, one can show that the chiral order parameter scales as

$$\langle \bar{\psi}\psi \rangle(t, h) = h^{1/\delta} F(z), \qquad (28)$$

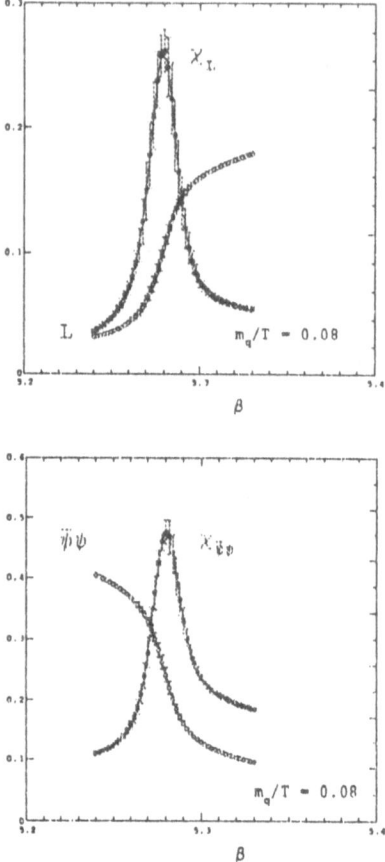

Fig. 1: Order parameters for the chiral and deconfinement transitions (lower and upper figures, respectively) are plotted as a function of the inverse QCD coupling $\beta = 6/g^2$. High (low) values of β correspond to high (low) values of the temperature. The associated susceptibilities are also plotted in each case. Courtesy of [20].

with $z = th^{1/\beta\delta}$, and the chiral susceptibility, defined as $\chi_m(t, h) = \partial\langle\bar{\psi}\psi\rangle/\partial m$, via

$$\chi_m(t, h) = \frac{1}{\delta}h^{\frac{1}{\delta}-1}[F(z) - \frac{z}{\beta}F'(z)]. \qquad (29)$$

The more familiar critical exponents δ and β are related to y_t and y_h via

$$\beta = \frac{(1 - y_h)}{y_t} \qquad \text{and} \qquad \delta = \frac{y_h}{1 - y_h}. \qquad (30)$$

The heights of the peaks of the susceptibilities scale themselves with the behavior

$$\chi_m^{\text{peak}} \sim m^{-z_m} \qquad \text{and} \qquad \chi_t^{\text{peak}} \sim m^{-z_m} \qquad (31)$$

with $z_m = 2 - 1/y_h$ and $z_t = (y_t - 1)/y_h + 1$.

	y_h	y_t	z_m	z_t
$O(4)$	0.83	0.45	0.79	0.34
$O(2)$	0.83	0.50	0.79	0.39
MF	3/4	1/2	2/3	1/3

Table 1: Critical exponents for $O(4)$, $O(2)$ and mean field theory (MF). Taken from [4].

The expected values for the critical exponents for the case of $O(4)$ symmetry, $O(2)$ symmetry, and mean field exponents (MF) are listed in Table I, in the form of y_h, y_t, and the corresponding values of z_m and z_t. The $O(2)$ symmetry exponents are also listed, because at finite lattice spacing, the exact chiral symmetry of the staggered fermion action is $U(1) \simeq O(2)$. Only sufficiently close to the continuum limit does one expect to find $O(4)$ exponents.

The calculated results for the exponents themselves, evaluated on different spatially sized lattices, are summarized in Table 2. Comparing Tables 1 and 2, one sees that at this stage, no definitive statement about the symmetry of the underlying Lagrangian can be made from lattice gauge theory. This is an indicator that vital study in this field is still necessary to determine the underlying symmetry group conclusively. It is probably necessary to increase the lattice sizes and move to smaller masses.

	8^3	12^3	16^3
z_m	0.84(5)	1.06(7)	0.93(8)
z_t	0.63(7)	0.94(12)	0.85(12)

Table 2: Critical exponents, as a function of the lattice size. Taken from Ref.[4].

Meson Screening Masses and $U_A(1)$ Restoration One of the questions that has raised some theoretical interest in the last few years is whether the $U_A(1)$ symmetry, i.e. the symmetry $\psi \to e^{i\alpha\gamma_5}\psi$, which leads to the *nonconserved current* A_0^μ that is given in Eq.(20) is also restored at finite temperature, at some point. For three flavors, this occurs trivially. A demonstration of this, following Ref.[23] is given.

In SU(3), the statement that $U_A(1)$ is restored, implies that $m_\pi = m_{\eta'}$. Since the masses of the particles are determined from the vacuum expectation values of the appropriate meson-meson correlators, we need to show only that

$$\langle \phi_3(x)\phi_3(0)\rangle = \langle \phi_0(x)\phi_0(0)\rangle, \tag{32}$$

where $\phi_3(x) = \bar{\psi}(x)i\gamma_5\lambda_3\psi(x)$ is the correlator for the π_0 and $\phi_0(x) = \bar{\psi}i\gamma_5\lambda_0\psi(x)$ is that for the η'. If one considers the specific axial transformation

$$\psi(x) \to \psi'(x) = e^{i\gamma_5(\sqrt{3}\lambda_8 - \lambda_3)\frac{\pi}{4}}\psi(x), \tag{33}$$

then, after a little algebra, one finds that the composite fields transform as

$$\phi_3(x) \to \phi'_3(x) = \sqrt{\frac{2}{3}}\phi_0(x) + \sqrt{\frac{1}{3}}\phi_8. \tag{34}$$

The correlator composed of these composite fields itself then transforms as

$$
\begin{aligned}
\langle \phi_3(x)\phi_3(0)\rangle &\to \langle \phi'_3(x)\phi'_3(0)\rangle \\
&= \frac{2}{3}\langle \phi_0(x)\phi_0(0)\rangle + \frac{1}{3}\langle \phi_8(x)\phi_8(0)\rangle \\
&\quad + \frac{\sqrt{2}}{3}\langle \phi_0(x)\phi_8(0)\rangle + \frac{\sqrt{2}}{3}\langle \phi_8(x)\phi_0(0)\rangle
\end{aligned}
\tag{35}
$$

The last two terms of this expression vanish, since the system is assumed to be $SU_V(3)$ symmetric. In addition, this implies that $\langle \phi_3(x)\phi_3(0)\rangle = \langle \phi_8(x)\phi_8(0)\rangle$, so that Eq.(35) implies that

$$\langle \phi_0(x)\phi_0(0)\rangle = \langle \phi_3(x)\phi_3(0)\rangle, \tag{36}$$

or that $m_{\eta'} = m_\pi$.

In retrospect, it is simple to understand why the symmetry must be restored. Noting that mathematical constructions containing traces of fields *preserve* the symmetry, while determinants or antisymmetric functions violate it, one sees that the lowest order combination of fields that would violate $U_A(1)$ would involve the completely antisymmetric tensor, and consequently contain *three* field combinations. Since one requires here only *two* field operator combinations in order to construct a meson-meson correlator, this must be $U_A(1)$ invariant in the chirally restored phase. This leads to the definitive statement: for $n < N_f$, all n-point functions in the chirally restored phase are $U_A(1)$ invariant.

From the previous argument, it is evident that in SU(2) the situation is more complicated. There are two independent chiral multiplets in this case: (σ, π) and (η', a_0). In lattice studies, the behavior of the masses of the π and the a_0 have been calculated. Here the integral over the correlators has been studied,

$$\chi_{M_i} = \int d^4x \langle \phi_i(x)\phi_i(0)\rangle \tag{37}$$

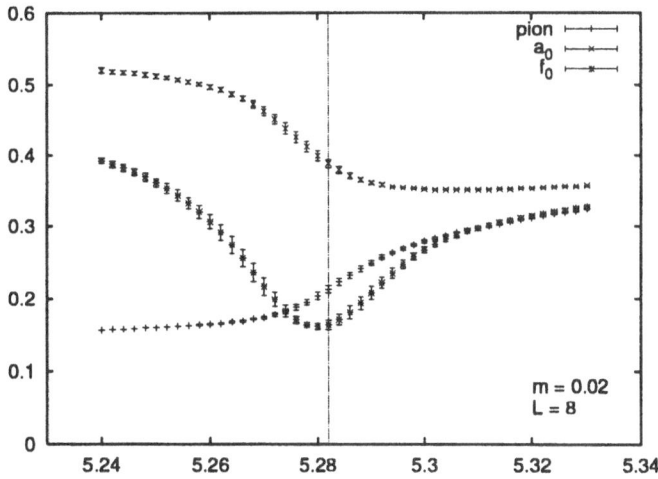

Fig. 2: Meson screening masses plotted as a function of $\beta = 6/g^2$. The critical value $\beta = \beta_c$ is indicated by the vertical line.

for $M_i = \pi$ or a_0, and the leading behavior of these correlators is assumed to be $\chi_{M_i} \sim m_{M_i}^{-2}$. A plot of the "screening masses" obtained in this fashion is shown in Fig. 2 as a function of $6/g^2$, with g the coupling in the QCD Lagrangian, which again represents increasing temperature over the region of the phase transition. It is interesting to note that the π and σ have become degenerate: in this picture, this occurs at some temperature slightly larger than T_c, with the σ undershooting the π curve and approaching it from below. That the σ meson undershoots the π curve is not expected from model calculations and may be a lattice artifact. This will be discussed in the following sections. One sees in Fig. 2 that the mass of the other scalar, the a_0, drops with temperature, but not as drastically as does the σ. One observes that it does *not* become degenerate with the π and σ over the temperature range indicated. Thus it does not appear from this particular calculation that $U_A(1)$ symmetry is restored in SU(2). An alternate approach, however, using the scaling arguments of Brown and Rho [24], *does* however indicate a degeneracy at the transition temperature [4]. Thus, in this section once again, the question of the restoration of $U_A(1)$ symmetry is not resolved.

Bulk Thermodynamic Quantities One of the most important contributions that lattice physics is able to provide are calculations of bulk thermodynamic quantities. In particular, the energy density and pressure densities are given by

$$e = \frac{T^2}{V} \frac{\partial}{\partial T} \ln Z \tag{38}$$

and

$$p = T \frac{\partial}{\partial V} \ln Z \tag{39}$$

in terms of the partition function Z. In practice [25], the pressure density is obtained from integration of the difference of the action densities at zero and finite temperature,

$$\frac{p}{T^4}\Big|_{\beta_0}^{\beta} = N_\tau^4 \int_{\beta_0}^{\beta} d\beta (S_0 - S_T). \tag{40}$$

Note that this quantity is defined in such a way that $p/T^4_{T=0} = 0$, in contrast to setting the usual thermodynamic limit of Nernst, i.e. the entropy $S(T = 0) = 0$ [26]. While this does not affect anything that follows, one should bear this in mind when making model comparisons, as will be done in Section 3.

In the following figures, we have chosen to illustrate the pressure and energy densities for lattice simulations that include quark degrees of freedom, rather than simply quenched QCD. In Fig. 3, we show the pressure density, plotted as a function of the scaled temperature, for four flavor QCD on a $16^3 \times 4$ lattice. A comparison is made on varying the quark masses, and using quenched QCD, in the latter case with appropriate scaling of the number of degrees of freedom. One sees that there is a *sharp* rise in the pressure density at $T = T_c$, and the curve tends to the Stefan-Boltzmann limit, but does not reach it over the temperature range $(3.5T_c)$ shown. The deviation from the ideal gas limit appears to be too large to be described by perturbation theory, suggesting here that the perturbative regime occurs for temperatures $T \gg T_c$.

The energy density of four flavor QCD on a $16^3 \times 4$ lattice is shown in Fig. 4. In this case, the energy density remains close to the ideal gas limit at temperatures of the order of $3T_c$, but overshoots it and approaches it from above for finite values of the quark mass. Whether this is a lattice artifact or not is presently unclear.

In concluding this section, we see that lattice gauge simulations are reaching a point where one may obtain "exact" results that stem directly from the discretized QCD Lagrangian. These can be used as a guide for constructing simple models, and conversely, simple models and simple predictions based solely on symmetry considerations such as discussed here[1] may be used as a

---

[1] Chiral random matrix theory [27] also falls into this category.

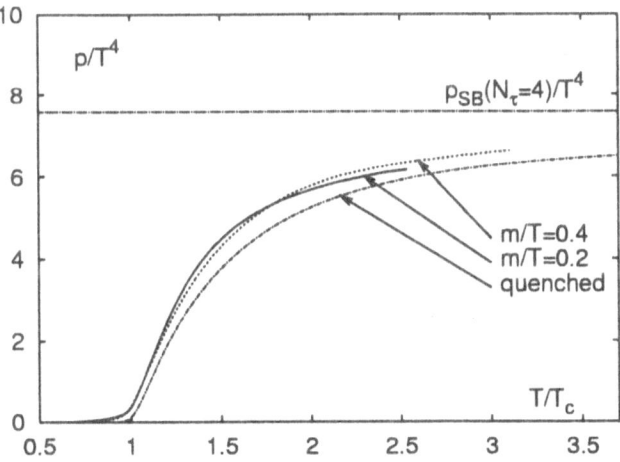

Fig. 3: Pressure density, plotted as a function of the scaled temperature T/T_c. (Taken from [3].)

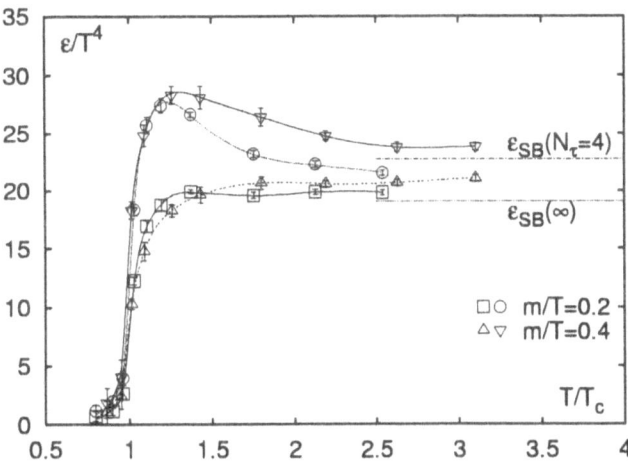

Fig. 4: Energy density plotted as a function of the scaled temperature, T/T_c. (Taken from [3].)

guide for the interpretation of the numerical results. As has emerged here, there are still many questions that are open for study.

With this, we turn to a different approach which is regarded by its protagonists as being an exact low energy representation of QCD, viz. chiral perturbation theory and investige what is known at finite temperature.

2.4   Chiral Perturbation Theory

A Brief Introduction Chiral perturbation theory starts with the premise that an effective Lagrangian for QCD at low temperatures can be written solely in terms of the observed baryonic (here mesonic) degrees of freedom, in such a way that global chiral symmetry is enforced. This is done in its most general form by collecting the mesonic degrees of freedom into the field

$$U(x) = \exp(i\pi^a \tau_a / F), \tag{41}$$

where π^a are the $SU(2)$ pion fields, τ_a the Pauli matrices, and F the pion decay constant, and contructing a Lagrangian density that is ordered in momenta. Such an expansion for the Lagrangian only starts at $O(p^2)$, and must contain an even number of derivatives in order to be Lorentz invariant. Writing

$$\mathcal{L}_{QCD} \to \mathcal{L}_{eff} = \mathcal{L}_{eff}^{(2)} + \mathcal{L}_{eff}^{(4)} + \mathcal{L}_{eff}^{(6)} + \ldots, \tag{42}$$

the lowest leading order term is

$$\mathcal{L}_{eff}^{(2)} = \frac{1}{4} F^2 \text{tr}(\partial_\mu U \partial^\mu U^\dagger), \tag{43}$$

which, taken on its own, is the (non-renormalizable) sigma model. QCD, as we have already discussed, is however *not* completely invariant under chiral symmetry. There is an explicit breaking of the symmetry due to the presence of the current quark mass matrix. The symmetry breaking term is in general given as

$$\mathcal{L}_{sb} = f(U, \partial U, \ldots) \times m^0, \tag{44}$$

where m^0 is the (real and diagonal) current quark mass matrix. One incorporates this into the effective Lagrangian by making not only an expansion in powers of the derivatives, but also in powers of m^0, i.e. $\mathcal{L}_{sb} \sim f(U) \times m^0$ to leading order. More precisely, this term takes the form (that is Lorentz and parity invariant)

$$\mathcal{L}_{sb} = \frac{1}{2} F^2 B \text{tr}(m^0(U + U^\dagger)), \tag{45}$$

introducing the new constant B. This is generally included in the definition of $\mathcal{L}_{eff}^{(2)}$, i.e.

$$\mathcal{L}_{eff}^{(2)} = \frac{1}{4} F^2 \text{tr}(\partial_\mu U \partial^\mu U^\dagger) + \frac{1}{2} F^2 B \text{tr}(m^0(U + U^\dagger)). \tag{46}$$

In this reckoning, one can thus regard m^0 as being of $O(p^2)$.

To make physical sense of the constant B, one may expand the field $U = \exp(i\pi \cdot \tau / F)$ in powers of the pion field π. The symmetry breaking part of the Lagrangian then becomes

$$\mathcal{L}_{sb} = (m_u^0 + m_d^0) B [F^2 - \frac{1}{2}\pi^2 + \frac{1}{24}\pi^4 F^{-2} + \ldots] \tag{47}$$

The first term in this expansion gives the vacuum energy generated by the symmetry breaking. The second term generates the pion mass, while the higher order terms describe further interactions of the π fields. By direct analogy with the QCD Hamiltonian, we know that the derivative of H_{QCD} with respect to m_u^0 generates the operator $\bar{u}u$. Thus the derivative of the vacuum energy with respect to the current quark mass gives the vacuum expectation value of this operator. Applying this to $\mathcal{L}_{eff}$, one has

$$\langle 0|\bar{u}u|0\rangle = \langle 0|\bar{d}d|0\rangle = -F^2 B\{1 + O(m)\}, \tag{48}$$

indicating that B is related to the condensate. Since the pion mass is given as

$$m_\pi^2 = (m_u^0 + m_u^0)B\{1 + O(m)\}, \tag{49}$$

one obtains the Gell-Mann-Oakes-Renner (GOR) relation [28],

$$F_\pi^2 M_\pi^2 = (m_u^0 + m_d^0)|\langle 0|\bar{u}u|0\rangle| \tag{50}$$

from Eqs.(48) and (49), on eliminating B.

To order p^4, the effective Lagrangian would contain two additional independent terms in the event that no current quark mass were present, i.e. one would include two new terms

$$\mathcal{L}_{eff}^{(4)} = \frac{1}{4}l_1(\text{tr}\{\partial_\mu U^\dagger \partial^\mu U\})^2 + \frac{1}{4}l_2\text{tr}(\partial_\mu U^\dagger \partial_\nu U)\text{tr}(\partial^\mu U^\dagger \partial^\nu U). \tag{51}$$

with new low energy constants l_1 and l_2. Including the current quark mass matrix again to construct an explicit symmetry breaking terms requires the inclusion of further additional terms, as was the case for $\mathcal{L}_{eff}^{(2)}$. For most purposes, this is sufficient. However, to obtain the most general form from which all propagators can be derived, it is useful to introduce *external fields* into the Lagrange density. Here the essential additions are $v_\mu(x)$ and $a_\mu(x)$ that are vector and axial vector in nature and which can be regarded as being of order p. Then, using the original notation of Ref.[1], the complete set of terms that contribute to $\mathcal{L}_{eff}^{(4)}$ were worked out by these authors and found to be, for SU(3)

$$\begin{aligned}
\mathcal{L}_{eff}^{(4)} = &\ L_1\langle \nabla_\mu U^\dagger \nabla^\mu U\rangle^2 + L_2\langle \nabla_\mu U^\dagger \nabla_\mu U\rangle\langle \nabla^\mu U^\dagger \nabla^\nu U\rangle \\
&+ L_4\langle \chi U^\dagger + \chi^\dagger U\rangle\langle \nabla U^\dagger \nabla^\mu U\rangle + L_6\langle \chi U^\dagger + \chi^\dagger U\rangle^2 \\
&+ L_8\langle \chi U^\dagger \chi U^\dagger + U\chi^\dagger U\chi^+\rangle + L_{10}\langle U^\dagger F_R^{\mu\nu} U F_{L\mu\nu}\rangle \\
&+ H_1\langle F_R^{\mu\nu} F_{R\mu\nu} + F_L^{\mu\nu} F_{L\mu\nu} + H_2\langle \chi^\dagger \chi\rangle + L_7\langle \chi U^\dagger - U\chi^\dagger\rangle^2 \\
&+ L_5\langle \nabla_\mu U\nabla^\mu U^\dagger(\chi U^\dagger + U\chi^\dagger)\rangle + iL_9\langle F_{\mu\nu}^L \nabla^\mu \nabla^\nu U^\dagger + F_{\mu\nu}^R \nabla^\mu \nabla^\nu U\rangle \\
&+ L_3\langle \nabla_\mu U\nabla^\mu U^\dagger \nabla_\nu U\nabla^\nu.U^\dagger\rangle
\end{aligned} \tag{52}$$

where using a different notation to Eq. (51) now, the low energy constants L_1 to L_{10}, and H_1 and H_2 have been introduced. The angular brackets are a shorthand notation for the trace. In this expression, one notes that the covariant derivative that is constructed using the external field must now appear,

$$\nabla_\mu = \partial_\mu - i\{a_\mu, U\}, \tag{53}$$

and $F^{\mu\nu}$ is the field strength tensor constructed from the *external* field, i.e.

$$F_{R,L}^{\mu\nu} = \pm\partial^\mu a^\nu \mp \partial^\nu a^\mu - i[a^\mu, a^\nu]. \tag{54}$$

Terms involving the current quark mass have been summarized into the field $\chi = 2B\hat{m}$, with $\hat{m} = (m_0^u + m_0^d)/2$. Note that the low energy constants L_i become renormalized when physical quantities are calculated, as this theory is perturbatively renormalizable *order by order*. A certain number of such physical quantities that are measured in experiment must then be used to fit the renormalized parameters at a given mass scale. Given definite values for these constants, predictions of other quantities can then be made.

Three ingredients are essential to any application that attempts to calculate quantities for chiral perturbation theory to a specific order. For example, should one wish to calculate to $O(p^4)$, the following steps must be taken: (1) The general $\mathcal{L}_{eff}^{(2)}$ of order p^2 is to be used at both the tree and one loop level. (2) The general $\mathcal{L}_{eff}^{(2)}$ of order p^4 is to be used only at tree level. (3) A renormalization program must be implemented to make physical predictions. The extension of this procedure to higher powers in p^2 is obvious.

Let us look at a standard example for the derivation of the pion mass [29]. In what follows, we denote the low energy constants appropriate to SU(2) [2] two flavors as being $L_i^{(2)}$. If one expands the Lagrangians $\mathcal{L}_{eff}^{(2)}$ and $\mathcal{L}_{eff}^{(4)}$ in terms of the pion fields, one finds

$$\mathcal{L}_{eff}^{(2)} = \frac{2}{3}[\partial^\mu\pi\partial_\mu\pi - m^2\pi\cdot\pi] + \frac{m^2}{6F^2}[(\pi\cdot\partial^\mu\pi)(\pi\cdot\partial_\mu\pi) - (\pi\cdot\pi)(\partial^\mu\pi\cdot\partial_\mu\pi)] + O(\pi^6), \tag{55}$$

while

$$\mathcal{L}_{eff}^{(4)} = \frac{m^2}{f^2}[16L_4^{(2)} + 8L_5^{(2)}]\frac{1}{2}\partial_\mu\pi \cdot \partial^\mu\pi$$

$$- \frac{m^2}{F^2}[32L_6^{(2)} + 16L_8^{(2)}]\frac{1}{2}\hat{m}^2\pi \cdot \pi + O(\phi^4). \tag{56}$$

The terms in $\mathcal{L}_{eff}^{(4)}$ that are of order π^4 contribute to physical quantities via one loop diagrams and one therefore does not need to consider these in a calculation to order $O(p^4)$. What is required however, are the one loop diagrams that are generated by $\mathcal{L}_{eff}^{(2)}$. For a calculation of the the renormalized pion mass, however, one can avoid evaluating any diagrams at all by simply

---

[2] These can be simply related to the l_i of Eq. (51), and the reader is referred to [2] for explicit details.

considering all possible contractions of two fields in these terms in $\mathcal{L}_{eff}^{(4)}$, to arrive at an "effective" effective Lagrangian, that takes the form

$$
\begin{aligned}
\mathcal{L}_{eff}^{(4)} &= \frac{1}{2}\partial^\mu \pi \cdot \partial_\mu \pi - \frac{1}{2}m^2 \pi \cdot \pi + \frac{5m_\pi^2}{12m^2}I(m_\pi^2)\pi \cdot \pi \\
&+ \frac{1}{6F^2}(\delta_{ik}\delta_{jl} - \delta_{il}\delta_{kl})I(m_\pi^2)(\delta_{ij}\partial^\mu \pi_k \partial_\mu \pi_l + \delta_{kl}m_\pi^2 \pi_i \pi_j) \\
&+ \frac{1}{2}\partial_\mu \pi \partial^\mu \pi \frac{m_\pi^2}{F_\pi^2}[16L_4^{(2)} + 8L_5^{(2)}] - \frac{1}{2}m_\pi^2 \pi \cdot \pi \frac{m_\pi^2}{F_\pi^2}[32L_6^{(2)} + 16L_8^{(2)}].
\end{aligned}
$$

(57)

In obtaining this result, the Feynman propagator

$$
i\Delta_{Fjk}(0) = \langle 0|\pi_j(x)\pi_k(x)|0\rangle = \delta_{jk}I(m_\pi^2)
$$

(58)

has been introduced and is written in terms of the integral

$$
I(m_\pi^2) = \mu^{4-d}\int \frac{d^d k}{(2\pi)^d}\frac{i}{k^2 - m_\pi^2} = \mu^{4-d}(4\pi)^{d/2}\Gamma(1 - \frac{d}{2})(m_\pi^2)^{\frac{d}{2}-1}
$$

(59)

that is treated with dimensional regularization, d being an arbitrary dimension. In addition, use has been made of the fact that derivatives of the Feynman propagator, defined as

$$
-\partial_\mu \partial_\nu i\Delta_{Fjk}(0) = \langle 0|\partial_\mu \pi_j(x)\partial_\nu \pi_k(x)|0\rangle = \delta_{ij}I_{\mu\nu}(m_\pi^2)
$$

(60)

can be expressed in terms of the integral $I(m_\pi^2)$ via

$$
I_{\mu\nu}(m_\pi^2) = \mu^{4-d}\int \frac{d^d k}{(2\pi)^d}k_\mu k_\nu \frac{i}{k^2 - m_\pi^2} = g_{\mu\nu}\frac{m_\pi^2}{d}I(m_\pi^2).
$$

(61)

Regrouping the kinetic and mass terms, Eq.(57) becomes

$$
\begin{aligned}
\mathcal{L}_{eff} &= \frac{1}{2}\partial^\mu \pi \cdot \partial_\mu \pi[1 + (16L_4^{(2)} + 8L_5^{(2)})\frac{m_\pi^2}{F_\pi^2} - \frac{2}{3F_\pi^2}I(m_\pi^2)] \\
&- \frac{1}{2}m^2 \pi \cdot \pi[1 + (32L_6^{(2)} + 16L_8^{(2)})\frac{m_\pi^2}{F_\pi^2} - \frac{1}{6F_\pi^2}I(m_\pi^2)].
\end{aligned}
$$

(62)

By expanding this expression in powers of $d - 4$ and renormalizing the pion field as $\pi_r = Z_\pi^{-1/2}\pi$, with

$$
Z_\pi = 1 - \frac{8m_\pi^2}{F_\pi^2}(2L_4^{(2)} + L_5^{(2)}) + \frac{m_\pi^2}{24\pi^2 F_\pi^2}[\frac{2}{d-4} + \gamma - 1 - \ln 4\pi + \ln \frac{m_\pi^2}{\mu^2}],
$$

(63)

one obtains the canonical form for the effective Lagrangian for pion fields,

$$
\mathcal{L}_{eff} = \frac{1}{2}\partial_\mu \pi_r \partial^\mu \pi_r - \frac{1}{2}M_\pi^2 \pi_r \cdot \pi_r,
$$

(64)

with the identification of the physical pion mass as

$$M_\pi^2 = m^2[1 + \frac{m_\pi^2}{32\pi^2 F_\pi^2} \ln \frac{m_\pi^2}{\mu^2} - \frac{8m_\pi^2}{F_\pi^2} L_{comb}]. \tag{65}$$

Here $L_{comb} = 2L_4^{(2)r} + L_5^{(2)r} - 4L_6^{(2)r} - 2L_8^{(2)r}$. In the original paper of Gasser and Leutwyler [1], M_π^2 was not obtained in this fashion, but rather from the expansion of the Fourier transform of the axial vector correlator, which has the form

$$J_{\mu\nu}^{ik}(p) = i \int d^4 s e^{ip(x-y)} \langle 0|T A_\mu^i(x) A_\nu^k(y)|0\rangle$$

$$= \delta^{ik}\{\frac{p_\mu p_\nu F_\pi^2}{M_\pi^2 - p^2} + \ldots\}, \tag{66}$$

where $A_\mu^i(x) = \bar{\psi}(x)\gamma_\mu\gamma_5\frac{\tau^i}{2}\psi(x)$. From this expression, the corresponding expansion for F_π has also been obtained.

Cool Chiral Perturbation Theory The evaluation of the condensate density at finite temperature was first carried out by Gerber and Leutwyler [5]. In their calculation, which involves $\mathcal{L}_{eff}^{(2)}$ and $\mathcal{L}_{eff}^{(4)}$, they find that the first term in the behavior of the condensate with temperature is quadratically decreasing, i.e.

$$\langle\bar{q}q\rangle = \langle 0|\bar{q}q|0\rangle_{T=0}[1 - \frac{T^2}{8F^2} - \frac{T^4}{384F^4} - \frac{T^6}{288F^6} \ln \frac{\Lambda_q}{T} + O(T^8)]. \tag{67}$$

This is a result that has been obtained under the assumption that quarks are massless, i.e. in the chiral limit. Λ_q is a scale factor constructed from the renomalized low energy constants, and is expected to be of the order of $\Lambda_q = 360..580$ MeV.

In a recent publication, Toublan [6] has investigated pion static properties with the aim of obtaining $O(p^6)$ accuracy in all quantities and to then verify the Gell-Mann–Oakes–Renner (GOR) [28] relation at finite temperature. To do so, the tree, one loop and two loop diagrams of $\mathcal{L}_{eff}^{(2)}$ are required, the tree and one loop graphs of $\mathcal{L}_{eff}^{(4)}$ are required plus the tree level graphs of $\mathcal{L}_{eff}^{(6)}$. In doing so, the result of Eq.(67) has been reconfirmed. In addition, the mass $M_\pi(T)$ and pion decay constant as a function of temperature are also evaluated, using the finite temperature axial vector correlator. In total, thirty-six Feynman graphs contribute to the correlator at this order! However, in the chiral limit, one is still lucky enough to have simple analytic forms for the temperature dependence. One finds

$$\frac{M_\pi^2(T)}{M_\pi} = 1 + \frac{T^2}{24F^2} - \frac{T^4}{36F^4} \ln \frac{\Lambda_M}{T} + O(T^6), \tag{68}$$

while

$$\frac{Re[F_\pi^t(T)]^2}{F_\pi^2}\Big|_{\hat{m}=0} = 1 - \frac{T^2}{6F^2} + \frac{T^4}{36F^4}\ln\frac{\Lambda_T}{T} + O(T^6). \tag{69}$$

and

$$\frac{Re[F_\pi^t(T) - F_\pi^s(T)]}{F_\pi}\Big|_{\hat{m}=0} = \frac{T^4}{27F^4}\ln\Lambda_\Delta T + O(T^6), \tag{70}$$

where $\Lambda_{M,T,\Delta}$ are various scales, whose sizes are determined by the renormalized couplings $L_1^r \ldots L_{10}^r$ that are a function of scale. They are determined numerically to be $\Lambda_M \simeq 1.9$ GeV, $\Lambda_T \simeq 2.3$ GeV, and $\Lambda_\Delta = 1.8$ GeV. Note that, at finite temperature, there is a separation of "temporal" and "spatial" pion decay constants. This comes about since Lorentz invariance is not maintained in a heat bath and the the singular part of the axial two point function takes the form

$$A_{\mu\nu}(q,T) = -\frac{f_\mu(q,T)f_\nu(q,T)}{q_0^2 - \Omega^2(q,T)} \tag{71}$$

where

$$f_0(q,T) = q_0 F_t(q,T) \qquad f_i(q,T) = q_i F_s(q,T), \tag{72}$$

with $i = 1..3$, and the decay constants $F_\pi^{s,t}$ are defined as

$$F_\pi^{s,t}(T) = F_{s,t}(q,T)|_{q=0}. \tag{73}$$

The GOR relation is modified so as to read [6]

$$\lim_{\hat{m}\to 0}\frac{M_\pi^2(T)Re[F_\pi^t(T)]^2}{\hat{m}\langle\bar{q}q\rangle_T} = -1 + O(T^6). \tag{74}$$

For this reason, we show graphs for $M_\pi^2(T)/M_\pi^2$ and $Re[F_\pi^t(T)]^2/F_\pi^2$, as a function of temperature in Figs. 5 and 6. The tree level result is given (dotted curve), together with the one loop computation (upper dashed line in Fig. 5, lower dashed line in Fig. 6) and the two loop approximation (solid curve). In both of these figures, a non-zero value of the quark mass has been assumed for these curves. In the chiral limit, one finds the lower (upper) dashed curve in Fig. 5 (6). What is evident from these two figures, is that chiral perturbation theory is not converging and appears to provide an *oscillating* series for these quantities. Thus for larger values of the temperature, $T > 100$ MeV say, one sees that the pion mass *decreases* with temperature in the two loop approximation, in contradistinction with the one loop result, the lattice results of the last section, and also in contradistinction with the model results obtained in the Nambu–Jona-Lasinio model, which will be presented later on in the following section. Convergence at temperatures in this range appears to be problematic, which is perhaps an indication that the series is at best asymptotic, or changes its nature due to the onset of the phase transition. In this range, one expects non-analytic behavior and it is unreasonable to expect a perturbation analysis to succeed. These curves clearly indicate that ChPT at finite temperature can at best be regarded as cool, so that the fundamental behavior at low temperatures sets a constraint on the finite temperature behavior of would-be effective models.

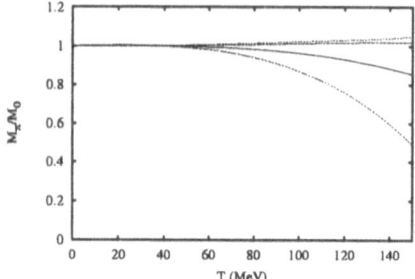

Fig. 5: The pion mass, scaled by its value at $T = 0$ is shown as a function of the temperature.

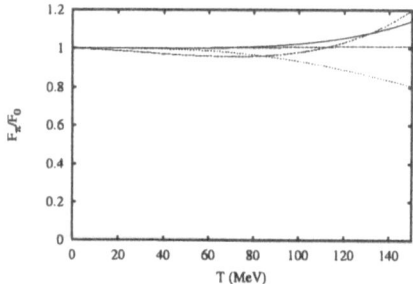

Fig. 6: The pion decay constant, scaled by its value at $T = 0$ is shown as a function of the temperature.

3   The Nambu–Jona-Lasinio Model

The Nambu–Jona-Lasinio (NJL) model has been reviewed in detail by several authors from different viewpoints [8–10], and consequently I do not wish to present any detail of this model other that a basic introduction here. Rather the purpose of this chapter is to illustrate that with the simple equations requiring little computational time, one can reproduce all the main features of the static properties that have been so arduously extracted from years of labor on the lattice. It is extremely encouraging to have a simple model that can be handled semi-analytically – one gains a tremendous amount of insight into the actual functioning of the mechanism of dynamical symmetry breaking and the consequences thereof.

Nevertheless, the NJL model is simply a model – in contrast to the results of the previous section, which are regarded as "factual", this section can only give model-dependent results. Accordingly it is only equitable to indicate, in addition to the successes provided by this approach, the failings also. These become obvious when examining bulk thermodynamic properties, such as pressure, energy and entropy densities, and will be discussed in what follows.

We shall then turn to dynamic properties, and examine the temperature dependence of scattering amplitudes in the quark-antiquark channel, which displays a divergence which we term critical scattering, in analogy to the phenomenon of critical opalescence that is observed in light scattering.

3.1   Order Parameter

We first consider the order parameter for the chiral transition that is obtained from the NJL Lagrangian, which, for two flavors of quarks, is taken to be

$$\mathcal{L}_{NJL} = \bar{\psi}(x)(i\,\not{\partial} - m_0)\psi(x) + G[(\bar{\psi}\psi)^2 + (\bar{\psi}i\gamma_5\tau\psi)^2], \qquad (75)$$

where G is a dimensionful coupling strength, and m_0 denotes the common current quark mass for u and d quarks. For three flavors of quarks, we use

$$\mathcal{L}_{NJL} = \bar{\psi}(x)(i\,\not{\partial} - m_f^0)\psi(x) + G\sum_{a=0}^{8}[(\bar{\psi}\lambda^a\psi)^2 + (\bar{\psi}\lambda^a\gamma_5\psi)^2]$$
$$- K\{\det\bar{\psi}(1+\gamma_5)\psi + \det\bar{\psi}(1-\gamma_5)\psi\}. \qquad (76)$$

Here G and K both are dimensionful coupling strengths and $m_f^0 = \text{diag}(m_0^u, m_0^d, m_0^s)$. The self-energy, in the mean field approximation, that corresponds to the lowest order term in an expansion in the inverse number of colors N_c [12,13], is given as[3]

$$m = m_0 - 2G\langle\langle\bar{\psi}\psi\rangle\rangle, \qquad (77)$$

and the condensate is given explicitly as

$$\langle\langle\bar{\psi}\psi\rangle\rangle = \frac{N_c N_f}{\pi^2}\int_0^\Lambda \frac{p^2}{E_p}[1 - f^-(p,\mu) - f^+(p,\mu)], \qquad (78)$$

---

[3] Since the coupling strengths turn out to be large, $G\Lambda^2 \sim 2$, an expansion in the number of couplings is inadmissable and an alternative expansion scheme must be used.

with

$$f^{\pm}(p,\mu) = \frac{1}{[1 + \exp \ \beta(E_p \pm \mu)]}. \tag{79}$$

One sees that the condensate is directly proportional to the value of the dynamically generated mass, in the event that the current quark mass is zero. Although the situation is more complicated in SU(3), where the dynamically generated quark masses satisfy coupled equations,

$$m_i = m_i^0 - \frac{GN_c}{\pi^2} m_i A(m_i, \mu_i) + \frac{KN_c^2}{\pi^2} m_j m_k A(m_j, \mu_j) A(m_k, \mu_k),$$
$$i \neq j \neq k \tag{80}$$

and the function

$$A(m_f, \mu_f) = \frac{16\pi^2}{\beta} \sum_n e^{i\omega_n \eta} \int_{|p|<\Lambda} \frac{d^3p}{(2\pi)^3} \frac{1}{(i\omega_n + \mu_f)^2 - E_f^2} \tag{81}$$

is proportional to the condensate density for a specific flavor,

$$A(m_f, \mu_f) \sim \langle\langle \bar{\psi}\psi \rangle\rangle_f, \tag{82}$$

the dynamically generated quark masses are equivalently order parameters of the phase transition, and we therefore plot these. They are shown here only for the SU(3) case, in Fig. 7, for a finite value of the current quark mass [30]. As expected, the phase transition that occurs in the chiral limit is washed out and becomes a cross over. Another feature that emerges in this model is that the strange quark mass remains large, even at temperatures $T \sim 300$ MeV, and does not reach its current mass value of 150 MeV until $T \gg T_c$.

3.2   Meson Masses

The meson masses for the scalar and pseudosalar sectors are determined via the well-known method of evaluating the quark-antiquark scattering amplitude in the random phase approximation, and searching for poles of this function. This involves knowing only the irreducible polarization function that one can construct from a single quark loop, the details of which can be found, for example, in [9,10,30]. One finds the masses that are shown in Figs. 8 and 9 for the pseudoscalar and scalar sectors, respectively. In Fig. 8, $2m_q$ is plotted in addition to m_π. The point at which these two curves cross is called the Mott temperature, T_{M_π}. For $T > T_{M_\pi}$, the pion is no longer a bound state, but is a resonance, with a finite width that is not shown here. Similarly we have plotted $m_q + m_s$, from which the kaonic Mott temperature T_{M_K} is defined. For $T \gg T_{M_K}$, the kaon is also a resonance with a finite width.

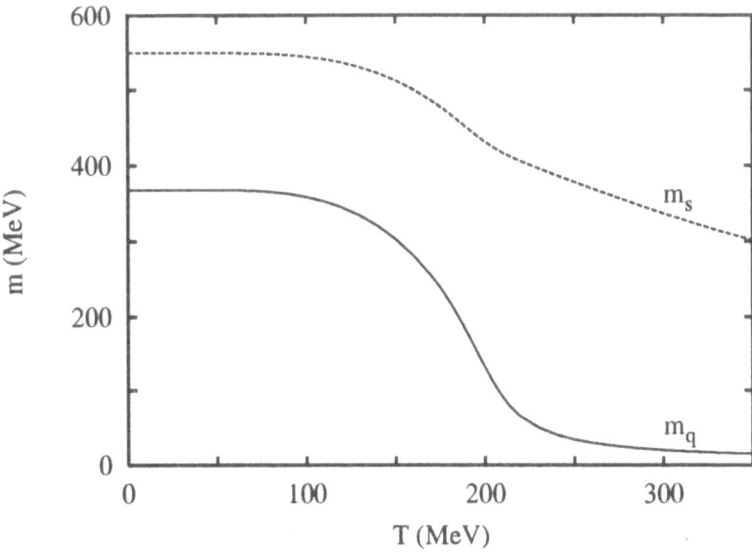

Fig. 7: Temperature dependence of the constituent quark masses. The solid line refers to the light quarks, the dashed to the strange quark [30].

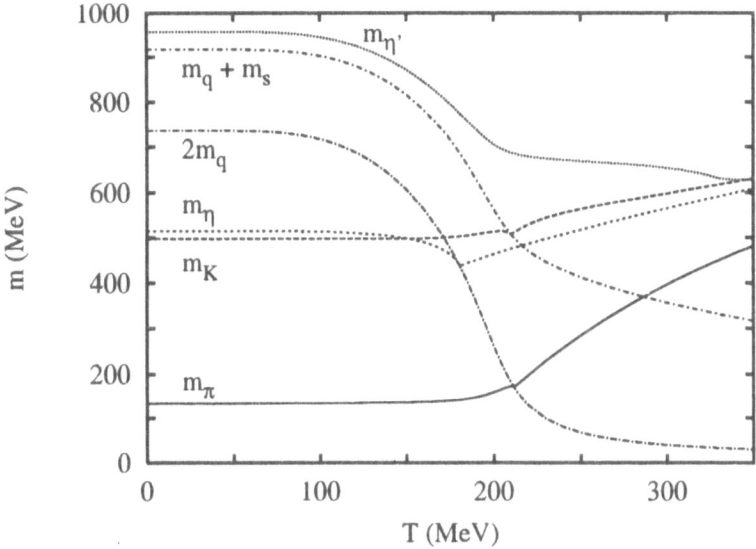

Fig. 8: Temperature dependence of the pseudoscalar meson masses, as well as that of $2m_q$ and $m_q + m_s$. [30].

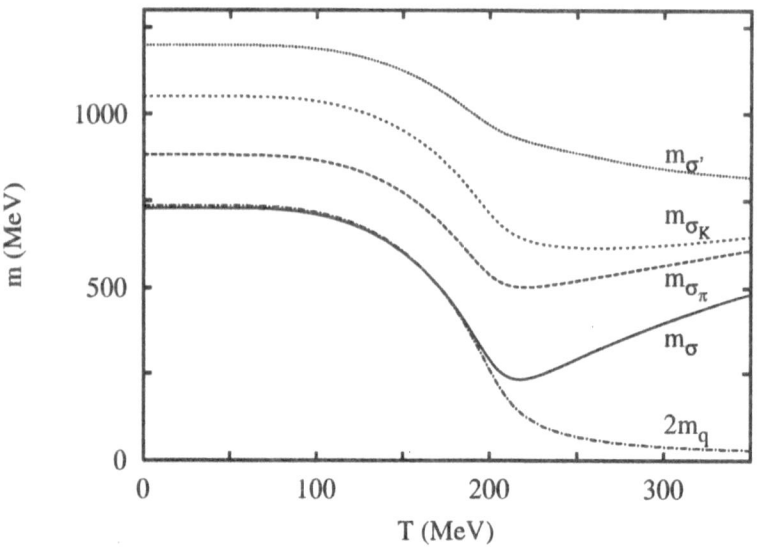

Fig. 9: Temperature dependence of the scalar meson masses and $2m_q$ [30].

These graphs deserve some comment. Firstly let us compare them with the figure showing the meson masses obtained via lattice gauge theory, Fig. 2. We note first that there are some fundamental differences in obtaining these graphs: (a) Figs. 8 and 9 show so-called *pole* masses, while Fig. 2 gives *screening* masses. Nevertheless, it has been shown that, in the NJL model, the temperature behavior of screening masses and pole masses is *qualitatively similar* [31], although quantitatively somewhat different. Since we cannot hope for any quantitative agreement at this stage, it is justifiable to make a comparison. (b) The NJL model calculation shown is for SU(3), while the lattice calculation is SU(2). With these points in mind, note that the σ and π mesons from figures 8 and 9 become degenerate at high temperatures, as observed also in Fig. 2. However, there is no undershooting of the σ meson. The meson labelled m_{σ_π} of Fig. 8 corresponds to the a_0 of Fig. 2. Here one observes qualitatively the same behavior, i.e. that this scalar meson also decreases strongly in the phase transition region. Thus one has an aesthetically pleasing agreement between the NJL model and the results obtained by lattice gauge theory for meson masses at this level.

A direct comparison of Fig. 7 with the results of chiral perturbation theory, i.e. with Fig. 5 is problematic. We simply make some comments: (a) the physics underlying Figs. 5 and 7 is completely different. Fig. 5 is obtained by constructing meson loops (the mesons are regarded as structureless point-like objects), while in Fig. 7, the pion is constructed from a quark-antiquark loop. Meson loops form corrections to this calculation and would be of the next order in $1/N_c$. Such corrections have in fact been evaluated, and it has been found that the leading order T^2 dependence of Eq.(68) is recovered [32]. The fact that in the final analysis the curve of M_π as a function of temperature is finally *decreasing* for chiral perturbation theory in Fig. 5, is in strong contradiction to both Figs. 2 and 8.

3.3   Bulk Thermodynamic Quantities

In the last subsection, we have indicated the successes of the NJL model in calculating the order parameter and masses as a function of temperature. In this subsection, we turn to bulk thermodynamic quantities. Here we will see that the model does not do as well, and that the lack of confinement makes itself strongly evident at low temperatures, while the cutoff of the model is a hindrance at high temperatures. We start with the thermodynamical potential Ω, calculated in the grand canonical ensemble. Given an interaction between fermions that is 4-point in nature such as in Eq.(75), Ω can be calculated quite generally as [10,33]

$$\Omega = \Omega_0 + \int_0^1 \frac{d\lambda}{\lambda} \frac{1}{2} \int \frac{d^3p}{(2\pi)^3} \frac{1}{\beta} \sum_n \exp(i\nu_n\eta) \mathrm{Tr}[\Sigma^\lambda(\nu_n,\boldsymbol{p})S^\lambda(\nu_n,\boldsymbol{p})], \quad (83)$$

where Ω_0 is the thermodynamic potential in the absence of interactions, and Σ^λ and S^λ designate the Matsubara self-energy and Green function associated with the system. The superscript λ refers to the fact that both S and Σ are to be evaluated with the introduction of an artifical coupling that multiplies the interaction Lagrangian $\mathcal{L}_{int}$. The Matsubara frequencies for fermions are, as required, odd, i.e. $\nu_n = (2n+1)\pi/\beta$, with $n = 0, \pm 1, \pm 2 \pm 3 \ldots$.

For the NJL Lagrangian of Eq.(75), in the mean field approximation, it is not necessary to apply Eq.(83). A straightforward calculation gives

$$\Omega = \Omega_q = -2N_cN_f \int \frac{d^3p}{(2\pi)^3} E_p$$
$$-\frac{2N_cN_f}{\beta} \int \frac{d^3p}{(2\pi)^3} \ln[1 + e^{-\beta(E_p+\mu)}][1 + e^{-\beta(E_p-\mu)}], \quad (84)$$

with $E_p^2 = p^2 + m^2$. As is evident from the label q, this appears to be a thermodynamic potential generated solely by the *quark* degrees of freedom.

We note also that the thermodyamical properties can only be measured relative to the physical vacuum,

$$\Omega_{vac}^{phys} = \Omega(T = 0, \mu = 0, m(0,0)), \quad (85)$$

which, for the mean field approximation, corresponds to

$$(\Omega_{vac}^{phys})_{mf} = \frac{(m - m_0)}{4G} - 2N_c N_f \int \frac{d^3 p}{(2\pi)^3} E_p. \tag{86}$$

To introduce *mesonic* degrees of freedom, it is necessary to go beyond the mean field approximation to include the next set of terms in the $1/N_c$ expansion. The self-energy in this case includes effective interactions in both the scalar and pseudoscalar channels [34],

$$\Sigma_{fl}^{\lambda}(\nu_n, p) = -\frac{1}{\beta} \sum_{n'} \int \frac{d^3 q}{(2\pi)^3} [S(\nu_{n'}, q) V_{\sigma}^{\lambda}(\nu_n - \nu_{n'}, p - q)$$
$$+ i\gamma_5 \tau S(\nu_{n'}, q) i\gamma_5 \tau V_{\pi}^{\lambda}(\nu_n - \nu_{n'}, p - q)], \tag{87}$$

and is constructed on summing the Fock and infinite RPA series that contribute to the self-energy in this order. Here

$$V_M^{\lambda}(\omega, q) = -2G\lambda[1 - 2G\lambda\Pi_M(\omega, q)]^{-1}, \tag{88}$$

and the irreducible polarization in the mesonic channel

$$\Pi_M(\omega_m, p) = \int \frac{d^3 q}{(2\pi)^3} \frac{1}{\beta} \sum_n \mathrm{Tr}\Gamma_M S(\nu_n + \omega_m, q + p)\Gamma_M S(\nu_n, q), \tag{89}$$

is determined by the vertex Γ_M for that channel. Inserting Eq.(87) into Eq.(83) yields the fluctuating part of the thermodynamic potential,

$$\Omega_{fl} = \sum_m \frac{N_M}{2} \int \frac{d^3 p}{(2\pi)^3} \frac{1}{\beta} \sum_n e^{i\omega_n \eta} \ln[1 - 2G\Pi_M(\omega_n, p)]. \tag{90}$$

The nature of this term is revealed on performing the frequency sum. One has

$$\Omega_{fl} = \Omega_{\pi} + \Omega_{\sigma}, \tag{91}$$

where, for each species $M = \pi$ or σ,

$$\Omega_M = -N_M \int \frac{d^3 p}{(2\pi)^3} \int_0^{\infty} d\omega [\frac{1}{2}\omega + \frac{1}{\beta} \ln(1 - e^{-\beta\omega})] \frac{1}{2\pi i}$$
$$\times \frac{d}{d\omega} \ln \frac{1 - 2G\Pi_M(\omega + i\epsilon, p)}{1 - 2G\Pi_M(\omega - i\epsilon, p)}. \tag{92}$$

Some analysis shows that a simple approximation for the polarization near the pole, *i.e.*

$$1 - 2G\Pi_M(\omega, p) = (\omega^2 - E_M^2) \times \mathrm{const} \tag{93}$$

leads to

$$\Omega_M = N_M \int \frac{d^3 p}{(2\pi)^3} [\frac{1}{2}E_M + \frac{1}{\beta} \ln(1 - e^{-\beta E_M})], \tag{94}$$

exactly as one would expect for the thermodynamic potential given bosonic degrees of freedom. The pole approximation is however insufficient, as one integrates over all energies, and in practice, the fact that the bound states also become delocalized resonances at the Mott point must also be accounted for. This has been done in introducing phase shifts in each channel [34].

In order to calculate the pressure, the physical vacuum given by Eq.(85) must be reevaluated to include a term from Ω_{fl}. One now has

$$\Omega_{vac}^{phys} = (\Omega_{vac}^{phys})_{mf} + (\Omega_{vac}^{phys})_{fl}. \tag{95}$$

and the pressure density is now

$$p = -\Omega_q - \Omega_{fl} + \Omega_{vac}^{phys}. \tag{96}$$

In Figs. 10 and 11, the pressure and associated energy densities evaluated from this thermodynamic potential are shown.

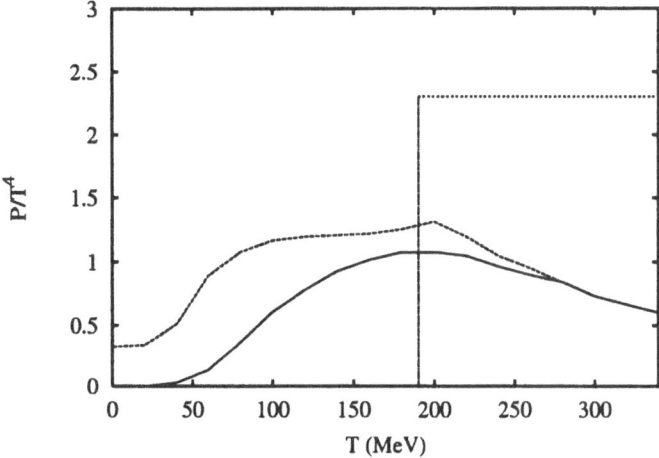

Fig. 10: The pressure density, scaled by T^4, is shown as a function of temperature. The lower curve is for the mean field case only, the upper includes fluctuations (mesons). The vertical line indicates the critical temperature and the horizontal one the Stefan-Boltzmann limit for an ideal quark gas [34].

In Fig. 10, one sees that the lower curve, corresponding to the mean field approximation calculation represents the quark degrees of freedom. There is an appreciable pressure that arises from this term, *i.e.* from the quark degrees of freedom, for temperatures $T < T_c$, which is indicated by the vertical line.

Including mesonic degrees of freedom rectifies the behavior at small temperatures, but still leaves a large intermediate range of temperatures $T < T_c$ that is dominated by these unphysical quark degrees of freedom. This is thus a direct consequence of the missing feature of confinement. The sharp rise in the pressure density shown in Fig. 3 cannot be modelled by a non-confining theory. At high temperatures, $T > T_c$, there is a small contribution from the mesonic degrees of freedom, that exist as correlated states with a finite width in the plasma. The main contribution arises however here from the quark degrees of freedom. The actual value obtained for the pressure density underestimates the Stefan-Boltzmann limit (shown as a horizontal line), since there is a finite cutoff on the quark momenta. Relaxing this constraint would lead to the pressure density approaching a constant.

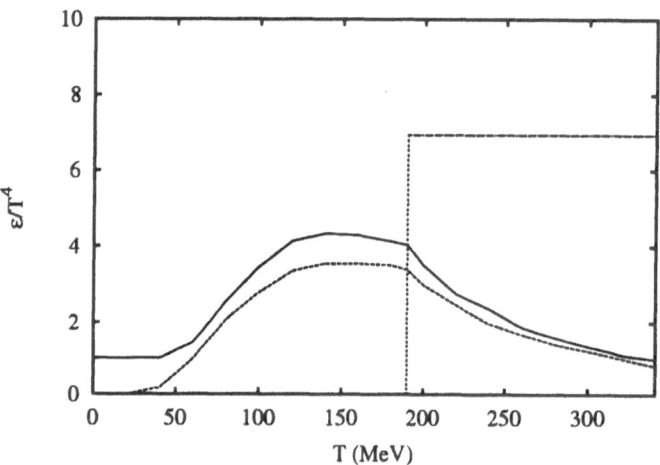

Fig. 11: The energy density, scaled by T^4, is shown as a function of temperature. The lower curve is for the mean field case only, the upper includes fluctuations (mesons). The vertical line indicates the critical temperature and the horizontal one the Stefan-Boltzmann limit for an ideal quark gas [34].

Similar comments can be made for the energy density: the intermediate temperature range 50MeV$< T < T_c$ is dominated by quark degrees of freedom, indicating the lack of confinement. The high temperature values $T > T_c$ do not approach the Stefan-Boltzmann limit, due to the cutoff.

   In concluding this subsection, one sees that one needs to include confinement in some fashion in order to be able to regain the lattice picture. From a thermodynamic point of view, the high temperature regime about $T \simeq T_c$ is probably best described by the model, in the sense that only quark degrees of freedom plus correlated mesonic states are present. In the next subsection, we thus study elastic quark-antiquark scattering about this point and indicate that a divergence occurs in the cross-section at $T = T_M$ and that the phenomenon of critical scattering as a consequence of the chiral phase transition is observed.

3.4   Critical Opalescence in the Quark-Antiquark Channel

In this section, we examine the behavior of the quark-antiquark scattering amplitude in the NJL model in the vicinity of the Mott temperature, which replaces the critical temperature when finite current quark masses are used. In SU(3), there are seven independent processes out of a total of fifteen for quark-antiquark scattering, taking isospin and charge conjugation symmetry into account. These are listed in Table 3. Mesons that can be exchanged in the s and t channels, as are given by the Feynman diagrams of Fig. 12 are also listed.

Process	Exchanged mesons (s channel)	Exchanged mesons (t channel)
$ud \to ud$	π, σ_π	$\pi, \eta, \eta', \sigma_\pi, \sigma, \sigma'$
$u\bar{s} \to u\bar{s}$	K, σ_K	$\eta, \eta', \sigma, \sigma'$
$u\bar{u} \to u\bar{u}$	$\pi, \eta, \eta', \sigma_\pi, \sigma, \sigma'$	$\pi, \eta, \eta', \sigma_\pi, \sigma, \sigma'$
$u\bar{u} \to d\bar{d}$	$\pi, \eta, \eta', \sigma_\pi, \sigma, \sigma'$	π, σ_π
$u\bar{u} \to s\bar{s}$	$\eta, \eta', \sigma, \sigma'$	K, σ_K
$s\bar{s} \to u\bar{u}$	$\eta, \eta', \sigma, \sigma'$	K, σ_K
$s\bar{s} \to s\bar{s}$	$\eta, \eta', \sigma, \sigma'$	K, σ_K

Table 3: Independent processes for $q\bar{q}$ scattering.

The transition amplitudes can be written as

$$-i\mathcal{M}_s = \delta_{c_1,c_2}\delta_{c_3,c_4}\bar{v}(p_2)Tu(p_1)[i\mathcal{D}_s^S(p_1+p_2)]\bar{u}(p_3)Tv(p_4)$$
$$+\delta_{c_1,c_2}\delta_{c_3,c_4}\bar{v}(p_2)(i\gamma_5 T)u(p_1)[i\mathcal{D}_s^P(p_1+p_2)]\bar{u}(p_3)(i\gamma_5 T)v(p_4),$$

$$(97)$$

and

$$-iM_t = \delta_{c_1,c_3}\delta_{c_2,c_4}\bar{u}(p_3)Tu(p_1)[i\mathcal{D}_t^S(p_1-p_3)]\bar{v}(p_2)Tv(p_4)$$
$$\delta_{c_1,c_3}\delta_{c_2,c_4}\bar{u}(p_3)(i\gamma_5 T)u(p_1)[i\mathcal{D}_t^P(p_1-p_3)]\bar{v}(p_2)(i\gamma_5 T)v(p_4),$$

$$(98)$$

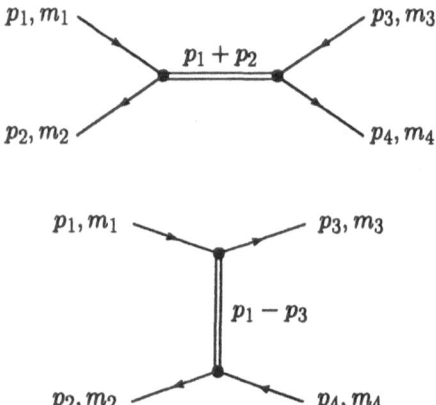

Fig. 12: Feynman diagrams for elastic $q\bar{q}$ scattering within the NJl model in an expansion to lowest order in $1/N_c$.

where T selects the isospin eigenvalue for a particular channel, and $D_{s,t}^{S,P}$ is the s or t channel, scalar or pseudoscalar quark-antiquark scattering amplitude, and which can be constructed from the corresponding polarization function. It has a simple form, for example [9]

$$\mathcal{D}_\pi(p_0, p) = \frac{2G^{eff}}{1 - 2G^{eff}\Pi_{u\bar{u}}^P(p_0, p)}, \qquad (99)$$

where G^{eff} is an effective SU(3) coupling strength in the pionic channel [9].

The differential cross section is constructed in the usual fashion as

$$\frac{d\sigma}{dt} = \frac{1}{16\pi[s - (m_u + m_s)^2][s - (m_u - m_s)^2]} \frac{1}{4N_c^2} \sum_{s,c} |\mathcal{M}_s - \mathcal{M}_t|^2, \quad (100)$$

while the total cross section is evaluated as

$$\sigma = \int dt \, \frac{d\sigma}{dt}[1 - f_F(\beta E_3)][1 - f_F(\beta E_4)], \qquad (101)$$

introducing a Fermi blocking factor for the final states. Here $E_i^2 = p_i^2 + m_i^2$, where $i = 3, 4$.

In Fig. 13, we show the total cross section for light quarks in the initial state, as a function of $\sqrt{s}$, at a temperature $T = 215$ MeV, which lies slightly higher than the pion and kaonic Mott temperatures, $T_{M_\pi} = 212$MeV and $T_{M_K} = 210$MeV. Both pions and kaons are sharp resonances now. At higher values of the temperature, these become broader resonances in the cross-section, as shown in Fig. 14. At the Mott temperature itself, when quarks bind into hadrons, the intermediate states in the s channel give rise to *infinite* cross sections at threshold. This feature, which also appears in other processes like $\pi\pi \to \pi\pi$ [35], $\pi\gamma \to \pi\gamma$ [36] or $q\bar{q} \to \gamma\gamma$ [37,38] is akin to the phenomenon of critical opalescence. This has been discussed in some detail in Ref.[39], and the interested reader is referred to this.

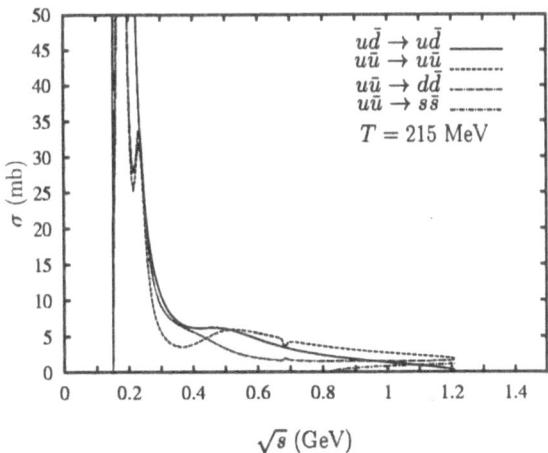

Fig. 13: Total cross section for $q\bar{q}$ scattering with only light quarks in the initial state, shown as a function of $\sqrt{s}$, at $T = 215$ MeV.

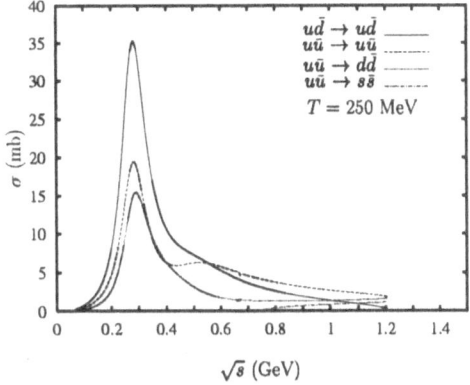

Fig. 14: Total cross section for $q\bar{q}$ scattering with only light quarks in the initial state, shown as a function of $\sqrt{s}$, at $T = 250$ MeV.

4   Non-equilibrium formulation and transport equation

The considerations of the first two sections discussed properties of chiral systems *in equilibrium*. If it were possible to measure any of the associated changes at the phase transition temperature, there would be no need for further discussion. However, because of the nature of confinement, we are unable to observe critical scattering directly, nor any of the other dramatic changes in pion properties. One tool for examining quark matter is via heavy ion collisions, and as such, over the short time scales over which collisions occur, it is unclear whether both thermal and chemical equilibrium can be reached during a collision. For this reason, we wish to investigate what the effects are of a condensate that changes with the medium, as well as medium dependent cross sections in a non equilibrium scenario.

There are several formal aspects that have to be understood before one can attempt actual collision simulations. Firstly one can set up an exact formal description of a relativistic fermionic system that is out of equilibrium via the method of Schwinger and Keldysh. From a heuristic point of view, however, we have a good understanding of the classical Boltzmann equation, so that it is important to establish a link between the two from which one can then go further. In doing so, one generally has a field theory with retarded and advanced Green functions. However, if we examine the collision term of the Boltzmann equation, we see that we require cross sections. However, we only know how to calculate these using causal Green functions. So we have to find a link telling us which level of approximation requires which Feynman graphs.

The content of this lecture is summarized briefly in the next paragraphs. (a) We wish to start from a chosen Lagrangian that gives a microscopically correct description of the world, and to formulate a non-equilibrium theory via a matrix of Green functions $S^{ij}(x_1, x_2)$ (i and j will be defined later!). This matrix of Green functions satisfies a matrix form of the Schwinger-Dyson equations, which as usual, cannot be solved exactly. (b) Some technical aid is required at this point. A centre of mass variable $X = (x_1 + x_2)/2$ and relative coordinate $u = x_1 - x_2$ are introduced, and one *Wigner transforms* the matrix of Green functions. This is simply a Fourier transform with respect to the relative coordinate u. At this point, the equations are still exact. (c) Now one seeks methods of solution. For a fermionic system, the exact method would involve making a spinor decomposition of the Green functions, and we would have 32 coupled equations to solve! This is simply too difficult, in particular for an expanding system, for which spatial gradients are important, and so we turn rather to making the *quasiparticle assumption*, which, coupled with an expansion in powers of $\hbar$, leads to the well-known kinetic theory of Boltzmann, here in relativistic form.

All that has been discussed is quite general for any fermionic theory. Using the Lagrange density of the Nambu–Jona-Lasinio Lagrangian with an expansion in $1/N_c$ illustrates how extensions to the standard binary collision forms in the Boltzmann equation come about, and clears the issue of the content of Feynman graphs for the cross sections that occur in the Boltzmann equation.

4.1   Closed time path – Schwinger-Keldysh formalism

There are several excellent texts that exist that cover the basics of the Schwinger-Keldysh formalism [40,41] for Green functions not in equilibrium. Detailed reviews using path integrals can be found in [42], while the more standard operator approach is to be found in [43–45]. Most confusing in this subject is simply notation: All the listed references use different ones. I shall conform to that of Landau[4], which is particularly transparent in setting up rules for a perturbative diagrammatic expansion.

Central to the problem of non-equilibrium systems is that the description via a single causal Green function alone, is inadequate. One requires the four Green functions,

$$i\hbar S^c(x,y) = \langle T\psi(x)\bar{\psi}(y)\rangle = i\hbar S^{--}(x,y)$$
$$i\hbar S^a(x,y) = \langle \tilde{T}\psi(x)\bar{\psi}(y)\rangle = i\hbar S^{++}(x,y)$$
$$i\hbar S^>(x,y) = \langle \psi(x)\bar{\psi}(y)\rangle = i\hbar S^{+-}(x,y)$$
$$i\hbar S^<(x,y) = -\langle \bar{\psi}(y)\psi(x)\rangle = i\hbar S^{-+}(x,y), \tag{102}$$

i.e. the causal and acausal propagators S^c and S^a, $S^>$ and $S^<$. In Eq.(102), T is the standard time ordering operator,

$$T(O(x)O(y)) = \theta(x_0 - y_0)O(x)O(y) - \theta(y_0 - x_0)O(y)O(x), \tag{103}$$

and $\tilde{T}$ the *anti*time ordering operator,

$$\tilde{T}(O(x)O(y)) = \theta(y_0 - x_0)O(x)O(y) - \theta(x_0 - y_0)O(y)O(x). \tag{104}$$

On the right hand side of Eq.(102), the superscripts $ij = +, -$ have been introduced (these were mentioned in the introduction to this section). This is an arbitrary but useful convention for constructing a matrix notation for summarizing the Green functions,

$$\underline{S} = \begin{pmatrix} S^{--} & S^{-+} \\ S^{+-} & S^{++} \end{pmatrix}. \tag{105}$$

It is automatically achieved by introducing the closed time path of Fig. 15, and setting the fields that occur in the Green function S^{ij} on the ith or jth branch respectively.

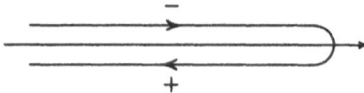

Fig. 15: Closed time path on which the Green functions are defined

There are many interlinking relationships that follow simply from the definition of the Green functions. For example, S^c and S^a are related to $S^>$ and $S^<$ via

$$S^{--}(x,y) = \theta(x_0 - y_0)S^{+-}(x,y) + \theta(y_0 - x_0)S^{-+}(x,y)$$
$$S^{++}(x,y) = \theta(y_0 - x_0)S^{+-}(x,y) + \theta(x_0 - y_0)S^{-+}(x,y). \tag{106}$$

All four Green functions are not independent, since

$$S^{--}(x,y) + S^{++}(x,y) = S^{-+}(x,y) + S^{+-}(x,y)$$
$$= S^K(x,y) = -\frac{i}{\hbar}\langle[\psi(x),\bar{\psi}(y)]\rangle, \tag{107}$$

defining the Keldysh Green function. In addition, one can define the retarded and advanced Green functions

$$i\hbar S^R(x,y) = \theta(x_0 - y_0)\langle\{\psi(x),\bar{\psi}(y)\}\rangle$$
$$i\hbar S^A(x,y) = -\theta(y_0 - x_0)\langle\{\psi(x),\bar{\psi}(y)\}\rangle, \tag{108}$$

which are also related to the S^{ij} via

$$S^R(x,y) = S^{--}(x,y) - S^{-+}(x,y) = S^{+-}(x,y) - S^{++}(x,y)$$
$$S^A(x,y) = S^{--}(x,y) - S^{-+}(x,y) = S^{-+}(x,y) - S^{++}(x,y), \tag{109}$$

which can also be verified directly from the definitions of these functions. One could consider working with the matrix of independent functions

$$\underline{S}' = \begin{pmatrix} 0 & S^A \\ S^R & S^K \end{pmatrix}, \tag{110}$$

but I will not do so in this chapter. Nevertheless, the retarded and advanced Green functions play a special role. Due to their simple analytic structure, plus the fact that the equations of motion that they satisfy (see Eq.(117) later!) are closed, means that one usually can find a simple analytic form for these functions.

---

[4] This differs from the labelling of [43] by a minus sign. Off-diagonal self-energies also differ by a minus sign.

The matrix of self-energies is defined now via the Dyson equation,

$$\underline{S}(x,y) = \underline{S}^0(x,y) + \int d^4z\, d^4w\, \underline{S}^0(x,w)\underline{\Sigma}(w,z)\underline{S}(z,y)$$

$$= \underline{S}^0(x,y) + \int d^4z\, d^4w\, \underline{S}(x,w)\underline{\Sigma}(w,z)\underline{S}^0(z,y). \qquad (111)$$

Pictorially, one can for example examine one element of this equation – say S^{++}. The equation that this function satisfies is given in Fig. 16, using an obvious notation. Thus one sees that all components of the self-energy are in fact required in order to evaluate one single component of $\underline{S}$.

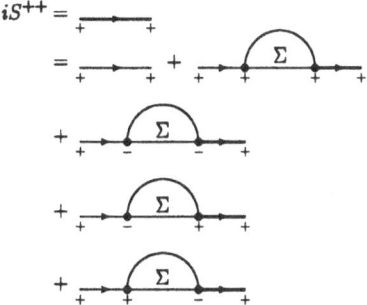

Fig. 16: Dyson equation for one component of the matrix of Green functions.

From the Dyson equation, one can derive the equations of motion for the components of $\underline{S}$, which are summarized as

$$(i\hbar\,\partial_x - m_0)\,\underline{S}(x,y) = \underline{\sigma}_z\delta^4(x-y) + \int d^4z\, \underline{\sigma}_z\underline{\Sigma}(x,z)\underline{S}(z,y), \qquad (112)$$

where

$$\underline{\sigma}_z = \begin{pmatrix} 1 & 0 \\ 0 & -1 \end{pmatrix}. \qquad (113)$$

By defining the retarded and advanced self-energies as

$$\Sigma^R = \Sigma^{--} + \Sigma^{-+}$$
$$\Sigma^A = \Sigma^{--} + \Sigma^{+-}, \qquad (114)$$

one finds the corresponding Dyson equations for $\underline{S}'$,

$$\underline{S}'(x,y) = \underline{S}'^{(0)}(x,y) + \int d^4z\, d^4w\, \underline{S}'^{(0)}(x,w)\underline{\Sigma}'(w,z)\underline{S}'(z,y), \qquad (115)$$

from which one sees that the Dyson equations for S^R and S^A are individually closed,

$$S_{\beta\gamma}^{R,A}(x,y) = S_{\beta\gamma}^{R,A(0)}(x,y) + \int d^4z d^4 w S_{\beta\mu}^{R,A(0)}(x,w) \Sigma_{\mu\nu}^{R,A}(w,z) S_{\nu\gamma}^{R,A}(z,y) \tag{116}$$

with corresponding equation of motion

$$(i\hbar \not{\partial}_x - m_0)_{\alpha\beta} S_{\beta\gamma}^{R,A}(x,y) = \delta_{\alpha\gamma}\delta^4(x-y) + \int d^4z \Sigma_{\alpha\beta}^{R,A}(x,z) S_{\beta\gamma}^{R,A}(z,y), \tag{117}$$

while the equation for the Keldysh function is integrodifferential,

$$(i\hbar \not{\partial}_x - m_0) S^K = \int (\Sigma^K S^A + \Sigma^R S^K) d^4 z. \tag{118}$$

For a free particle, it is useful to note that the solution for the retarded and advanced functions follows immediately as

$$S^{R,A}(p) = \frac{\not{p}+m_0}{p^2 - m_0^2 \pm i\epsilon p_0}. \tag{119}$$

4.2   Transport and Constraint Equations

Of the matrix of Green functions, consider only the equation of motion for $S^{-+}(x,y)$ that follows from Eq.(112). This is

$$(i\hbar \not{\partial}_x - m_0)_{\alpha\beta} S_{\beta\gamma}^{-+}(x,y) = \int d^4z \left\{ \Sigma_{\alpha\beta}^{--}(x,z) S_{\beta\gamma}^{-+}(z,y) \right.$$
$$\left. + \Sigma_{\alpha\beta}^{-+}(x,z) S_{\beta\gamma}^{++}(z,y) \right\} \tag{120}$$

In a similar fashion, one can derive the equation of motion

$$S_{\alpha\beta}^{-+}(x,y)\left(-i\hbar \overleftarrow{\not{\partial}_y} - m_0\right)_{\beta\gamma} = \int d^4z \{ -S_{\alpha\beta}^{--}(x,z) \Sigma_{\beta\gamma}^{-+}(z,y)$$
$$-S_{\alpha\beta}^{-+}(x,z) \Sigma_{\beta\gamma}^{++}(z,y) \}. \tag{121}$$

It turns out to be slightly more convenient to cast these equations in an alternative form, using the relations Eq.(109) between the Green functions, and similar ones for the self-energies. We write

$$(i\hbar \not{\partial}_x - m_0)_{\alpha\beta} S_{\beta\gamma}^{-+}(x,y) = \int d^4z \{ \Sigma_{\alpha\beta}^{-+}(x,z) S_{\beta\gamma}^{+-}(z,y) - \Sigma_{\alpha\beta}^{+-}(x,z) S_{\beta\gamma}^{-+}(z,y)$$
$$+\Sigma_{\alpha\beta}^A(x,z) S_{\beta\gamma}^{-+}(z,y) - \Sigma_{\alpha\beta}^{-+}(x,z) S_{\beta\gamma}^R(z,y) \} \tag{122}$$

and

$$S_{\alpha\beta}^{-+}(x,y)\left(-i\hbar\overleftarrow{\partial_y}-m_0\right)_{\beta\gamma} = \int d^4z\{ -S_{\alpha\beta}^{R}(x,z)\Sigma_{\beta\gamma}^{-+}(z,y)$$

$$+ S_{\alpha\beta}^{-+}(x,z)\Sigma_{\beta\gamma}^{A}(z,y)\}, \quad (123)$$

It is now a tedious technical task to Wigner transform Eqs.(122) and (123). We illustrate this on a simple example and then simply give the final result. Introducing relative and centre of mass variables $u = x-y$ and $X = \frac{1}{2}(x+y)$, the Wigner transform of $S(x,y)$ is defined to be

$$S(X,p) = \int d^4u\, e^{ip\cdot u/\hbar} S(X+\frac{u}{2}, X-\frac{u}{2}). \quad (124)$$

To Wigner transform say the first term on the left hand side of Eq.(122) requires an integral of the form

$$\int d^4u\, e^{ipu/\hbar}\partial_y^\mu f(x,y)$$

$$= \int d^4u\, e^{ipu/\hbar}(\frac{1}{2}\frac{\partial}{\partial X_\mu} - \frac{\partial}{\partial u_\mu})f(X+\frac{1}{2}u, X-\frac{1}{2}u)$$

$$= \frac{1}{2}\frac{\partial}{\partial X_\mu}\int d^4u\, e^{ipu/\hbar} f(X+\frac{1}{2}u, X-\frac{1}{2}u)$$

$$+ \int d^4u(\frac{\partial}{\partial u_\mu}e^{ipu/\hbar})f(X+\frac{1}{2}u, X-\frac{1}{2}u)$$

$$= (\frac{1}{2}\partial_X^\mu + \frac{ip^\mu}{\hbar})f(X,p). \quad (125)$$

Similarly one can show that

$$\partial_x^\mu f(x,y) \rightarrow (-i\frac{p^\mu}{\hbar} + \frac{1}{2}\partial_X^\mu)f(X,p) \quad (126)$$

$$f(y)g(x,y) \rightarrow f(X)\exp\left(\frac{i\hbar}{2}\frac{\overleftarrow{\partial}}{\partial X^\mu}\frac{\overrightarrow{\partial}}{\partial p_\mu}\right)g(X,p) \quad (127)$$

$$f(x)g(x,y) \rightarrow f(X)\exp\left(-\frac{i\hbar}{2}\frac{\overleftarrow{\partial}}{\partial X^\mu}\frac{\overrightarrow{\partial}}{\partial p_\mu}\right)g(X,p) \quad (128)$$

$$\int d^4z f(x,z)g(z,y) \rightarrow f(X,p)\exp\left(-\frac{i\hbar}{2}(\frac{\overleftarrow{\partial}}{\partial X^\mu}\frac{\overrightarrow{\partial}}{\partial p_\mu} - \frac{\overleftarrow{\partial}}{\partial p_\mu}\frac{\overrightarrow{\partial}}{\partial X^\mu})\right)g(X,p)$$

$$(129)$$

need be made on Wigner transforming the product functions on the left hand side of the last equations. Applying these relations to Eqs.(122) and (123) leads to the rather complex forms for the equations of motion,

$$\{i\hbar\gamma^\mu(\frac{1}{2}\frac{\partial}{\partial X^\mu} - \frac{ip_\mu}{\hbar}) - m_0\}S^{-+}(X,p) =$$

$$\Sigma^{-+}(X,p)\hat{\Lambda}S^{+-}(X,p) - \Sigma^{+-}(X,p)\hat{\Lambda}S^{-+}(X,p)$$
$$+ \Sigma^{A}(X,p)\hat{\Lambda}S^{-+}(X,p) - \Sigma^{-+}(X,p)\hat{\Lambda}S^{R}(X,p)$$

$$(130)$$

and

$$S^{-+}(X,p)\{-i\hbar\gamma^{\mu}(\frac{1}{2}\frac{\overleftarrow{\partial}}{\partial X^{\mu}} + \frac{ip_{\mu}}{\hbar}) - m_0\} = -S^{R}(X,p)\hat{\Lambda}\Sigma^{-+}(X,p)$$
$$+ S^{-+}(X,p)\hat{\Lambda}\Sigma^{A}(X,p),$$

$$(131)$$

$$\hat{\Lambda} = \exp\left(-\frac{i\hbar}{2}(\frac{\overleftarrow{\partial}}{\partial X^{\mu}}\frac{\overrightarrow{\partial}}{\partial p_{\mu}} - \frac{\overleftarrow{\partial}}{\partial p_{\mu}}\frac{\overrightarrow{\partial}}{\partial X^{\mu}})\right).$$

$$(132)$$

Now subtracting and adding these resulting equations, one arrives at two futher equations, which we identify as the transport and constraint equations respectively:

$$\frac{i\hbar}{2}\{\gamma^{\mu}, \frac{\partial S^{-+}}{\partial X^{\mu}}\} + [\not{p}, S^{-+}(X,p)] = I_{-}$$

$$(133)$$

and

$$\frac{i\hbar}{2}[\gamma^{\mu}, \frac{\partial S^{-+}}{\partial X^{\mu}}] + \{\not{p} - m_0, S^{-+}\} = I_{+}.$$

$$(134)$$

In these equations, the terms that occur on the right hand side are decomposed into three types of contribution, one containing at least one retarded function, one with at least one advanced function and a further term with neither, which in the semi-classical limit is the origin of the collision integral. Explicitly, one has

$$I_{\mp} = I_{\text{coll}} + I_{\mp}^{A} + I_{\mp}^{R},$$

$$(135)$$

with

$$I_{\text{coll}} = \Sigma^{-+}(X,p)\hat{\Lambda}S^{+-}(X,p) - \Sigma^{+-}(X,p)\hat{\Lambda}S^{-+}(X,p)$$
$$= I_{\text{coll}}^{\text{gain}} - I_{\text{coll}}^{\text{loss}},$$

$$(136)$$

$$I_{\mp}^{R} = -\Sigma^{-+}(X,p)\hat{\Lambda}S^{R}(X,p) \pm S^{R}(X,p)\hat{\Lambda}\Sigma^{-+}(X,p)$$

$$(137)$$

and

$$I_{\mp}^{A} = \Sigma^{A}(X,p)\hat{\Lambda}S^{-+}(X,p) \mp S^{-+}(X,p)\hat{\Lambda}\Sigma^{A}(X,p).$$

$$(138)$$

Equations (133) and (134) are the central, exact equations that describe the non-equilibrium evolution of a system of interacting quarks. To actually see that these are in fact transport and constrint equations known from Vlasov of Boltmann theory requires some (hard) work. This follows only under certain approximations, and of course one needs some model in order to specify the interactions. For this purpose, we will use the Nambu–Jona-Lasinio model.

Before doing this however, note that an exact solution of Eqs.(133) and (134) follows formally on making a spinor decomposition,

$$-i\hbar S^{-+} = F + i\gamma_5 P + \gamma^\mu V_\mu + \gamma^\mu \gamma_5 A_\mu = \frac{1}{2}\sigma^{\mu\nu} S_{\mu\nu}. \qquad (139)$$

The equations for the projected functions $F \sim \mathrm{tr}S^{-+}$, $P \sim \mathrm{tr}\gamma_5 S^{-+}$, ..., form a set of 16 times 2 coupled equations that need to be solved simultaneously. This is not only a formidable task from the computational point of view, it also offers at present little physical insight.

For reasons of simplicity, therefore, we introduce the quasiparticle ansatz that contains the quark and antiquark distribution functions $f_q(X,p)$ and $f_{\bar{q}}(X,p)$, and which puts these on their mass shell,

$$S^{-+}(X,p) = 2\pi i \frac{\not{p} + m}{2E_p}[\delta(p_0 - E_p)f_q(X,p) - \delta(p_0 + E_p)\bar{f}_{\bar{q}}(X,-p)] \quad (140)$$

with $\bar{f}_{q,\bar{q}} = 1 - f_{q,\bar{q}}$. Similar expressions can also be easily written down for the remaining components of the matrix S^{ij}.

4.3   The Vlasov equation for the NJL model

At this point, one cannot go futher unless one specifies a theory or model from which the self-energy can be calculated. A four point interaction like that of the SU(2) NJL model is particularly simple to handle because the Feynman rules are particularly simple: (a) a directed line represents a fermion. The signs attributed to the beginning (i) and end (j) of the line reflect in the Green function iS^{ji} to be associated with the line. (b) an interaction line can have only a single sign on both of its ends. If the sign is $\pm$, it is to be translated as $\pm iV$, with V being the interaction strength. In the NJL model, this is $V = -2G$.

According to these rules, in the Hartree approximation, it follows immediately that $\Sigma^{+-} = \Sigma^{-+} = 0$, so that $I_{\mathrm{coll}} = I_R = 0$. Only $I_A \neq 0$. Furthermore $\Sigma^A(X,p) = \Sigma^A(X) = m(X)$ alone, so that

$$I_-^A = \Sigma^A(X)[1 - \frac{i\hbar}{2}(\overleftarrow{\partial}_x\overrightarrow{\partial}_p - \overleftarrow{\partial}_p\overrightarrow{\partial}_x)]S^{-+}$$
$$- S^{-+}[1 - \frac{i\hbar}{2}(\overleftarrow{\partial}_x\overrightarrow{\partial}_p - \overleftarrow{\partial}_p\overrightarrow{\partial}_x)]\Sigma^A(X) + O(\hbar^2)$$

$$(141)$$

or

$$I_-^A = -i\hbar\partial_\mu\Sigma^A\partial_p^\mu S^{-+}, \qquad (142)$$

and the transport equation becomes

$$\frac{i\hbar}{2}\{\gamma^\mu \frac{\partial S^{-+}(X,p)}{\partial X^\mu}\} + [\not{p}, S^{-+}(X,p)] = -i\hbar\partial_\mu\Sigma^A\partial_p^\mu S^{-+}. \qquad (143)$$

Assuming that the quasiparticle ansatz for $S^{-+}(X,p)$ of Eq.(140) holds and that the mass is to be considered as the dynamically generated Hartree mass that is to be self-consistently determined, one can insert Eq.(140) into Eq.(143), take the trace over spinor indices and integrate over a positive energy interval Δ_+ that contains E_p, to arrive at an equation for the quark distribution function,

$$\frac{\partial}{\partial X^0} f_q(X,p) = p_i \partial^i \frac{f(X,p)}{E_p} + m(X)\partial_i m(X)\partial_p^i \left(\frac{f_q(X,p)}{E_p}\right) = 0. \quad (144)$$

On performing the derivatives and extracting a factor of $1/E_p$, one can write this as

$$\frac{1}{E_p}(p^\mu \partial_\mu f_q(X,p) + m(X)\partial_\mu m(X)\partial_p^\mu f_q(X,p)) = 0, \quad (145)$$

which is the Vlasov equation for the model. It must be solved concurrently with the gap equation for $m(X)$,

$$m(X) = m_0 + 4GN_c m(X) \int \frac{d^3 p}{(2\pi\hbar)^3} \frac{1}{E_p(X)} [1 - f_q(X,p) - f_{\bar{q}}(X,p)], \quad (146)$$

that is derived directly from the Hartree self-energy. The constraint equation, in this same approximation in the expansion in $\hbar$, is

$$(p^2 - m^2(X))f_q(X,p) = 0, \quad (147)$$

which validates our use of the quasiparticle assumption as being exact. Equation (145) indiates that chiral symmetry breaking enters via the condensate or mass already as a spatially varying potential in the Vlasov equation.

4.4   The Boltzmann equation for the NJL model

In principle, the next step from a physical point of view would be to incorporate all self-energy diagrams of the next order in $1/N_c$. This would correspond to meson exchange [12]. This has not been done yet formally [47] and we will touch on this briefly in the following subsection. Here we shall rather examine the simpler problem of considering our self-energy with at least two interaction vertices, such as shown in Figs. 17 and 18 for the NJL model. These are the minimal types of diagram that can possibly give rise to an off-diagonal self-energy Σ^{+-} say, and therefore to a non-vanishing contribution to the gain and loss terms that comprise I_{coll} in the transport equation, Eq.(133). We will not give details here, but just note the salient features [46].

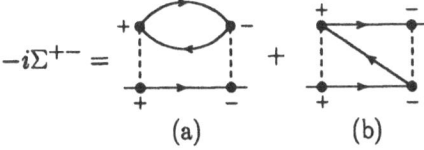

$$-i\Sigma^{+-} =$$
(a)   (b)

Fig. 17: Direct (a) and exchange (b) graphs that contribute to Σ^{-+} and which contain two interaction lines. The vertices can be either all scalar or all pseudoscalar in nature.

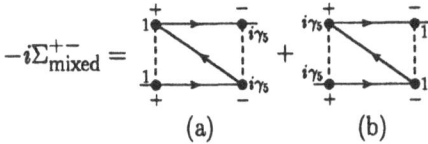

$$-i\Sigma^{+-}_{\text{mixed}} =$$
(a)   (b)

Fig. 18: Mixed graphs that contribute to $\Sigma^{+-}_{\text{mixed}}$ and which contain two interaction lines.

Firstly, a direct translation of the off-diagonal graphs, in the scalar channel say,

$$\Sigma^{+-}_\sigma(X,p) = -4G^2\hbar^2 \int \frac{d^4p_1}{(2\pi\hbar)^4} \frac{d^4p_2}{(2\pi\hbar)^4} \frac{d^4p_3}{(2\pi\hbar)^4} (2\pi\hbar)^4 \delta(p - p_1 + p_2 - p_3)$$
$$\times [S^{+-}(X,p_1)\text{tr}\left(S^{-+}(X,p_2)S^{+-}(X,p_3)\right)$$
$$-S^{+-}(X,p_1)S^{-+}(X,p_2)S^{+-}(X,p_3)], \tag{148}$$

contains a product of three Green functions. Recalling that this will be multiplied by S^{-+} in the collision integral and also that we must trace and integrate the result first over a positive energy interval, we can easily see that such a procedure will lead to eight terms that each contain some product of four quark or antiquark distribution functions, such as for example

$$\text{coefficient} \times \bar{f}_q(p)f_q(p_2)\bar{f}_q(p_3)f_q(p_4) \tag{149}$$

In a loose sense, if one designates $f_{q,\bar{q}}(p)$ to represent an incoming quark (antiquark) and $\bar{f}_{q\bar{q}}$ to represent an outgoing quark (antiquark), then one can draw diagrams associated with each process. For example, the products listed in Eq.(149) would represent quark-quark scattering. A similar term of the eight possible leads to quark-antiquark scattering, while the remaining six that are not listed (but which are easily worked out), are shown in Fig. 19. Some of these look like the typical vacuum fluctuation processes that would occur in any relativistic theory and in addition to these, there are others that give rise to pair creation and annihilation. All six graphs of this figure can be shown to vanish from energy-momentum conservation due to the quasiparticle assumption! This gives us an indication of the complexity and richness of the theory that would go beyond the standard collision scenario if one relaxes this assumption.

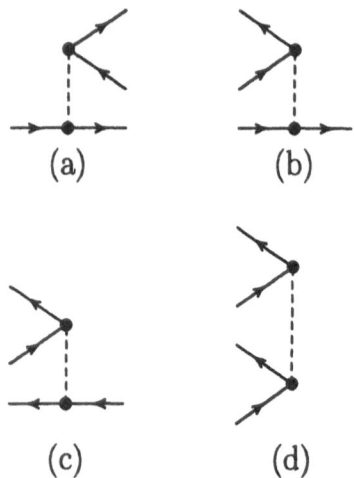

Fig. 19: Six graphs that arize from the term $\sim \Sigma^{+-}S^{-+}$. These are heuristic graphs are are not Feynman diagrams.

Secondly, it is important to verify that the coefficient functions in the term of (149) in fact truly give rise to the differential cross section for elastic quark-quark scattering as would be calculated from real Feynman diagrams (and not heuristic graphs of Fig. 16) such as are displayed in Fig. 20. In fact, this has been explicitly demonstrated to be the case [46]. One finds that the contribution from Fig. 17(a) gives rise to the amplitudes squared of both the s or u channels for qq scattering (or s or t channels for $q\bar{q}$ scattering), while Fig. 17(b) is required to produce the interference terms between them. It appears that evaluating nonequilibrium self-energies for the Boltzmann equation leads to scatering processes that can be obtained from all possible combinations of cutting the slef-energy grphas of Fig. 17 vertically, reminiscent of the Wick-Cutkowsky rules [17].

Finally, one arrives at a Boltzmann equation from Eq.(133). It reads

$$p^\mu \partial_\mu f_q(X, \boldsymbol{p}) + m(X) \partial_\mu m(X) \partial_p^\mu f_q(X, \boldsymbol{p}) =$$
$$N_c \int d\Omega \int \frac{d^3 p_2}{(2\pi\hbar)^3 2E_{p_2}} |v_p - v_2| 2E_p 2E_{p_2}$$

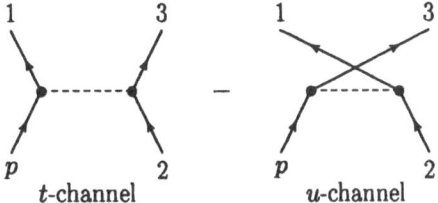

Fig. 20: t and u channel Feynman graphs for elastic quark-quark scattering, to lowest order in $1/N_c$.

$$\times \{\frac{1}{2}\frac{d\sigma}{d\Omega}|_{qq \to qq}(p2 \to 13)(f_q(p_1)\bar{f}_q((p_3)\bar{f}_q(p) - \bar{f}(p_1)f_q(p_2)\bar{f}_q(p_3)f_q(p))$$

$$+\frac{d\sigma}{d\Omega}|_{q\bar{q} \to q\bar{q}}(p2 \to 13)(f_q(p_1)\bar{f}_{\bar{q}}(p_2)f_{\bar{q}}(p_3)\bar{f}_q(p) - \bar{f}_q(p_1)\bar{f}_{\bar{q}}(p_3)f_{\bar{q}}(p_2)f_q(p))\},$$

$$(150)$$

The constraint derived earlier, Eq.(147), however, remains unaltered. From the Boltzmann equation, it is apparent that the changes in the condensate with the medium affect the equation in two possible places: (a) As with the Vlasov equation, a medium dependent potential occurs on the left hand side that is related to the effective quark mass in medium and (b) the cross-sections occurring on the right hand side are medium dependent, and also depend on changes of the quark and meson masses in the medium. As we have seen in the preceding section, the cross section for quark-antiquark scattering diverges at the phase transition.

The actual answer as to what one should expect from numerical simulations is however unclear: since the differential cross-sections are averaged over, one may lose the sharp signal of the divergence. However, the force term on the left hand side may still play an essential role. At this stage also, too many physical features are still lacking, in particular, the coupling of the quark degrees of freedom to mesons and their coupling back to the quarks. This must lead to a hadronization scenario. In the final subsection of this chapter, we briefly sketch how this might occur. For numerical simulations thus far, we refer the reader to [48] and other references cited therein.

4.5   Higher orders in $1/N_c$ and meson production

As already pointed out earlier, the expansion in the coupling strength that was used for selecting the diagrams of the last section is inadmissable, because $G\Lambda \sim 2$. Going to higher orders in the $1/N_c$ expansion is however non-trivial, as a *symmetry conserving* set of graphs must be chosen. From [12,13], we know that this comprises firstly the set of graphs of Fig. 21 for the self-energy, where the "F" denotes the new full Green function that must be newly determined in a self consistent fashion.

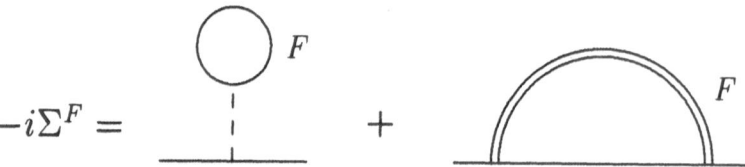

$$-i\Sigma^F =$$

Fig. 21: Self-consistent self-energy that includes meson exchange.

Denoting the two terms in the self-energy as Σ_1 and Σ_2 respectively, one can make an expansion of the new full Green function about that governed by Σ_1,

$$\Sigma = \Sigma_1 + \Sigma_2(k) \tag{151}$$

$$
\begin{aligned}
S^F &= S_{\Sigma_1} + S_{\Sigma_1}\Sigma_2 S^F \\
&= \frac{1}{\not{p} - \Sigma_1} + \frac{1}{\not{p} - \Sigma_1}\Sigma_2\frac{1}{\not{p} - \Sigma_1} + \dots
\end{aligned}
\tag{152}
$$

Concomitantly, the irreducible polarization $\Pi^F(k)$ now occurring in the quark antiquark scattering amplitude

$$-iD^F(k) = \frac{2iG}{1 - 2G\Pi^F(k)} \tag{153}$$

must contain further terms,

$$\Pi^F = \Pi^0 + \delta\Pi \tag{154}$$

where Π^0 is the simple quark loop, in order to be symmetry conserving. The graphs required for Π^F are shown in Fig. 22.

Inserting the expansion of the Green function and the irreducible polarization into the full self-energy of Fig. 21, leads to graphs that contain *inter alia* diagrams of the form shown in Fig. 23.

This gives us an intuitive understanding that, on evaluating these diagrams in the non-equilibrium scenario, we should no longer simply obtain a cross-section for elastic quark-quark and quark-antiquark scattering, but also the hadronization process of $q\bar{q} \to MM'$, where M and M' are mesons. Much work however, remains to be done in this regard.

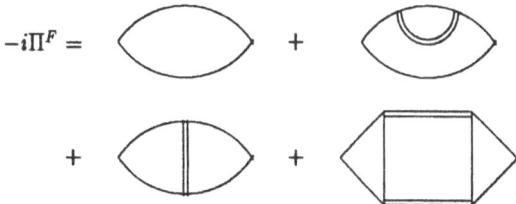

$$-i\Pi^F =$$

Fig. 22: Contributions to the irreducible polarization to next to leading order in the $1/N_c$ expansion.

Fig. 23: Diagrams occurring in Σ, that lead to the hadronization of a quark into two mesons.

5   Concluding comments

In this series of lectures, we have investigated some aspects of chiral symmetry breaking at finite temperatures. We have seen that in the last few years, much information is emerging from the lattice gauge community that tells us about the transition region itself. Chiral perturbation theory, on the other hand, while being excellent in the low temperature regime, cannot adequately describe a phase transition.

In the following section, we have investigated the Nambu–Jona-Lasinio model at finite temperatures. It gives a remarkably good qualitative agreement with the lattice data in the realm of static properties. It fails, however, to describe the bulk thermodynamic properties well, primarily due to the fact that confinement is lacking. The NJL model gives a simple picture for a delocalization rather than a deconfinement transition. Associated with this (physically appealing) picture that bound mesons become delocalized at the transition temperature – now the Mott temperature – and are still correlated states with a finite width in the quark medium, are marked divergences in many functions, such as the pion radius, π-π and π-K scattering lengths (not discussed here), as well as the phenomenon of critical scattering, observed in the quark-antiquark channel.

Due to the fact that none of the apparent singularities are directly observable experimentally, we have turned to transport theory, in order to investigate what effects are to be expected from a condensate density that is medium dependent. Calculations at this stage indicate that a Boltzmann equation is dependent on the condensate through a force term, and also via the cross-sections that arize from binary collisions among the quarks and antiquarks. Howver, the stage of calculation is still primitive: a consistent physical theory that includes mesons and which overcomes the problems associated with the lack of confinement is required before one can expect to obtain credible results. This, of course, leaves the path open for future research.

6   Acknowledgments

I would like to thank Jean Cleymans for the opportunity of being able to speak in Cape Town and for providing a comfortable and stimulating scientific atmosphere. In preparing this manuscript, I am indebted to both E. Laermann and P. Rehberg for providing some of the figures in postscript form. A hearty thanks also goes to G. Papp for his concerted efforts and timely thinking in nursing a collapsing computer. This work has been supported in part by the Deutsche Forschungsgemeinschaft DFG under the contract number Hu 233/4-4, and by the German Ministry for Education and Research (BMBF) under contract number 06 HD 742.

References

1. J. Gasser and H. Leutwyler, Ann. Phys. (N.Y.) **158** (1984) 142.
2. J. Gasser and H. Leutwyler, Nucl. Phys. **B250** (1985) 517.
3. F. Karsch, Nucl. Phys. B (Proc. Suppl.) **60A** (1988) 169.
4. E. Laerman, Nucl. Phys. B (Proc. Suppl.) **60A** (1988) 180.
5. P. Gerber and H. Leutwyler, Nucl. Phys. **B321** (1989) 387.
6. D. Toublan, Phys. Rev. **D 56** (1997) 5629.
7. I.M. Barbour, S.E. Morrison, E.G. Klepfish, J.B. Kogut and M.-P. Lombardo, Nucl. Phys. B. (Proc. Suppl.) **60A** (1998) 220.
8. Y. Nambu and G. Jona-Lasinio, Phys. Rev. **122** (1961) 345; *ibid.* **124** (1961) 246.
9. S.P. Klevansky, Rev. Mod. Phys. **64** (1992) 649.
10. U. Vogl and W. Weise, Prog. Part. Nucl. Phys. **27** (1991) 195; T. Hatsuda and T. Kunihiro, Phys. Rep. **247** (1994) 221.
11. See, for example, F. Karsch, in *Quark Gluon Plasma*, edited by R.C. Hwa (World Scientific, Singapore, 1990).
12. E. Quack and S.P. Klevansky, Phys. Rev. **C49** (1994) 3283.
13. V. Dmitrasinović, H.-J. Schulze, R. Tegen and R.H. Lemmer, Ann. Phys. (N.Y.) **238** (1995) 332.
14. B.-J. Schaefer and H.-J. Pirner, Nucl. Phys. **A627** (1997) 481.
15. Chr. V. Christov, A. Blotz, H.C. Kim, P. Pobylitsa, T. Watabe, T. Meissner, E. Ruiz Arriola, Prog. Part. Nucl. Phys. **37** (1996) 91.

16. J. Goldstone, Nuovo Cimento, **19** (1961) 154. See also S. Coleman, Erice Lectures 1973, Laws of Hadronic Matter (Academic Press, New York, Edited by A. Zichichi), 1975, p139.
17. C. Itzykson and J.-B. Zuber, *Quantum Field Theory* (McGraw-Hill, New York, 1980).
18. G. 't Hooft, Phys. Rev. Lett. **37** (1976) 8; Phys. Rev. **D14** (1976) 3432.
19. A. Ukawa, Nucl. Phys. B (Proc. Suppl.) **17** (1990) 118 and references cited therein.
20. E. Laermann, private communication.
21. R. Pisarski and F. Wilczek, Phys. Rev. **D29** (1994) 338.
22. See, for example, N. Goldenfield, *Lectures on Phase Transitions and the Renormalization Group*, (Addison-Wesley, USA, 1992).
23. M.C. Birse, T.D. Cohen and J.A. McGovern, Phys. Lett. **B388** (1996) 137.
24. G.E. Brown and M. Rho, Phys. Rev. Lett. **66** (1991) 2720; *ibid.* Nucl. Phys. **A590** (1995) 527c.
25. J. Engels, J. Fingberg, F. Karsch, D. Miller and M. Weber, Phys. Lett. **B 252** (1990) 625.
26. M. Asakawa and T. Hatsuda, Phys. Rev. **D55** (1997) 4488.
27. M.A. Novak, M. Rho and I. Zahed, Chiral Nuclear Dynamics, (World Scientific, Singapore, 1996)
28. M. Gell-Mann, R. Oakes and B. Renner, Phys. Rev. **175** (1968) 2195.
29. J.F. Donoghue, E. Golowich and B.R. Holstein, *Dynamics of the Standard Model*, (Cambridge UP, USA, 1992).
30. P. Rehberg, S.P. Klevansky and J. Hüfner, Phys. Rev. C **53** (1996) 410.
31. W. Florkowski and B.L. Friman, Acta Phys. Pol. **B25** (1994) 271; *ibid.*, Z. Phys. **C61** (1994) 171.
32. W. Florkowski and W. Broniowski, Phys. Lett. **B386** (1996) 62.
33. A.L. Fetter and J.D. Walecka, *Quantum Theory of Many-particle Systems* (McGraw-Hill, New York, 1971).
34. P. Zhuang, J. Hüfner and S.P. Klevansky, Nucl. Phys. **A 576** (1994) 525.
35. E. Quack, P. Zhuang, Y. Kalinovsky, S.P. Klevansky and J. Hüfner, Phys. Lett. **B348** (1995) 1.
36. A.E. Dorokhov, J. Hüfner, S.P. Klevansky, P. Rehberg and M.K. Volkov, Z. f. Physik **C75** (1997) 127.
37. S.P. Klevansky, Nucl. Phys. **A575** (1994) 605.
38. P. Rehberg, Y. Kalinovsky and D. Blaschke, Nucl. Phys. **A622** (1997) 478.
39. J. Hüfner, S.P. Klevansky and P. Rehberg, Nucl. Phys. **A606** (1996) 260.
40. J. Schwinger, J. Math. Phys. **2** (1961) 407.
41. L.V. Keldysh, JETP **20** (1965) 1018.
42. K.-C. Chou, Z.-B. Su, B.-L. Hao and L. Yu, Phys. Rep. **118** (1985) 1.
43. W. Botermans and R. Malfliet, Phys. Rep. **198** (1990) 115.
44. S.R. de Groot, W.A. van Leeuwen and Ch. G. van Weert, *Relativistic Kinetic Theory* ,(North Holland, 1980)
45. L.D. Landau and E.M. Lifschitz, *Physikalische Kinetik* , vol 10, (Akademie Verlag, Berlin, 1986).
46. S.P. Klevansky, A. Ogura and J. Hüfner, Ann. Phys. (N.Y.) **261** (1997) 37.
47. S.P. Klevansky, P. Rehberg, A. Ogura and J. Hüfner, Hirschegg Conference 1997, (GSI, Darmstadt, 1997) p397.
48. P. Rehberg and J. Hüfner, Nucl. Phys. **A 635** (1998) 511.

Physics and Astrophysics of Strange Quark Matter

Jes Madsen[1]

Institute of Physics and Astronomy, University of Aarhus,
DK-8000 Århus C Denmark

Abstract. 3-flavor quark matter (strange quark matter; SQM) can be stable or metastable for a wide range of strong interaction parameters. If so, SQM can play an important role in cosmology, neutron stars, cosmic ray physics, and relativistic heavy-ion collisions. As an example of the intimate connections between astrophysics and heavy-ion collision physics, this Chapter gives an overview of the physical properties of SQM in bulk and of small-baryon number strangelets; discusses the possible formation, destruction, and implications of lumps of SQM (quark nuggets) in the early Universe; and describes the structure and signature of strange stars, as well as the formation and detection of strangelets in cosmic rays. It is concluded, that astrophysical and laboratory searches are complementary in many respects, and that both should be pursued to test the intriguing possibility of a strange ground state for hadronic matter, and (more generally) to improve our knowledge of the strong interactions.

1 Introduction

Hadronic matter is expected to undergo a transition to quark-gluon plasma under conditions of high temperature and/or baryon chemical potential. These conditions may be achieved for a brief moment in ultrarelativistic heavy-ion collisions, but they are also likely to appear in Nature. A very high density (and comparatively low temperature) environment exists in the interior of neutron stars, which may actually contain significant amounts of quark matter in the interior. High temperatures (but rather low baryon chemical potential) were realized in the first 10^{-4} seconds after the Big Bang, and here a hot quark-gluon plasma state must have existed until the temperature dropped to 100–200 MeV due to the adiabatic expansion of the Universe.

This Chapter will outline some of the possible ways in which astrophysics may teach us about the existence and properties of quark-gluon plasmas. The advantage relative to laboratory searches is, that truly bulk systems can be studied, and that the timescales involved are much longer than those relevant to collisions. Disadvantages are that astrophysicists (with possible exceptions if strange quark matter is absolutely stable) can only observe indirect consequences of the plasma state for example in the properties of pulsars or in the distribution of light nuclei produced a few minutes after the Big Bang. It will be shown, however, that astrophysics arguments in many cases can be used to constrain parameters significantly relative to direct experimental approaches because of the large volumes and timescales involved.

The implications of quark-gluon plasmas in astrophysics and cosmology are many-fold, and I shall focus on aspects related to the idea of (meta)stability of strange quark matter through discussions partly biased by my own research interests.

Lumps of up, down, and strange quarks (strange quark matter, SQM), with masses ranging from small nuclei to neutron stars, rather than ^{56}Fe, could be the ground state of hadronic matter even at zero temperature and pressure. This possibility, first noted by Bodmer in 1971 [1], has attracted much attention since Witten resurrected the idea in 1984 [2]. The existence of stable or metastable SQM would have numerous consequences for physics and astrophysics, and testing some of these consequences should ultimately tell us whether SQM really exists.

First it was believed that SQM might give a natural explanation of the cosmological dark matter problem. While not ruled out, this idea is now less popular, but strange quark matter may still be important in astrophysical settings, such as strange stars. Numerous investigations have searched for deposits of SQM on the Earth and in meteorites, so far unsuccessfully, and recently relativistic heavy-ion collision experiments have been performed and/or proposed to test the idea. Cosmic ray searches have come up with a few potential candidates for small SQM-lumps (strangelets), but at present no compelling evidence for stable SQM has been presented. This, however, does not rule it out. Most searches for SQM are sensitive to strangelets with very low baryon number, A, and as discussed later, finite size effects have a significant destabilizing effect on such objects, even if SQM is stable in bulk.

There is a significant range of strong interaction parameters for which SQM in bulk is stable. But even if it is not, many of the (astro)physical implications are more or less unchanged in the case of metastable SQM. In neutron stars, for instance, the high pressure brings SQM closer to stability relative to hadronic matter, and it is quite likely, that neutron stars contain cores of strange quark matter, even if SQM is unstable at zero pressure. In relativistic heavy-ion collision experiments, strangelets need "only" survive for 10^{-8} seconds to be of interest. In fact, (meta)stable strangelets may be one of the "cleanest" signatures for formation of a quark-gluon plasma in such collisions.

The present review tries to give an account of the status of strange quark matter physics and astrophysics, as of early 1998, but of course not all aspects are covered in equal detail. In particular, nothing is said about the heroic experimental efforts to produce strangelets in heavy-ion collisions. A collection of papers describing all aspects of SQM and a list of references to the field through mid-1991 can be found in [3]. An earlier review was given in [4]. Recent reviews include [5–12], and the thorough reader will notice, that some parts of the present Chapter borrows from my own papers among these since the physics discussed therein remains more or less unchanged. Refs. [11,12] also discuss the related issue of lumps of metastable strange hadronic matter, which will not be dealt with here.

Section 2 discusses the physics of SQM, starting out with simple estimates of why 3-flavor quark-matter is likely to be more bound than the 2-flavor alternative, proceeding with more detailed descriptions of SQM in bulk. Smaller systems (strangelets), for which finite-size effects are crucial, are described in Section 3. Most of the results are based on the MIT bag-model, but it is worth stressing from the outset that this should only be viewed as a crude approximation to reality, ultimately to be surpassed by direct QCD-calculations.

Section 4 deals with the possible production of lumps of SQM (often called quark nuggets) in the cosmological quark-hadron phase transition, and the struggle of quark nuggets to survive evaporation and boiling in a hostile environment. It turns out, that only large nuggets are likely to survive, but the physics involved in the destruction process is illuminating as it resembles the (time-reversed) physics involved in strangelet production in heavy-ion collisions. Implications of surviving quark nuggets for Big Bang nucleosynthesis and the dark matter problem are also discussed.

Perhaps the most likely place to discover SQM (even if it is not absolutely stable) is in neutron stars. These could be "hybrid", "strange", or even "mixed" (the first term conventionally used for neutron stars with quark cores; the second for "true" quark stars in case of SQM stability, and the latter for objects with mixed phases of quark matter and nuclear matter). Section 5 describes these stars, their implications for our understanding of pulsars, and the possible connection to the energetic gamma-ray bursters.

Strangelets surviving from the early Universe *or* released from strange stars in binary systems have been searched for in cosmic ray detectors and in meteorites and mineral deposits. So far there are only a few potential candidates, but more sensitive experiments will soon be carried out. Section 6 discusses some of the limits obtained. It also presents an astrophysical argument which either improves the Earth-based flux-limits by many orders of magnitude (almost excluding absolutely stable SQM), or predicts that all neutron stars are strange stars, if SQM is stable (the prediction to choose depends on whether any pulsars can be proven to be ordinary neutron stars).

Conclusions and a brief outlook are provided in Section 7.

2   Physics of SQM in Bulk

2.1   Does Strange Matter Conflict with Experience?

At first sight, the possibility that quark matter could be absolutely stable seems to contradict daily life experiences (and experiments) showing that nuclei consist of neutrons and protons, rather than a soup of quarks. If a lower energy state exists, then why are we here? Why have we not decayed into strange quark matter?

The answer to this obvious question is, that (meta)stability of strange quark matter requires a significant fraction of strange quarks to be present. Conversion of an iron nucleus into an $A = 56$ strangelet thus demands a very high order weak interaction to change dozens of u- and d-quarks into s-quarks at the same time. Such a process has negligible probability of happening. For lower A the conversion requires a lower order weak interaction, but as demonstrated later, finite-size effects destabilize small strangelets so that they become unstable or only weakly metastable even if strange quark matter is stable in bulk.

Therefore (meta)stability of strange quark matter does not conflict with the existence of ordinary nuclei. On the other hand, the existence of ordinary nuclei shows, that quark matter composed of u- and d-quarks alone is unstable, a fact that will be used later on to place constraints on model parameters.

Another constraint from our mere existence can be placed on the **electrical charge** of strangelets. If energy is gained by converting ordinary matter into strange quark matter, strangelets with negative quark charge, even if globally neutral due to a cloud of positrons, would have devastating consequences, eating up the nuclei they would encounter. Even a small stable component in the cosmos would be intolerable (but they could still appear as metastable products in heavy-ion collisions, like the recent charge -1, mass 7.4 GeV event in NA52 at CERN [13]). A positive charge on the quark surface (neutralized by surrounding electrons) is less problematic, because ordinary nuclei will be electrostatically repelled. The barrier has to be of a certain height, though, in order not to impact stellar evolution (see below). Note that **neutrons** are easily absorbed. As demonstrated later, this has important consequences for quark star formation and can be used to constrain strange matter properties using several astrophysical lines of reasoning. It may even lead to practical applications in energy production, etc. [14].

2.2   Simple Arguments for (Meta)Stability

As argued above, quark matter composed of u and d-quarks is expected to be unstable (except from 3-quark baryons). Introducing a third flavor makes it possible to reduce the energy relative to a two-flavor system, because an extra Fermi-well is available. The introduction of an extra fermion-flavor makes it possible to increase the spatial concentration of quarks, thereby reducing the total energy. A penalty is paid because the mass of the s-quark is high compared to that of u and d, so stability is most likely for low s-quark mass.

To make the argument slightly more quantitative, consider non-interacting, massless quarks[1] inside a confining bag at temperature $T = 0$, without external pressure. For a massless quark-flavor, i, the Fermi momentum, p_{Fi}, equals the chemical potential, μ_i (throughout the chapter, unless otherwise noted, $\hbar = c = k_B = 1$; for an introduction to Fermi-gas thermodynamics, see for instance Ref. [15]). Thus the number densities are $n_i = \mu_i^3/\pi^2$, the energy densities $\epsilon_i = 3\mu_i^4/(4\pi^2)$, and the pressures $P_i = \mu_i^4/(4\pi^2)$. The sum of the quark pressures is balanced by the confining bag pressure, B; $\sum_i P_i = B$; the total energy density is $\epsilon = \sum_i \epsilon_i + B = 3\sum_i P_i + B = 4B$, and the density of baryon number is $n_B = \sum_i n_i/3$. Notice that the sum of the constituents pressures, as well as the total energy density are given solely in terms of the bag constant, B.

For a gas of u and d-quarks charge neutrality requires $n_d = 2n_u$, or $\mu_2 \equiv \mu_u = 2^{-1/3}\mu_d$. The corresponding two-flavor quark pressure is $P_2 = P_u + P_d = (1 + 2^{4/3})\mu_2^4/(4\pi^2) = B$, the total energy density $\epsilon_2 = 3P_2 + B = 4B$, and the baryon number density $n_{B2} = (n_u + n_d)/3 = \mu_2^3/\pi^2$, giving an energy per baryon of

$$\epsilon_2/n_{B2} = (1 + 2^{4/3})^{3/4}(4\pi^2)^{1/4}B^{1/4} = 6.441B^{1/4} \approx 934\text{MeV}B_{145}^{1/4}, \quad (1)$$

where $B_{145}^{1/4} \equiv B^{1/4}/145\text{MeV}$; 145MeV being the lowest possible choice for reasons discussed below.

A three-flavor quark gas is electrically neutral for $n_u = n_d = n_s$, i. e. $\mu_3 \equiv \mu_u = \mu_d = \mu_s$. For fixed bag constant the three-quark gas should exert the same pressure as the two-quark gas (leaving also the energy density, $\epsilon_3 = 3P_3 + B = 4B$, unchanged). That happens when $\mu_3 = [(1+2^{4/3})/3]^{1/4}\mu_2$, giving a baryon number density of $n_{B3} = \mu_3^3/\pi^2 = [(1 + 2^{4/3})/3]^{3/4}n_{B2}$. The energy per baryon is then

$$\epsilon_3/n_{B3} = 3\mu_3 = 3^{3/4}(4\pi^2)^{1/4}B^{1/4} = 5.714B^{1/4} \approx 829\text{MeV}B_{145}^{1/4}; \quad (2)$$

lower than in the two-quark case by a factor $n_{B2}/n_{B3} = (3/(1 + 2^{4/3}))^{3/4} \approx 0.89$.

The possible presence of electrons was neglected in the calculations above. For two-flavor quark matter, including electrons in chemical equilibrium via $u + e^- \leftrightarrow d + \nu_e$, so that $\mu_u + \mu_e = \mu_d$, gives more cumbersome equations, but only changes ϵ_2/n_{B2} to $6.445B^{1/4}$, since μ_e turns out to be rather small. Three-flavor quark matter does not contain electrons for non-interacting, massless quarks.

One may therefore gain of order 100 MeV per baryon by introducing an extra flavor. At fixed confining bag pressure the extra Fermi-well allows one

---

[1] Since current quark masses rather than constituent quark masses enter in the MIT bag model used to describe SQM, this is a very good approximation for u and d-quarks with $5\,\text{MeV} \approx m_u < m_d \approx 10\,\text{MeV} \ll 300\,\text{MeV} \approx \mu_u, \mu_d$.

to pack the baryon number denser into the system, thereby gaining in binding energy.

The energy per baryon in a free gas of neutrons is the neutron mass, $m_n = 939.6$MeV; in a gas of ^{56}Fe it is 930 MeV. Naively, stability of ud-quark matter relative to neutrons thus corresponds to $\epsilon_2/n_{B2} < m_n$, or $B^{1/4} < 145.9$MeV ($B^{1/4} < 144.4$MeV for stability relative to iron). The argument can be turned around: Since one observes neutrons and ^{56}Fe in Nature, rather than ud-quark matter, it is concluded that $B^{1/4}$ must be larger than the numbers just quoted. More detailed calculations including finite-size effects and Coulomb-forces do not change these numbers much, so we shall assume for the present purpose that $B^{1/4} = 145$MeV is an experimental lower limit for $\alpha_s = 0$. (Here α_s denotes the strong "fine-structure" constant; $\alpha_s = 0$ corresponding to non-interacting quarks except for the confinement given by B).

Bulk strange quark matter is absolutely stable relative to a gas of iron for $B^{1/4} < 162.8$MeV, metastable relative to a neutron gas for $B^{1/4} < 164.4$MeV, and relative to a gas of Λ-particles (the ultimate production limit in heavy-ion collisions) for $B^{1/4} < 195.2$MeV. These numbers are upper limits. As demonstrated below, a finite s-quark mass as well as a non-zero strong coupling constant decreases the limit on $B^{1/4}$.

The presence of ordinary nuclei in Nature *cannot* be used to turn the values of $B^{1/4}$ just quoted for SQM into lower limits. Conversion of a nucleus into a lump of SQM requires simultaneous transformation of roughly A u- and d-quarks into s-quarks. The probability for this to happen involves a weak interaction coupling to the power A, i. e. it does not happen. This leads to the conclusion, that even if SQM is the lowest energy state for hadronic matter in bulk, its formation requires a strangeness-rich environment or formation via a "normal" quark-gluon plasma in relativistic heavy-ion collisions, the early Universe, or a neutron star interior. All of these possibilities will be explored in the following.

2.3   SQM in Bulk at $T = 0$

The estimates above assumed $m_s = \alpha_s = 0$. Non-zero α_s was found by Farhi and Jaffe [16] to correspond effectively to a reduction in B. In the interest of simplicity I will therefore set $\alpha_s = 0$ in most of the following. The energy "penalty" paid by having to form s-quarks at a finite mass of 50–300 MeV calls for more detailed calculations, however. Such calculations are usually performed within the MIT bag model [17,18].

Strange quark matter contains degenerate Fermi gases of u, d, and s quarks, and e^- or e^+. Chemical equilibrium is maintained by weak interactions,

$$d \leftrightarrow u + e^- + \bar{\nu}_e \qquad (3)$$

$$s \leftrightarrow u + e^- + \bar{\nu}_e \qquad (4)$$

$$u + s \leftrightarrow d + u, \tag{5}$$

where the first two reactions should be understood to include also the various permutations of the involved particles.

Neutrinos generally escape the system, so we shall ascribe to them no chemical potential. Thus the chemical potentials in equilibrium are given by

$$\mu_d = \mu_s = \mu_u + \mu_e. \tag{6}$$

Knowing the chemical potentials one can calculate the thermodynamic potentials.

$$\Omega_{e,V} = -\frac{\mu_e^4}{12\pi^2} \tag{7}$$

$$\Omega_{u,V} = -\frac{\mu_u^4}{4\pi^2} \tag{8}$$

$$\Omega_{d,V} = -\frac{\mu_d^4}{4\pi^2} \tag{9}$$

$$\Omega_{s,V} = -\frac{\mu_s^4}{4\pi^2} \left((1-\lambda^2)^{1/2}(1 - \frac{5}{2}\lambda^2) + \frac{3}{2}\lambda^4 \ln \frac{1 + (1-\lambda^2)^{1/2}}{\lambda} \right), \tag{10}$$

defining $\lambda \equiv m_s/\mu_s$.

Number densities are given by

$$n_{i,V} = -\partial\Omega_{i,V}/\partial\mu_i; \tag{11}$$

i. e. $n_{e,V} = \mu_e^3/3\pi^2$, $n_{u,V} = \mu_u^3/\pi^2$, $n_{d,V} = \mu_d^3/\pi^2$, $n_{s,V} = \mu_s^3(1-\lambda^2)^{3/2}/\pi^2$. The total pressure is

$$P = \sum_i P_i - B = -\sum_i \Omega_{i,V} - B = 0, \tag{12}$$

and charge neutrality requires

$$\frac{2}{3}n_{u,V} - \frac{1}{3}n_{d,V} - \frac{1}{3}n_{s,V} - n_{e,V} = 0. \tag{13}$$

The total energy density is

$$\epsilon = \sum_i (\Omega_{i,V} + n_{i,V}\mu_i) + B, \tag{14}$$

and the density of baryon number

$$n_B = \frac{1}{3}(n_{u,V} + n_{d,V} + n_{s,V}). \tag{15}$$

Combining Eqs. (6) and (13) leaves only one independent chemical potential, which can be determined from the pressure balance, Eq. (12). Thus

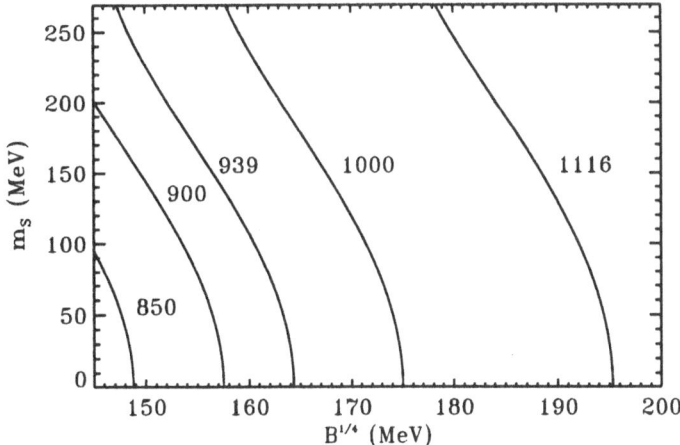

Fig. 1: Energy per baryon for bulk strange quark matter as a function of bag constant and strange quark mass.

all parameters can be calculated for a given choice of m_s and B. Results of such calculations are shown in Figure 1. Similar calculations were originally done by Farhi and Jaffe [16].

The calculations above assumed zero temperature and external pressure. Finite temperature and external pressure can be relevant in connection with cosmology (Section 4) and strange stars (Section 5) respectively, and also for strangelet creation in collision experiments. The relevant extensions of the formulae above will be given in Section 3.4.

3   Strangelets

So far the treatment of SQM has focused on the bulk properties. This approximation is generally valid for large baryon numbers. For $A \ll 10^7$ the quark part of SQM is smaller than the Compton wavelength of electrons, so electrons no longer ensure local charge neutrality. Therefore Coulomb energy has to be taken into account, though the fortuitous cancellation of $q_u + q_d + q_s = \frac{2}{3} - \frac{1}{3} - \frac{1}{3} = 0$ means that Coulomb energy is much less important for strangelets than for nuclei. For even smaller baryon numbers (in practice $A < 10^3$) other finite size effects such as surface tension and curvature have to be taken into account.

Several strangelet searches with relativistic heavy-ion collisions as well as cosmic ray searches have been carried out, and others are planned for the future. Most of these searches are sensitive only to low A-values, so it is important to know the properties of small lumps of strange quark matter (strangelets).

In the following I will describe the physical properties of strangelets in the language of the MIT-bag model (only limited work has been performed using other models—qualitatively confirming the MIT-bag results, though quantitative details can differ). First, I will discuss results obtained from direct solution of the Dirac equation with MIT-bag boundary conditions; such mode filling calculations correspond to a nuclear shell model. Then I will show how the mean behavior of the shell model results can be understood physically in terms of a liquid drop model calculation based on a smoothed density of states, and how approximations to the liquid drop results give simple formulae for strangelet masses etc. Finally I discuss the changes introduced if strangelets are at finite rather than zero temperature.

3.1   Shell Model

Mode-filling for large numbers of quarks in a spherical MIT-bag was performed for ud-systems by Vasak, Greiner and Neise [19], and for 2- and 3-flavor systems by Farhi and Jaffe [16], and Greiner et $al.$ [20] (see also [21]). Gilson and Jaffe [22] published an investigation of low-mass strangelets for 4 different combinations of s-quark mass and bag constant with particular emphasis on metastability against strong decays. Further parameter ranges were studied and compared to liquid drop model calculations by Madsen [23], and recently new shell-model studies were published by Schaffner-Bielich et $al.$ [24]. All of these calculations were performed for $\alpha_s = 0$, which will also be assumed in the following.

In the MIT bag model noninteracting quarks are confined to a spherical cavity of radius R. They satisfy the free Dirac equation inside the cavity and obey a boundary condition at the surface, which corresponds to no current flow across the surface. The bag itself has an energy of BV. In the simplest version the energy (mass) of the system is given by the sum of the bag energy and the energies of individual quarks,

$$E = \sum_{i=u,d,s} \sum_{\kappa} N_{\kappa,i}(m_i^2 + k_{\kappa,i}^2)^{1/2} + B4\pi R^3/3. \qquad (16)$$

Here $k_{\kappa,i} \equiv x_{\kappa,i}/R$, where $x_{\kappa,i}$ are eigenvalues of the equation

$$f_\kappa(x_{\kappa,i}) = \frac{-x_{\kappa,i}}{(x_{\kappa,i}^2 + m_i^2 R^2)^{1/2} + m_i R} f_{\kappa-1}(x_{\kappa,i}). \qquad (17)$$

f_κ are regular Bessel functions of order κ ,

$$f_\kappa(x) = \begin{cases} j_\kappa(x) & \kappa \geq 0 \\ y_\kappa(x) = (-1)^{\kappa+1} j_{-\kappa-1}(x) & \kappa < 0 \end{cases} \qquad (18)$$

For states with quantum numbers (j,l) κ takes the values $\kappa = \pm(j + \frac{1}{2})$ for $l = j \pm \frac{1}{2}$. For a given quark flavor each level has a degeneracy of $N_{\kappa,i} = 3(2j+1)$ (the factor 3 from color degrees of freedom). For example, the $1S_{1/2}$ ground-state $(j = 1/2, l = 0, \kappa = -1)$ for a massless quark corresponds to solving the equation $\tan x = x/(1 - x)$, giving $x \simeq 2.0428$. The ground state has a degeneracy of 6 per flavor.

For massless quarks (finding the equilibrium radius from $\partial E/\partial R = 0$) one gets

$$E = 364.00\text{MeV}B_{145}^{1/4}\left(\sum x_{\kappa,i}\right)^{3/4} \tag{19}$$

where the sum is to be taken over all $3A$ quark-levels, and the numbers $x_{\kappa,i}$ for massless quarks are tabulated in [19].

For massive quarks the level filling scheme is more cumbersome (see e.g. Refs. [22,23]). Fixing bag constant and quark-masses, for each baryon number one must fill up the lowest energy levels for a choice of radius; then vary the radius until a minimum energy is found ($\partial E/\partial R = 0$). Since levels cross, the order of levels is changing as a function of R. This is easily seen in the Figures, where one notices discontinuous changes in the position of shells.

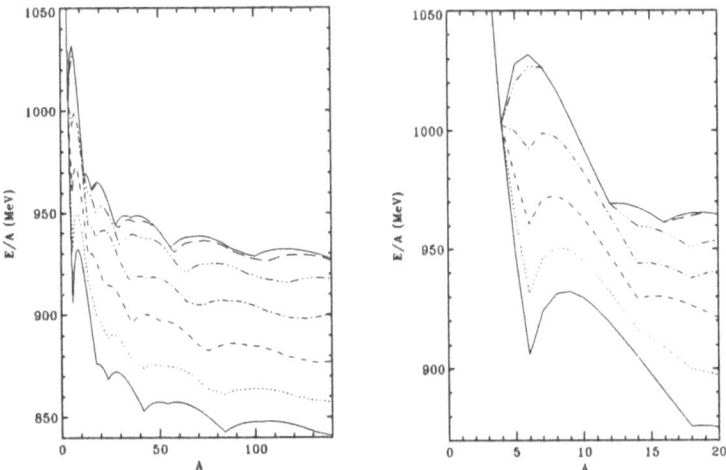

Fig. 2: Energy per baryon (in MeV) for strangelets with $B^{1/4} = 145\text{MeV}$ and m_s from 0–300 MeV in steps of 50 MeV (m_s increases upward). The figure on the right shows an expanded view of the low-mass region to highlight the change of "magic numbers" with changing m_s.

One notices that the energy per baryon smoothly approaches the bulk limit for $A \to \infty$, whereas the energy grows significantly for low A. For low s-quark mass shells are recognized for $A = 6$ (3 colors and 2 spin orientations per flavor), and less conspicuous ones for $A = 18, 24, 42, 54, 60, 84, 102$ etc. As m_s increases it becomes more and more favorable to use u and d rather than s-quarks, and the "magic numbers" change; for instance the first closed shell is seen for $A = 4$ rather than 6.

Equation (16) can be modified by inclusion of Coulomb energy and zero-point fluctuation energy. As already discussed the Coulomb energy is generally small. The zero-point energy is normally included as a phenomenological term of the form $-Z_0/R$, where fits to light hadron spectra indicate the choice $Z_0 = 1.84$. This was used, for instance, by Gilson and Jaffe [22]. Roughly half of this phenomenological term is due to center-of-mass motion, which can be included more explicitly by substituting $\left[\left(\sum x_{\kappa,i}\right)^2 - \sum x_{\kappa,i}^2\right]^{3/8}$ instead of $\left(\sum x_{\kappa,i}\right)^{3/4}$ in Eq. (19). The proper choice of α_s and Z_0 is a tricky question. As discussed by Farhi and Jaffe [16] the values are intimately coupled to B and m_s, and it is not obvious that values deduced from bag model fits to ordinary hadrons are to be preferred. This uncertainty may have an important effect for $A < 5$–10, but the zero-point energy quickly becomes negligible for increasing A for reasons explained in Section 3.2. It means, however, that it is difficult to match strangelet calculations to experimental data concerning ordinary hadrons or limits on the putative $A = 2$ H-dibaryon.

3.2   Liquid Drop Model

Mode-filling calculations are rather tedious but do of course give the "correct" results as far as the model can be trusted. But for many applications a global mass-formula analogous to the liquid drop model for nuclei is of great use and also gives further physical insight.

A phenomenological approach to a strangelet mass-formula was undertaken by Crawford et al. [25,26] whereas Berger and Jaffe [27] made a detailed analysis within the MIT bag model. They included Coulomb corrections and surface tension effects stemming from the depletion in the surface density of states due to the mass of the strange quark. Both effects were treated as perturbations added to a bulk solution with the surface contribution derived from a multiple reflection expansion. Madsen [23,28,29] gave a self-consistent treatment including also the very important curvature energy.

The following discussion closely follows [23]. All calculations are done for zero temperature and strong coupling constant, α_s. As argued by Farhi and Jaffe [16] the latter assumption can be relaxed by a re-scaling of the bag constant. Also, I shall concentrate on systems small enough ($A < 10^7$) to justify neglect of electrons. Strangelets with $A \ll 10^7$ are smaller than the electron Compton wavelength, and electrons are therefore mainly localized outside the quark phase. Thus strangelets do not obey a requirement of local charge neutrality, as was the case for SQM in bulk. This leads to a small Coulomb energy, which is rather negligible for the mass-formula (less than a few MeV per baryon), but which is decisive for the charge-to-mass ratio of the strangelet. A characteristic of strangelets, which is perhaps the best experimental signature, is that this ratio is very small compared to ordinary nuclei. Finally, I neglect charge screening, an issue of negligible importance for the mass formula, but of some importance for the charge-to-mass ratio for systems of radii above 5–10 fm ($A > 10^2$–10^3) [30].

In the ideal Fermi-gas approximation the energy of a system composed of quark flavors i is given by

$$E = \sum_i (\Omega_i + N_i \mu_i) + BV + E_{\text{Coul}}. \tag{20}$$

Here Ω_i, N_i and μ_i denote thermodynamic potentials, total number of quarks, and chemical potentials, respectively. B is the bag constant, V is the bag volume, and E_{Coul} is the Coulomb energy.

In the multiple reflection expansion framework of Balian and Bloch [31], the thermodynamical quantities can be derived from a density of states of the form

$$\frac{dN_i}{dk} = 6\left\{ \frac{k^2 V}{2\pi^2} + f_S\left(\frac{m_i}{k}\right) kS + f_C\left(\frac{m_i}{k}\right) C + \ldots \right\}, \tag{21}$$

where area $S = \oint dS$ ($= 4\pi R^2$ for a sphere) and extrinsic curvature $C = \oint \left(\frac{1}{R_1} + \frac{1}{R_2}\right) dS$ ($= 8\pi R$ for a sphere). Curvature radii are denoted R_1 and R_2. For a spherical system $R_1 = R_2 = R$. The functions f_S and f_C will be discussed below.

In terms of volume-, surface-, and curvature-densities, $n_{i,V}$, $n_{i,S}$, and $n_{i,C}$, the number of quarks of flavor i is

$$N_i = \int_0^{k_{Fi}} \frac{dN_i}{dk} dk = n_{i,V} V + n_{i,S} S + n_{i,C} C, \tag{22}$$

with Fermi momentum $k_{Fi} = (\mu_i^2 - m_i^2)^{1/2} = \mu_i (1 - \lambda_i^2)^{1/2}$; $\lambda_i \equiv m_i/\mu_i$.

The corresponding thermodynamic potentials are related by

$$\Omega_i = \Omega_{i,V} V + \Omega_{i,S} S + \Omega_{i,C} C, \tag{23}$$

where $\partial \Omega_i / \partial \mu_i = -N_i$, and $\partial \Omega_{i,j} / \partial \mu_i = -n_{i,j}$. The volume terms are given by

$$\Omega_{i,V} = -\frac{\mu_i^4}{4\pi^2} \left[(1 - \lambda_i^2)^{1/2} \left(1 - \frac{5}{2}\lambda_i^2\right) + \frac{3}{2}\lambda_i^4 \ln \frac{1 + (1 - \lambda_i^2)^{1/2}}{\lambda_i} \right], \tag{24}$$

$$n_{i,V} = \frac{\mu_i^3}{\pi^2} (1 - \lambda_i^2)^{3/2}. \tag{25}$$

The surface contribution from massive quarks is derived from

$$f_S\left(\frac{m}{k}\right) = -\frac{1}{8\pi} \left\{ 1 - \left(\frac{2}{\pi}\right) \tan^{-1} \frac{k}{m} \right\} \tag{26}$$

as [27]

$$\Omega_{i,S} = \frac{3}{4\pi}\mu_i^3 \left[\frac{(1-\lambda_i^2)}{6} - \frac{\lambda_i^2(1-\lambda_i)}{3} \right. \tag{27}$$

$$-\frac{1}{3\pi}\left(\tan^{-1}\left[\frac{(1-\lambda_i^2)^{1/2}}{\lambda_i}\right] - 2\lambda_i(1-\lambda_i^2)^{1/2}\right.$$

$$\left.\left. + \lambda_i^3 \ln\left[\frac{1+(1-\lambda_i^2)^{1/2}}{\lambda_i}\right]\right)\right];$$

$$n_{i,S} = -\frac{3}{4\pi}\mu_i^2 \left[\frac{(1-\lambda_i^2)}{2} - \frac{1}{\pi}\left(\tan^{-1}\left[\frac{(1-\lambda_i^2)^{1/2}}{\lambda_i}\right] - \lambda_i(1-\lambda_i^2)^{1/2}\right)\right] \tag{28}$$

For massless quarks $\Omega_{i,S} = n_{i,S} = 0$, whereas $f_C(0) = -1/24\pi^2$ gives [16,28,29] $\Omega_{i,C} = \mu_i^2/8\pi^2$; $n_{i,C} = -\mu_i/4\pi^2$.

The curvature terms have never been derived for massive quarks, but as shown by Madsen [23], the following *Ansatz* (found from analogies with the surface term and other known cases) works:

$$f_C\left(\frac{m}{k}\right) = \frac{1}{12\pi^2}\left\{1 - \frac{3}{2}\frac{k}{m}\left(\frac{\pi}{2} - \tan^{-1}\frac{k}{m}\right)\right\}. \tag{29}$$

This expression has the right limit for massless quarks ($f_C = -1/24\pi^2$) and for infinite mass, which corresponds to the Dirichlet boundary conditions studied by Balian and Bloch [31] ($f_C = 1/12\pi^2$). Furthermore, the expression gives perfect fits to mode-filling calculations (see the Figures and discussion below). From this *Ansatz* one derives the following thermodynamical potential and density:

$$\Omega_{i,C} = \frac{\mu_i^2}{8\pi^2}\left[\lambda_i^2 \log\frac{1+(1-\lambda_i^2)^{1/2}}{\lambda_i} + \frac{\pi}{2\lambda_i} - \frac{3\pi\lambda_i}{2} + \pi\lambda_i^2\right.$$

$$\left. - \frac{1}{\lambda_i}\tan^{-1}\frac{(1-\lambda_i^2)^{1/2}}{\lambda_i}\right]; \tag{30}$$

$$n_{i,C} = \frac{\mu_i}{8\pi^2}\left[(1-\lambda_i^2)^{1/2} - \frac{3\pi}{2}\frac{(1-\lambda_i^2)}{\lambda_i} + \frac{3}{\lambda_i}\tan^{-1}\frac{(1-\lambda_i^2)^{1/2}}{\lambda_i}\right]. \tag{31}$$

With these prescriptions the differential of $E(V, S, C, N_i)$ is given by

$$dE = \sum_i \left(\Omega_{i,V}dV + \Omega_{i,S}dS + \Omega_{i,C}dC + \mu_i dN_i\right) + BdV + dE_{\text{Coul}}. \tag{32}$$

Minimizing the total energy at fixed N_i by taking $dE = 0$ for a sphere gives the pressure equilibrium constraint

$$B = -\sum_i \Omega_{i,V} - \frac{2}{R}\sum_i \Omega_{i,S} - \frac{2}{R^2}\sum_i \Omega_{i,C} - \frac{dE_{\text{Coul}}}{dV}, \tag{33}$$

with

$$E_{\text{Coul}} = \frac{\alpha Z_V^2}{10R} + \frac{\alpha Z^2}{2R}, \tag{34}$$

$$\frac{dE_{\text{Coul}}}{dV} = -\frac{\alpha Z_V^2}{40\pi R^4} - \frac{\alpha Z^2}{8\pi R^4}, \tag{35}$$

where $Z_V = \sum_i q_i n_{i,V} V$ is the volume part of the total charge, Z, whereas charge $Z - Z_V = \sum_i q_i(n_{i,S}S + n_{i,C}C)$ is distributed on the surface. The quark charges are $q_u = 2/3$, $q_d = q_s = -1/3$. Eliminating B from Eq. (20) then gives the energy for a spherical quark lump as

$$E = \sum_i (N_i\mu_i + \frac{1}{3}\Omega_{i,S}S + \frac{2}{3}\Omega_{i,C}C) + \frac{4}{3}E_{\text{Coul}}. \tag{36}$$

The optimal composition for fixed baryon number, A, can be found by minimizing the energy with respect to N_i at fixed V, S, and C giving

$$0 = dE = \sum_i \left(\mu_i + \frac{\partial E_{\text{Coul}}}{\partial N_i}\right) dN_i. \tag{37}$$

Massless Quarks—Bulk Limit For uncharged bulk quark matter Eq. (36) reduces to the usual result for the energy per baryon

$$\epsilon^0 = A^{-1} \sum_i N_i^0 \mu_i^0, \tag{38}$$

where superscript 0 denotes bulk values. The energy minimization, Eq. (33), corresponds to

$$B = -\sum_i \Omega_{i,V}^0 = \sum_i \frac{(\mu_i^0)^4}{4\pi^2}. \tag{39}$$

The last equality assumes massless quarks. In the bulk limit the baryon number density is given by

$$n_A^0 = \frac{1}{3} \sum_i \frac{(\mu_i^0)^3}{\pi^2}, \tag{40}$$

and one may define a bulk radius per baryon as

$$R^0 = (3/4\pi n_A^0)^{1/3}. \tag{41}$$

For quark matter composed of massless u, d, and s-quarks, the Coulomb energy vanishes at equal number densities due to the fact that the sum of the quark charges is zero. Thus it is energetically most favorable to have equal chemical potentials for the three flavors. From the equations above one may derive the following bulk expressions for 3-flavor quark matter:

$$\mu_i^0 = \left(\frac{4\pi^2 B}{3}\right)^{1/4} = 1.905 B^{1/4} = 276.2\text{MeV} B_{145}^{1/4}; \tag{42}$$

$$n_A^0 = (\mu_i^0)^3/\pi^2 = 0.700B^{3/4} \tag{43}$$

$$R^0 = (3/4\pi n_A^0)^{1/3} = 0.699B^{-1/4}. \tag{44}$$

And the energy per baryon is

$$\epsilon^0 = 3\mu_i^0 = 5.714B^{1/4}, \tag{45}$$

in agreement with Eq. (2).

Following Berger and Jaffe [27] one may to first order regard Coulomb, surface (and here correspondingly curvature) energies as perturbations on top of the bulk solution. In this approach one gets

$$\begin{aligned}
\frac{E}{A} &= \epsilon^0 + A^{-1}\sum_i \Omega_{i,C}^0 C^0 = \epsilon^0 + \frac{3^{13/12}B^{1/4}}{\pi^{1/6}2^{1/6}A^{2/3}} \\
&\approx \left[829\text{MeV} + 351\text{MeV}A^{-2/3}\right]B_{145}^{1/4}.
\end{aligned} \tag{46}$$

The corresponding result for 2-flavor quark matter (c.f. [29]) is

$$\frac{E}{A} = \epsilon^0 + A^{-1}\sum_i \Omega_{i,C}^0 C^0 \approx \left[934\text{MeV} + 291\text{MeV}A^{-2/3}\right]B_{145}^{1/4}. \tag{47}$$

Massive s-Quarks—Bulk Limit For $m_s > 0$ the energy minimization, Eq. (39), changes to

$$\begin{aligned}
B &= -\sum_i \Omega_{i,V}^0 \\
&= \sum_{i=u,d} \frac{(\mu_i^0)^4}{4\pi^2} + \frac{(\mu_s^0)^4}{4\pi^2}\left[(1-\lambda^2)^{1/2}\left(1 - \frac{5}{2}\lambda^2\right)\right. \\
&\quad \left. + \frac{3}{2}\lambda^4 \ln\frac{1 + (1-\lambda^2)^{1/2}}{\lambda}\right],
\end{aligned} \tag{48}$$

and the baryon number density is now given by

$$n_A^0 = \frac{1}{3}\left[\sum_{i=u,d} \frac{(\mu_i^0)^3}{\pi^2} + \frac{(\mu_s^0)^3}{\pi^2}(1-\lambda^2)^{3/2}\right]. \tag{49}$$

A bulk radius per baryon is still defined by Eq. (41).

In bulk equilibrium the chemical potentials of the three quark flavors are equal, $\mu_u^0 = \mu_d^0 = \mu_s^0 \equiv \mu^0 = \epsilon^0/3$. Neglecting Coulomb energy one may approximate the energy per baryon of small strangelets as a sum of bulk, surface and curvature terms, using the chemical potential calculated in bulk:

$$\frac{E}{A} = \epsilon^0 + A^{-1}\sum_i \Omega_{i,S}^0 S^0 + A^{-1}\sum_i \Omega_{i,C}^0 C^0, \tag{50}$$

where $S^0 = 4\pi(R^0)^2 A^{2/3}$ and $C^0 = 8\pi(R^0)A^{1/3}$. Examples for $B^{1/4} = 145\text{MeV}$ are (with s-quark mass in MeV given in parenthesis)

$$\epsilon(0) = 829\text{MeV} + 0\text{MeV}A^{-1/3} + 351\text{MeV}A^{-2/3} \tag{51}$$

$$\epsilon(50) = 835\text{MeV} + 61\text{MeV}A^{-1/3} + 277\text{MeV}A^{-2/3} \tag{52}$$

$$\epsilon(150) = 874\text{MeV} + 77\text{MeV}A^{-1/3} + 232\text{MeV}A^{-2/3} \tag{53}$$

$$\epsilon(200) = 896\text{MeV} + 53\text{MeV}A^{-1/3} + 242\text{MeV}A^{-2/3} \tag{54}$$

$$\epsilon(250) = 911\text{MeV} + 22\text{MeV}A^{-1/3} + 266\text{MeV}A^{-2/3} \tag{55}$$

$$\epsilon(300) = 917\text{MeV} + 0.3\text{MeV}A^{-1/3} + 295\text{MeV}A^{-2/3} \tag{56}$$

$$\epsilon(350) = 917\text{MeV} + 0\text{MeV}A^{-1/3} + 296\text{MeV}A^{-2/3} \tag{57}$$

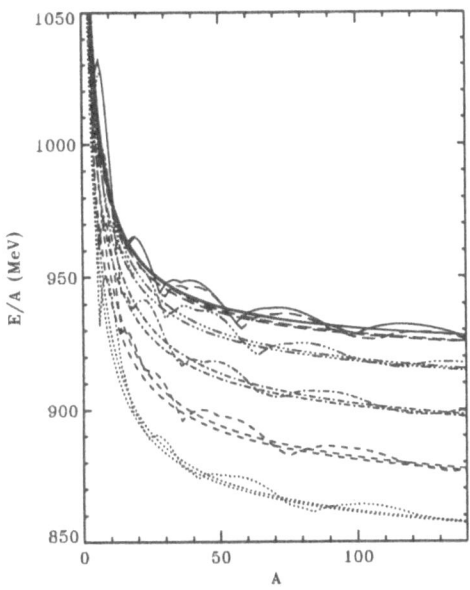

Fig. 3: Shell-model and liquid drop model results compared for $B^{1/4} = 145\text{MeV}$ with massless u and d quarks, and with m_s in the range 50–300 MeV in steps of 50 MeV. For each value of m_s the upper smooth curve is the full liquid drop model result, whereas the lower smooth curve is the bulk approximation.

The bulk approximations above generally undershoot the correct solution with properly smoothed density of states by 2MeV for $A > 100$, 5MeV for $A \approx 50$, 10MeV for $A \approx 10$ and 20MeV for $A \approx 5$ (Figure 3). This is because the actual chemical potentials of the quarks increase when A

decreases, whereas the bulk approximations use constant μ. For massless
s-quarks the expression for $\epsilon(0)$ scales simply as $B^{1/4}$. The same scaling applies for $m_s > \epsilon^0/3$, where no s-quarks are present; in the example above
the scaling can be applied to $\epsilon(350)$. For intermediate s-quark masses both
ϵ and m_s should be multiplied by $B_{145}^{1/4}$ to scale the results. For instance, if
$B^{1/4} = 165\text{MeV}$ one finds $\epsilon(150) = 985\text{MeV} + 93\text{MeV}A^{-1/3} + 265\text{MeV}A^{-2/3}$;
$\epsilon(250) = 1027\text{MeV} + 46\text{MeV}A^{-1/3} + 284\text{MeV}A^{-2/3}$. Coulomb effects were
not included above. Their inclusion would have no influence for $m_s \to 0$, but
would change the results by a few MeV for large m_s. In particular charge
neutral ud-quark matter has $\epsilon = \left[934\text{MeV} + 291\text{MeV}A^{-2/3}\right] B_{145}^{1/4}$ (Eq. (47))
rather than the $\left[917\text{MeV} + 296\text{MeV}A^{-2/3}\right] B_{145}^{1/4}$ found above (Eq. (57)).

In connection with the shell-model calculations I described the effects of a
zero-point energy of the form $-Z_0/R$, and claimed that it was important only
for $A < 10$. This can be understood in the bulk approximation of constant μ,
because the zero-point term per baryon is proportional to $A^{-4/3}$ compared
to $A^{-1/3}$ and $A^{-2/3}$ for surface and curvature energies. The full term to be
added to the bulk approximation expressions for a given ϵ^0 is:

$$\epsilon_{\text{zero}} = -Z_0(4/243\pi)^{1/3} \left[2 + [1 - (3m_s/\epsilon^0)]^{3/2}\right]^{1/3} \epsilon^0 A^{-4/3}, \qquad (58)$$

typically of order $-200Z_0\text{MeV}A^{-4/3}$.

3.3   Shell Model versus Liquid Drop Model

Self-consistent solutions can be obtained from Eq. (36). These solutions are
compared to the shell-model calculations and the bulk approximations in the
Figures. The fits are very good, showing that inclusion of surface tension and
curvature energy via the multiple reflection expansion explains the overall
behavior of the results.

3-flavor quark matter is energetically favored in bulk, and could be absolutely stable relative to ^{56}Fe for $144\text{MeV} < B^{1/4} < 163\text{MeV}$. The lower limit
corresponds to experimentally excluded stability of ud quark matter, whereas
the upper limit corresponds to a bulk energy per baryon of uds-matter of 930
MeV for $m_s = 0$.

Finite-size systems are strongly destabilized by the curvature energy, with
a magnitude of about $300\text{MeV}A^{-2/3}B_{145}^{1/4}$ for 3 quark flavors. This may pose
problems for the experimental attempts of producing strange quark matter,
since these experiments so far can only hope to create quark lumps with
baryon number $A < 20$–30, and observe lifetimes exceeding 10^{-8} seconds.
Further destabilization occurs for finite-mass s-quarks, where the surface
tension (exactly zero for massless quarks) adds up to $90\text{MeV}A^{-1/3}$ to the
energy.

Writing $E/A = \epsilon^0 + c_{\text{surf}}A^{-1/3} + c_{\text{curv}}A^{-2/3}$, with $c_{\text{surf}} \approx 100\text{MeV}$ and
$c_{\text{curv}} \approx 300\text{MeV}$, the stability condition $E/A < m_n$ may be written as $A >$

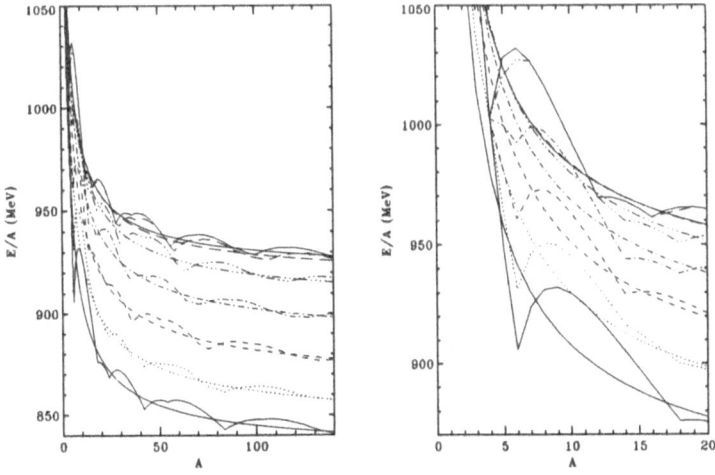

Fig. 4: As Figure 2 but showing also the liquid drop results.

$A_{\min}^{\mathrm{abs}}$, where

$$A_{\min}^{\mathrm{abs}} = \left(\frac{c_{\mathrm{surf}} + [c_{\mathrm{surf}}^2 + 4c_{\mathrm{curv}}(m_n - \epsilon^0)]^{1/2}}{2(m_n - \epsilon^0)} \right)^3 . \qquad (59)$$

Stability at baryon number 30 requires a bulk binding energy in excess of 65 MeV, which is barely within reach for $m_s > 100\mathrm{MeV}$ if, at the same time, ud-quark matter shall be unstable. The proposed cosmic ray strangelet-candidates with baryon number 370 [32] would for stability require a bulk binding energy per baryon exceeding 20 MeV to overcome the combined curvature and surface energies. Absolute stability relative to a gas of $^{56}\mathrm{Fe}$ corresponds to furthermore using 930 MeV instead of m_n, whereas stability relative to a gas of Λ-particles (the ultimate limit for formation of short-lived strangelets) would correspond to substitution of $m_\Lambda = 1116\mathrm{MeV}$.

Another way of stating the results is to calculate the minimum baryon number for which long-lived metastability with respect to neutron emission is possible. This requires $dE_{\mathrm{curv}}/dA + dE_{\mathrm{surf}}/dA < m_n - \epsilon^0$, or

$$A_{\min}^{\mathrm{meta}} = \left(\frac{c_{\mathrm{surf}} + [c_{\mathrm{surf}}^2 + 3c_{\mathrm{curv}}(m_n - \epsilon^0)]^{1/2}}{3(m_n - \epsilon^0)} \right)^3 . \qquad (60)$$

To have $A_{\min}^{\mathrm{meta}} < 30$ requires $m_n - \epsilon^0 > 30\mathrm{MeV}$, which is possible, but only for a narrow range of parameters.

This should not, however, defer experimentalists from pursuing the proposed searches. After all, the MIT bag model is only an approximation, and in particular shell effects can have a stabilizing effect. As stressed by Gilson and

Jaffe [22] the fact that the slope of E/A versus A becomes very steep near magic numbers can lead to strangelets that are metastable (stable against single baryon emission) even for $\epsilon^0 > 930\mathrm{MeV}$. Also, the time-scale for energetically allowed decays has not been calculated. Pauli-blocking is known to delay weak quark conversion in strangelets [33–36], and this will probably have a significant influence on the lifetimes. The existence of small baryon number strangelets is ultimately an experimental issue.

3.4   Strangelets at Finite Temperature

Whereas the calculations above deal with strangelets at zero temperature, the environment in heavy ion collisions is expected to be hot. An advantage of the asymptotic mass formula compared to the shell-model calculations is, that it can fairly easily be generalized to non-zero temperature.

The general expression for the thermodynamic potential, Ω_i, is

$$\Omega_i = \mp g_i T \int_0^\infty dk \, \frac{dN_i}{dk} \ln\left[1 \pm \exp(-(\epsilon(k) - \mu)/T)\right] \tag{61}$$

where the upper sign is for fermions, the lower for bosons, and the density of states, $\frac{dN_i}{dk}$, is given by Eq. (21). For *massless* quarks (including antiquarks) an integration gives, per flavor,

$$\Omega_q = -\left(\frac{7\pi^2}{60}T^4 + \frac{\mu^2 T^2}{2} + \frac{\mu^4}{4\pi^2}\right)V + \left(\frac{T^2}{24} + \frac{\mu^2}{8\pi^2}\right)C, \tag{62}$$

with a corresponding quark number

$$N_q = -\frac{\partial \Omega_q}{\partial \mu} = \left(\mu T^2 + \frac{\mu^3}{\pi^2}\right)V - \frac{\mu}{4\pi^2}C. \tag{63}$$

For gluons

$$\Omega_g = -\frac{8\pi^2}{45}T^4 V + \frac{4}{9}T^2 C. \tag{64}$$

The total Ω can be found from summing the terms above, and other thermodynamical quantities like the free energy and the internal energy can be derived. For 3 massless quark flavors of equal chemical potential one finds

$$\Omega = \left(-\frac{19\pi^2}{36}T^4 - \frac{3}{2}\mu^2 T^2 - \frac{3}{4\pi^2}\mu^4 + B\right)V + \left(\frac{41}{72}T^2 + \frac{3}{8\pi^2}\mu^2\right)C \tag{65}$$

$$F = \left(-\frac{19\pi^2}{36}T^4 + \frac{3}{2}\mu^2 T^2 + \frac{9}{4\pi^2}\mu^4 + B\right)V + \left(\frac{41}{72}T^2 - \frac{3}{8\pi^2}\mu^2\right)C \tag{66}$$

$$E = \left(\frac{19\pi^2}{12}T^4 + \frac{9}{2}\mu^2 T^2 + \frac{9}{4\pi^2}\mu^4 + B\right)V - \left(\frac{41}{72}T^2 + \frac{3}{8\pi^2}\mu^2\right)C. \tag{67}$$

Strangelets are in mechanical equilibrium at fixed temperature and baryon number when $dF = 0$, corresponding to

$$BV = \left(\frac{19\pi^2}{36}T^4 + \frac{3}{2}\mu^2 T^2 + \frac{3}{4\pi^2}\mu^4\right)V - \left(\frac{41}{216}T^2 + \frac{1}{8\pi^2}\mu^2\right)C \quad (68)$$

In this case one gets the following expressions for the thermodynamic potential, free energy, internal energy and baryon number:

$$\Omega = \left(\frac{41}{108}T^2 + \frac{1}{4\pi^2}\mu^2\right)C \quad (69)$$

$$F = \left(3\mu^2 T^2 + \frac{3}{\pi^2}\mu^4\right)V + \left(\frac{41}{108}T^2 - \frac{1}{2\pi^2}\mu^2\right)C \quad (70)$$

$$E = 4BV \quad (71)$$

$$A = \left(\mu T^2 + \frac{1}{\pi^2}\mu^3\right)V - \frac{\mu}{4\pi^2}C. \quad (72)$$

Notice that the equations above can also be used in connection with bulk SQM, for instance in an astrophysical context, by simply putting $C = 0$. An external pressure can be accommodated by substituting $B + P_{\text{external}}$ in place of B.

Dotted curves in Figure 5 shows the energy per baryon for finite temperature strangelets according to the formulae above. Results are given for fixed entropy per baryon, where the entropy is calculated from $S \equiv -\partial\Omega/\partial T|_{V,\mu}$. These results were first presented in [37]. A similar treatment, including finite m_s, was published in [38], whereas Ref. [39] shows results for a corresponding finite temperature shell model calculation, finding that shell structures are washed away at $T > 10\,\text{MeV}$, which means that liquid drop model and shell model results become indistinguishable at high T (S/A).

As discussed in more detail in [37] further complications arise from the fact, that strangelets must be color singlets. This has no influence on the ground state energy for $T = 0$, but for $T > 0$ quarks are statistically distributed over energy levels, and the color singlet constraint reduces the number of possible configurations, forcing the energy up for fixed entropy (see also [40]). The effect is important for $A < 100$ as illustrated in Figure 5. Similar effects result from insisting that strangelets shall have a definite momentum. These destabilising effects can be important in connection with experiments, which inevitably create strangelets with rather high entropies. A tremendous job remains to be done in calculating the details of strangelet formation, evolution, and decay modes, including realistic non-equilibrium effects, etc.!

4 SQM in Cosmology

4.1 Formation, Evaporation and Boiling of Quark Nuggets

If the cosmological quark-hadron phase transition was first order, supercooling may result in concentration of baryon number inside shrinking bubbles of

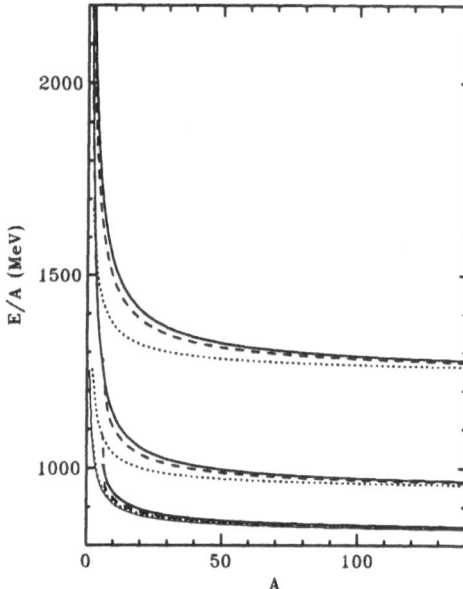

Fig. 5: E/A as a function of A for strangelets with equal numbers of massless u, d, and s quarks for entropy per baryon of 0, 1, 5 and 10, and $B^{1/4} = 145$MeV. Solid curves include color singlet and zero momentum constraints, dashed curves only the color singlet constraint, and dotted curves are without constraints. Entropy increases upward. For $S = 0$ $(T = 0)$ the three curves completely overlap (lowest solid curve).

quark phase. The amount of baryon concentration depends on the permeability of the "membrane" separating the phases and on the turbulent removal of quarks from the phase boundary. If a quark bubble is able to get rid of entropy fast enough (primarily in the form of neutrinos and photons) relative to the rate of baryon number removal, there is a chance of reaching baryon number densities in the quark bubbles approaching nuclear matter density. In other words, a quark nugget may form. Whether or not this actually happens, or whether one is left with the less extreme, but also interesting scenario where all of the quarks end up in inhomogeneously distributed neutrons and protons, giving non-standard Big Bang nucleosynthesis, has been a topic of much debate [2,41–44], and the final word has probably not been said.

But even if cosmological quark nuggets do form, they find themselves in a very hostile environment with a temperature of order 100 MeV. Under such conditions the nuggets are unstable against surface evaporation [45–48] and boiling [49–51]; but the crucial question from a cosmological point of view is whether some nuggets may survive due to the relatively short time-scale for cooling the Universe (the age of the Universe at temperature T being roughly $t_{\text{sec}} = T_{\text{MeV}}^{-2}$).

Alcock and Farhi [45] showed that the timescale for complete evaporation of a quark nugget was smaller than the age of the Universe at temperature T for baryon numbers below

$$A_{\text{evap}} \approx 2 \times 10^{56} \exp(-3I_n/T) f_n^3, \tag{73}$$

where f_n (the phase boundary penetrability of neutrons) was assumed to be close to unity. For a homogeneous quark nugget the neutron binding energy $I_n = m_n - \mu_u - 2\mu_d$ was estimated to be of order 20 MeV. For such a binding, primordial nuggets with baryon number $A < 10^{55}$ evaporate almost instantly when neutrino heating becomes possible at $T \approx 50$ MeV [45].

However, the surface evaporation of neutrons and protons reduces μ_u and μ_d, and leads to an increase in μ_s. Weak decays, diffusion and convection work to counteract this, but the net result is an s-quark enriched layer near the surface. (Small nuggets are s-quark enriched throughout their interior). The most efficient way to remove the s-quarks is then to emit them in kaons ($\overline{K^0}$, K^-) along with thermal $\bar{u}$ and $\bar{d}$. A quasi-equilibrium situation arises with an effective $I_n \approx 350$ MeV [46]. Thereby the baryon number of nuggets surviving evaporation is reduced to 10^{46}, and a proper inclusion of reabsorption of emitted hadrons (a calculation that has so far not been done) may reduce the number somewhat.

Cosmological nugget evaporation (time-reversed) is closely related to the distillation mechanism proposed for strangelet production in relativistic heavy-ion collisions [52,53,20,14,54–57]. There strangeness enhancement occurs due to emission of K^+ and K^0.

The calculations described above assume that the penetrability of the phase boundary is near 100%. It has been argued that the penetrability may be reduced by a few orders of magnitude in a chromoelectric flux tube model. This would decrease A_{evap} by a factor f_n^3, permitting smaller nuggets (possibly down to $A = 10^{39}$) to survive [47,48]. Again, the limit on A may be further reduced by reabsorption.

Primordial nuggets are superheated, and may therefore boil by forming bubbles of hadronic gas in their interiors [49]. However, even though boiling is thermodynamically allowed, it probably does not play an important role for primordial nuggets (or in heavy-ion collisions for that matter), since the time-scale is too short for bubble-nucleation to take place [50,51]. The surface evaporation described above is thus the decisive mechanism.

Some authors have argued [58,59], that boiling will take place unless a large external pressure (e.g. due to a gravitationally bound shell of nucleons) is there to prevent it. Such gravitational stabilization only works for masses close to those of stars ($A \approx 10^{57}$). However, the authors discuss only whether boiling is thermodynamically possible, but neglect that there is not enough time for the bubbles to nucleate.

Apart from trace abundances, one should not expect nuggets smaller than $10^{30} - 10^{40}$ to survive from the early Universe. This however brings one well within the causality limit set by the baryon number inside the horizon during the cosmic quark-hadron phase transition,

$$A_{hor} \approx 10^{49} \left(\frac{100\text{MeV}}{T} \right)^2 , \qquad (74)$$

and includes the "most probable" range of baryon numbers originally predicted by Witten [2]. It also leaves open the possibility that SQM may explain the dark matter problem, and if we understood the details of the quark-hadron phase transition, we could even calculate the relative abundances of dark and ordinary matter from first principles.

There is a possibility, that also small traces of primordial nuggets with low baryon numbers are left over from the early Universe. Even such traces may in fact be "observed" using the astrophysical detectors discussed in Section 6, or via Big Bang nucleosynthesis, as explained in Section 4.2.

4.2   Quark Nuggets and Big Bang Nucleosynthesis

A crucial property of quark nuggets is the positive electrostatic surface potential of the quark phase, which is due to the quarks being stronger bound than the electrons (electrostatic forces are weaker than strong forces). For typical nugget parameters the electrostatic potential can be several MeV, so except at very high temperatures, protons and nuclei are repelled from nuggets, whereas neutrons are absorbed, adding one unit of baryon number.

This opens the intriguing possibility of using SQM as an energy source [14], at least in principle. It also makes it possible to use Big Bang nucleosynthesis as well as the properties of pulsars to place very stringent limits on the abundance of quark nuggets in the Universe.

During Big Bang nucleosynthesis ($T \leq 1$ MeV), nuggets absorb neutrons but not protons. This means that the presence of quark nuggets reduces the neutron-to-proton ratio, thereby lowering the production of ^{4}He. The helium-production is very sensitive to the total amount of nugget-area present, and in order not to ruin the concordance with observations, one finds [60] that only nuggets with $A > A_{BBN} \approx 10^{23} \Omega_{nug}^3 h^6 f_n^3$ are allowed during nucleosynthesis. Here Ω_{nug} is the present-day nugget contribution to the cosmic density (in units of the critical density), h is the Hubble parameter in units of 100 km sec^{-1} Mpc^{-1}, and $f_n \leq 1$ is the penetrability of the nugget surface. Slightly stronger limits can be obtained from inclusion of all light nuclei instead of ^{4}He only [61]. (Ref. [62] found good correspondence with nucleosynthesis for a nugget-dominated, $\Omega = 1$ Universe if $A \approx 10^{17}$, but as shown in [46], this was due to an erroneous emission rate for nucleons.)

The nucleosynthesis calculations leading to A_{BBN} neglected inhomogeneities in the nucleon distribution, and all nuggets were assumed to have the same baryon number. However, the formation of ^{4}He is an on-off process over a limited range of A, so the detailed behavior of the inhomogeneities may not be so important.

Note that SQM, in spite of it carrying baryon number, does *not* contribute to the usual nucleosynthesis limit on Ω_{baryon}. The SQM baryon number is "hidden" in quark nuggets long before Big Bang nucleosynthesis begins, and the nuggets only influence nucleosynthesis if they have a big total surface area, as described above.

Evaporating nuggets would lead to strongly inhomogeneous nucleosynthesis with enhanced heavy-element formation. This aspect has recently been studied in [63].

4.3   Quark Nuggets as Dark Matter

Witten [2] argued that quark nuggets might be a natural explanation of the cosmological dark matter problem, in principle allowing a calculation of the relative amount of dark matter and ordinary baryons. In view of the evaporation discussed above, this idea now seems less likely, but is certainly not ruled out for $A > 10^{30}$. Massive quark nuggets decouple from thermal equilibrium with the radiation bath very early in the history of the Universe, quickly slow down, and behave as cold dark matter in the context of galaxy formation.

Of course it should again be noted, that all of the interesting cosmological consequences of the quark-hadron phase transition require the transition to be first order, in agreement with recent lattice QCD calculations.

5   SQM in Neutron Stars; Strange Stars

It has been known for many years, that neutron stars may in fact be "hybrid stars" consisting of "ordinary" nuclear matter in the outer parts and quark matter in the central regions. This will be the case if SQM is metastable at zero pressure, being stabilized relative to hadronic matter by the high pressure within a neutron star [64–66].

If SQM is absolutely stable at zero pressure, an even more intriguing possibility opens up, namely the existence of "strange stars" [2,67–69] consisting completely of SQM (perhaps apart from a minor crust to be discussed below). Such strange stars behave quite differently from neutron stars due to the unusual equation of state. For massless quarks the total energy density is given by $\rho = \rho_q + B$, and the total pressure by $P = P_q - B$, where ρ_q is the energy density of quarks, and the pressure of the quarks is $P_q = \rho_q/3$, since massless quarks are relativistic. The equation of state is thus given by

$$P = \frac{1}{3}\left(\rho - 4B\right). \tag{75}$$

The exact equation of state taking into account $m_s \neq 0$ is very similar [68] since s-quarks are relativistic for low m_s and not present for high m_s. (I here assume that $\alpha_s = 0$, but recall from Section 2.3, that a non-zero α_s effectively corresponds to a reduction of B).

The structure of a strange star is calculated from the Oppenheimer-Volkoff equation, describing the balance between gravity and pressure gradient, using the equation of state given above. The surface of the star corresponds to $P = 0$, a condition fulfilled for $\rho = 4B$, which for typical values of B is somewhat more than the density of ordinary nuclear matter! For stellar masses below $1M_\odot$ ($M_\odot$ is the solar mass) this density is almost constant throughout the star, so to a good approximation total mass and radius are related by $M \propto R^3$, a relation in striking contrast to ordinary neutron stars, where $M \propto R^{-3}$. This means that low-mass neutron stars and strange stars have widely different radii, possibly allowing observational distinction. Unfortunately Nature prefers to form these compact objects with masses near $1.4M_\odot$, according to stellar evolution models. For such a mass gravity rather than bag pressure plays the dominant stabilizing role, and there is no significant difference between neutron star and strange star radii. Also the maximum mass given by gravitational instability (the Chandrasekhar limit) is similar, of order $2M_\odot$. In contrast to ordinary neutron stars, which are unstable for masses below $0.1\,M_\odot$, strange stars have no minimum mass; the sequence continues smoothly to the domain of strangelets.

For the simple equation of state discussed above, the only natural energy scale in the problem is $B^{1/4}$. Thus there exists a homology transformation between strange star models for different values of B. In particular, the maximum mass of a strange star is given by

$$M_{\max} = 2.006 B_{145}^{-1/2} M_\odot. \tag{76}$$

The corresponding minimal radius, maximal moment of inertia, maximal central density, surface density, and minimal rotation period (the so-called Kepler period corresponding to mass-shedding at the equator), are given by

$$R_{\min} = 10.94 B_{145}^{-1/2} \text{km}, \tag{77}$$

$$I_{\max} = 2.256 \times 10^{45} B_{145}^{-3/2} \text{g cm}^2, \tag{78}$$

$$\rho_{\max} = 1.97 \times 10^{15} B_{145} \text{g cm}^{-3}, \tag{79}$$

$$\rho_{\text{surf}} = 4.102 \times 10^{14} B_{145} \text{g cm}^{-3}, \tag{80}$$

$$P_{\min} = 0.66 B_{145}^{-1/2} \text{ms}. \tag{81}$$

Bare strange stars (strange stars with quark matter all the way to the surface) have quite unusual properties. The density abruptly jumps from 0 to ρ_{surf} (Eq. (80)), and the density is almost constant through the interior (except when the mass is close to $M_{\max}$). The plasma frequency of the star is huge, meaning that photons with energies below 20MeV are reflected from the surface, whereas the star itself can only emit photons with higher energies [68,70]. Even more important, because of the strong interaction binding of the surface material, the star is not subject to the "Eddington limit", which for ordinary neutron stars limits the luminosity to be below 10^{38} erg/s (for higher luminosities the radiation pressure would exceed the gravitational attraction and expel the surface layers). As discussed below, this could lead to important "applications" of strange stars.

This approach may however be oversimplified because real strange stars may have surfaces more like ordinary neutron stars. In particular, a solid crust of ordinary material may form from accretion by the strange star after formation, or from material that was not converted during neutron star burning (see Section 5.5). Such a crust may be held up by the extreme, outward directed electrostatic potential of 10^{17}–10^{18}V/cm, created by the electron atmosphere with a thickness of a few hundred Fermi. This atmosphere merely expresses that the electrostatic binding of electrons is weaker than the strong binding of quarks; therefore the electron distribution does not end abruptly like that of quarks (the detailed structure was found from a Thomas-Fermi calculation by Alcock, Farhi and Olinto [68]; see also [71]).

The electrostatic potential can sustain a significant crust of ordinary neutron star material. The limit is given by the neutron drip density (4×10^{11} gcm^{-3}), above which neutrons drip out of nuclei and would be swallowed by the quark phase. This crust may be decisive for interpretation of pulsar behavior (Section 5.2).

As emphasized by Glendenning, Kettner and Weber [72,73], the existence of crusts not only changes the mass-radius relation for strange stars, but also opens a rich plethora of new stellar configurations. In particular, one may have a sequence of "strange dwarfs", much like white dwarfs except for an SQM core. At present there is no well-studied model for formation of such strange dwarfs.

Another possibility for formation of a (solid?) crust has been suggested in [74]. This mechanism relies on the existence of stable, low-baryon number strangelets (in this context sometimes denoted "quark-alphas" for the $A = 6$ strangelet analog of a helium nucleus [75]) which could act as "nuclei" in the surface region. Whereas this possibility may seem less likely from the discussion in Section 3, it can not be entirely ruled out.

Finally, it is worth noticing that Glendenning [76] has argued that neutron stars may contain regions with mixed quark and hadron phases. (This possibility was missed in earlier studies due to an erroneous assumption of *local* rather than *global* charge neutrality). Depending on parameters the mixed phase region can occupy a significant fraction of the star, and may show unusual topologies (plate-like or cylinder-like structures, rather than just spherical quark bubbles embedded in hadrons or vice versa [77–79]).

Studies of strange stars have not been pursued to the degree of detail known for ordinary neutron stars, and it is premature to draw any detailed conclusions. However, in the following, I shall look at some of the properties expected and emphasize the possible observable differences between neutron stars and strange stars.

5.1   Neutrino Cooling

A distinction between strange stars and neutron stars was for a long time believed to be a much more rapid cooling of SQM due to neutrino emitting weak interactions involving the quarks [68]. Thus a strange star was presumed to be much colder than a neutron star of similar age, a signature potentially observable from x-ray satellites. Only a few speculative mechanisms, such as the existence of kaon condensates might mimic the speed of quark matter neutrino cooling. Recently the story has been complicated considerably by the finding that ordinary neutron β-decay may be energetically allowed in nuclear matter [80], so that the cooling rate can be comparable to that of SQM. For this reason I shall not discuss the issue here, but refer the reader to an excellent review of neutron star cooling by Pethick [81], and a recent reinvestigation of strange star cooling by Schaab et al. [82].

5.2   Pulsar Glitches

One important feature seems to distinguish strange stars from neutron stars in a manner with observable consequences, and that is the distribution of the moment of inertia inside the star. Ordinary neutron stars older than a few months have a crust made of a crystal lattice or an ordered inhomogeneous medium reaching from the surface down to regions with density $2 \times 10^{14}\,\mathrm{g\,cm^{-3}}$. This crust contains about 1% of the total moment of inertia. Strange stars in contrast can only support a crust with density below the neutron drip density ($4.3 \times 10^{11}\,\mathrm{g\,cm^{-3}}$). This is because free neutrons would be absorbed and converted by the strange matter. Such a strange star crust contains at most a few times 10^{-5} of the total moment of inertia. This is an upper bound, since the strange star may have no crust at all, depending on its prior evolution. And recent studies of the mechanical balance between electric and gravitational forces on the crust indicate, that only densities up to perhaps $10^{11}\mathrm{g\,cm^{-3}}$ may be achieved [83,84].

As stressed by Alpar [85], and also pointed out by Haensel, Zdunik, and Schaeffer [67], and by Alcock, Farhi, and Olinto [68], this difference in the moment of inertia stored in the crust of neutron stars and strange stars seems to pose significant difficulties for explaining the glitch-phenomenon observed in radio pulsars with models based on strange stars. Glitches are observed as a sudden speed-up in the rotation rate of pulsars. The fractional change in rotation rate Ω is $\Delta\Omega/\Omega \approx 10^{-6}$—$10^{-9}$, and the corresponding fractional change in the spin-down rate $\dot{\Omega}$ is of order $\Delta\dot{\Omega}/\dot{\Omega} \approx 10^{-2}$—$10^{-3}$. Regardless of the detailed model for the glitch phenomenon these jumps must involve the decoupling and recoupling of a component in the star containing a significant fraction, I_i/I, of the total moment of inertia; $fI_i/I = \Delta I/I \approx \Delta\Omega/\Omega \approx 10^{-6}$—$10^{-9}$ (Alpar actually argued that $fI_i/I \approx \Delta\dot{\Omega}/\dot{\Omega} \approx 10^{-2}$—$10^{-3}$, where f is the fractional change in I_i, but this is not necessary [86]). This role is played by the inner crust of an ordinary neutron star, but the crust around a strange star is smaller; less than a few times $10^{-5}M_\odot$ with I_{crust}/I around a few times 10^{-5} for ordinary neutron star masses of $1.4M_\odot$ (higher for less massive stars). These numbers are based on models by Glendenning and Weber [86] assuming a maximum mass crust, i.e. a crust reaching neutron drip density at the base, so it seems fair to conclude, that strange stars in fact may have sufficiently massive crusts to account for glitches, but that parameters in that case are fairly tightly constrained.

Other possibilities for glitches in strange stars could involve a crust composed of strangelets (cf. the "quark-alpha" scenario in [74]), not to mention the possibility of a quark-hadron mixed phase [76–79]. There is still a lack of any detailed model for how the magnetic field structure and other crucial aspects of a pulsar can be modeled for strange stars. Presumably a strange star cannot do the job without significant structure, such as a crust and/or superfluidity/superconductivity in certain regions. These issues have only been very superficially studied and need further consideration. The present lack of such models should not be used to dismiss the possibility of strange stars.

5.3   Strange Star Oscillation and Maximum Rotation Rate

One of the most interesting differences between neutron stars and strange stars is related to the damping of instabilities.

First it should be noticed that a strange star is a very stable system. Strange stars may have radial oscillations with a fundamental period of 0.06–0.3 ms [87], but these are characterized by rapid damping in a matter of seconds [88–91]. This is due to the extremely high viscosity of SQM.

The large viscosity also plays a role in setting the maximum rotation limit for strange pulsars (or hybrid stars with SQM cores). The ultimate rotation limit corresponds to mass-shedding from the equator of the star (this is called the Kepler limit and is of order 0.6 msec for a strange star, Eq. (81); see Zdunik [92] for a review). But before reaching such rotation rates, the pulsars become unstable to non-radial deformations and are slowed down by emission of gravitational radiation. Shear and bulk viscosities tend to stabilize the star against these instabilities [89,93], and the high value for the bulk viscosity may mean that strange pulsars in contrast to ordinary pulsars can reach submillisecond periods [91]. Thus the discovery of very fast pulsars may be an indication favoring the existence of strange stars.

And even more exciting, it has been shown over the last few months, that ordinary neutron stars when they are young and hot are subject to a new class of instabilities, called r-mode instabilities [94–96], which during their first year of existence slows the rotation rate to only a few per cent of the Kepler limit. Rotation periods faster than 10 msec are unlikely after that, until some pulsars at a much later age may be spun-up by angular momentum transfer in binary systems, and thereby explain the rapid old pulsars with periods down to 1.56 msec. In contrast, strange stars are not subject to these instabilities until they are thousands of years old, and even then only for periods faster than 2-3 msec [97]. This seems to imply, that the most robust signature for the existence of strange stars (or neutron stars with a substantial fraction of high viscosity quark matter in the interior) is to search for young pulsars with rotation periods below, say, 5 msec (even stars with longer periods may candidate). These can not be ordinary neutron stars, whereas quark matter is the only substance known to have a bulk viscosity high enough to offer an explanation.

The bulk viscosity of strange quark matter depends on the rate of the non-leptonic interaction

$$u + d \leftrightarrow s + u. \tag{82}$$

(The rate for this reaction has recently been calculated by Madsen [35], and Heiselberg [36]; earlier studies, including that of Ref. [33] are incorrect). This reaction changes the concentrations of down and strange quarks in response to the density changes involved in vibration or rotational instabilities, thereby causing dissipation. This dissipation is most efficient if the rate of reaction (82) is comparable to the frequency of the density change. If the weak rate is very small, the quark concentrations keep their original values in spite of a periodic density fluctuation, whereas a very high weak rate means that the matter immediately adjusts to follow the true equilibrium values reversibly. But in the intermediate range dissipation due to PdV-work is important.

The importance of dissipation due to Eq. (82) was first stressed by Wang and Lu [88] in the case of neutron stars with quark cores. These authors made a numerical study of the evolution of the vibrational energy of a neutron star with an $0.2M_\odot$ quark core, governed by the energy dissipation due to Eq. (82). Sawyer [89] expressed the damping in terms of the bulk viscosity, a function of temperature and oscillation frequency, which he tabulated for a range of densities and strange quark masses. Sawyer's tabulation has later been used in studies of quark star vibration [90], and of the gravitational radiation reaction instability determining the maximum rotation rate of pulsars [93]. The latter study concluded, that the bulk viscosity is large enough to be important for temperatures exceeding 0.01 MeV, but that it should be a few orders of magnitude larger to generally dominate the stability properties.

However, as has been pointed out in [91], the bulk viscosities calculated in [89] depend on the assumption, that the rate of Eq. (82) can be expanded to first order in $\delta\mu = \mu_s - \mu_d$, where $\mu_i \approx 300\text{MeV}$ are the quark chemical potentials. This assumption is not correct at low temperatures ($2\pi T \ll \delta\mu$), where the dominating term in the rate is proportional to $\delta\mu^3$. Furthermore, the rate in [89] is too small by an overall factor of 3, and a discrepancy of 2-3 orders of magnitude, perhaps due to unit conversions, appears as well. Taken together, these effects lead to an upward correction of the bulk viscosity by several orders of magnitude, and thereby increases the importance for the astrophysical applications. The non-linearity of the rate also means, that the bulk viscosity is no longer independent of the amplitude of the density variations. The resulting bulk viscosity is (in cgs-units, with m_s, T, and $\mu_d \approx 235 \text{MeV}(\rho/\rho_{\text{nuc}})^{1/3}$ in MeV, and the oscillation frequency ω in s^{-1})

$$\zeta \approx 3.09 \times 10^{28} m_s^4 \omega^{-2} \left(\frac{\rho}{\rho_{\text{nuc}}}\right) \left[\frac{3}{4}\left(\frac{m_s^2}{3\mu_d}\frac{\Delta v}{v_0}\right)^2 + 4\pi^2 T^2\right] \text{g cm}^{-1}\text{s}^{-1}. \quad (83)$$

For typical values ($m_s = 100$ MeV, $\mu_d = 300$MeV, $\omega = 2\times10^4$ s^{-1}) this is $\zeta \approx 1.6 \times 10^{28} \left[93(\Delta v/v_0)^2 + 39T^2\right] \text{g cm}^{-1}\text{s}^{-1}$, where $\Delta v/v_0$ is the perturbation amplitude.

For a star of constant density (an excellent approximation for a strange star, except very close to the gravitational instability limit) Sawyer [89] estimated the damping time as

$$\tau_D \approx 1.5 \times 10^{25} \zeta^{-1} \text{s}. \quad (84)$$

Thus, even at very low temperatures, high amplitude oscillations are damped in fractions of a second, and those of low amplitude in a matter of minutes, if one takes into account, that the temperature of the star increases due to the heat released by viscous dissipation, which can speed up the damping of vibrations.

The discussion above was based on rather crude estimates [91]. A detailed, general relativistic, numerical treatment along the lines of Cutler *et al.* [90] is clearly needed.

As mentioned previously, viscosity also plays an important role in setting the maximum rotation rate of pulsars. Gravitational radiation reaction instabilities (as opposed to "Keplerian mass-shedding") is supposed to set the ultimate rotation rate limit, but the larger the damping by shear and bulk viscosity is, the closer the rate can get to the Keplerian limit given in Eq. (81).

The shear viscosity of SQM due to quark scattering has recently been recalculated by Heiselberg and Pethick [98]. Their results for $T \ll \mu$ can be written as

$$\eta \approx 4.0 \times 10^{15} \left(\frac{0.1}{\alpha_S}\right)^{5/3} \left(\frac{\rho}{\rho_{nuc}}\right)^{14/9} T^{-5/3} \text{gcm}^{-1}\text{s}^{-1}. \quad (85)$$

Investigations by Colpi and Miller [93] based on the older viscosities in [89,99] indicated, that the minimal rotation period of strange stars might be set by the gravitational radiation reaction instability of $m = 2$ or $m = 3$ modes at or just below 1 millisecond. With the new, much larger, viscosities, the non-axisymmetric instabilities will be suppressed, and it is not unreasonable to expect, that the maximum rotation frequency of strange stars will be close to the Keplerian limit. Detailed numerical calculations like those in Colpi and Miller [93], including the new viscosities and effects of dissipative heating, are required to settle the issue, but they are complicated by the non-linear behavior of the new bulk viscosity.

Whether or not the ultimate rotation period of strange stars can be significantly smaller than for neutron stars is of importance for old pulsars spun-up by accretion. But perhaps the most clear-cut signature for the existence of strange stars would be the (almost) lack of sensitivity to r-mode instabilities, which as mentioned earlier allows young strange stars to rotate much faster than young neutron stars [97].

5.4   Gamma-Ray Bursters

Strange stars because of their high surface density, strong binding (making it possible to circumvent the Eddington limit), and special emission properties have been suggested as explanations for some of the more mysterious cosmic events, namely γ-ray bursters. These are bursts of γ-rays of a few seconds duration, coming from unidentified sources which are presumably at extragalactic distances.

No consensus exists concerning the nature of these bursts, but Alcock, Farhi and Olinto [100] suggested a detailed model for the most prominent of the bursters, the one on 5 March 1979. Their model is based on an impact of a $10^{-8} M_\odot$ lump of SQM on a rotating strange star, and the authors are able to explain most of the observations concerning energetics and time-scales under the assumption that the burster is located in a supernova remnant in the Large Magellanic Cloud, as position measurements seem to indicate. An alternative model for this source and for soft γ-repeaters in the framework of strange stars with "quark-alpha" surface properties was suggested in [101]. Other strange star models for soft γ-repeaters and x-ray bursters include [102,103].

γ-ray bursters at truly cosmological distances could be due to collisions of two strange stars in binary systems [104], each collision releasing 10^{50} ergs in the form of gamma rays over a time-scale of 0.2 s.

There are, however, literally hundreds of different models for γ-bursts, and in spite of improved observational data the interpretation is at present unclear.

A recent identification of the x-ray source Her X-1 as a strange star [105] was unfortunately based on incorrect use of bag model parameters [106].

5.5   Formation of Strange Stars

If strange quark matter is stable, strange stars may be formed during supernova-explosions, and neutron stars can be converted to strange stars by a number of different mechanisms, such as pressure-induced transformation to uds-quark matter via ud-quark matter, sparking by high-energy neutrinos, or triggering due to the intrusion of a quark nugget. These and other possibilities were described by Alcock, Farhi, and Olinto [68].

As soon as a lump of strange matter comes in contact with free neutrons it starts converting them into strange matter. The burning of a neutron star into a strange star was discussed by Baym et al. [107] and Olinto [108], and it was shown that the star would be converted on a rather small time-scale set by quark diffusion and flavor-changing weak interactions. (The huge difference in the speed of the conversion front found in these papers is partly due to the omission of a factor $c^{1/2}$, where c is the speed of light, in equation (6) of Baym et al.) Later studies [109,110] found burning times in the range of 1–10^3 seconds under various parameter assumptions (see also Olinto [111] for a review). For the fastest burning times, the energy liberated may be important for the supernova mechanism and supernova neutrino bursts. Horvath and Benvenuto [112] have questioned the stability of "slow" neutron combustion and suggested that the conversion takes place much faster as a detonation. So far, the investigations of neutron star burning have been rather crude, neglecting many aspects of transport theory, heat conduction etc. A detailed study of this phenomenon would be interesting.

Perhaps the most likely mechanisms for initiating the formation of a strange star involves either a seed of SQM in the star (see Section 6), or thermal formation of quark matter bubbles. Thermal triggering of neutron star transformation may be understood qualitatively in terms of simple boiling theory. Before considering a more realistic equation of state it is instructive to study the boiling of a pure neutron gas into quarks. The quark bubbles formed consist of u and d quarks in the ratio 1:2; only later weak interactions may change the composition to an energetically more favorable state. Thus quark chemical potentials are related by $\mu_d = 2^{1/3}\mu_u$, and $\mu_n = \mu_u + 2\mu_d = (1 + 2^{4/3})\mu_u$, assuming chemical equilibrium across the phase boundary.

The free energy involved in formation of a spherical quark bubble of radius R and volume V is given by

$$F = -\Delta PV + 8\pi\gamma R \tag{86}$$

where

$$\Delta P = P_{ud} - P_n = \frac{\mu_u^4 + \mu_d^4}{4\pi^2} - B - P_n \tag{87}$$

and the curvature energy coefficient

$$\gamma = \frac{\mu_u^2 + \mu_d^2}{8\pi^2}. \tag{88}$$

The free energy has a maximum at the critical radius

$$r_c = (2\gamma/\Delta P)^{1/2} \tag{89}$$

and the corresponding free energy

$$W_c \equiv F(r_c) = 16\pi\gamma r_c/3 \tag{90}$$

is the work required to form a bubble of this radius which is the smallest bubble capable of growing. It is a standard assumption in the theory of bubble nucleation in first order phase transitions that bubbles form at this particular radius at a rate given by[2]

$$\mathcal{R} \approx T^4 \exp(-W_c/T). \tag{91}$$

The simplest possible equation of state for the neutron gas is that of a zero temperature, nonrelativistic degenerate Fermi-gas, where

$$P_n = \frac{(\mu_n^2 - m_n^2)^{5/2}}{15\pi^2 m_n} \tag{92}$$

and the baryon density

$$n_B = \frac{(\mu_n^2 - m_n^2)^{3/2}}{3\pi^2} \tag{93}$$

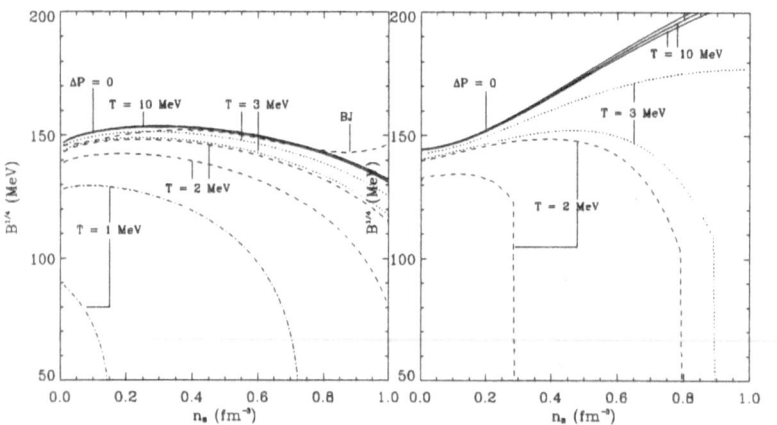

Fig. 6: Upper limits on the bag constant allowing thermal nucleation of quark matter bubbles in neutron stars as a function of baryon number density in the hadron phase. To the left are shown results for the simple neutron gas model discussed in the text. To the right for a more realistic mean field approximation. See text for further explanations.

A necessary condition for boiling is that $\Delta P > 0$. This leads to an upper limit on the bag constant, B_{max}, from Eq. (87) as illustrated in Fig. 6 (from

---

[2] The prefactor may differ from T^4, but this is of minor practical importance due to the dominant exponential.

[113]—the corresponding limit for the Bethe-Johnson equation of state is shown for comparison; it is seen to be very similar). This was also used as a criterion for neutron star stability by Krivoruchenko and Martemyanov [114].

Also shown in Fig. 6 is the limit on the bag constant below which bubble nucleation takes place at rates exceeding $1\,\mathrm{km}^{-3}\mathrm{Gyr}^{-1}$ and $1\,\mathrm{cm}^{-3}\mathrm{s}^{-1}$, respectively, for temperatures of 1, 2, 3 and 10 MeV (B_{max} can be considered as the limit for infinite temperature). One notes that the possibility of bubble nucleation is fairly insensitive to the temperature as soon as T exceeds a few MeV, whereas thermally induced bubble nucleation is impossible for $T < 2\mathrm{MeV}$ (recall from Section 2.2 that the stability of ordinary nuclei against decay into quark matter requires that $B > (145\mathrm{MeV})^4$). This confirms an estimate in [115] (see also [116,117]). The range of bag constants for which a hot neutron star may transform into quark matter is thus roughly $145\mathrm{MeV} < B^{1/4} < 155\mathrm{MeV}$.

Results for a more realistic mean field equation of state are also shown in the Figure (see [113] for further details). While the detailed numbers change, the overall conclusion does not. Quark matter bubbles may nucleate (possibly followed by burning of the star into SQM) in neutron stars/supernovae if the bag constant is low, and if the temperature exceeds a few MeV (thus the process is most likely during the supernova explosion itself). Should thermal nucleation not take place, one of the other mechanisms mentioned above must be relied on. Apart from seed-induced burning, all of these are likely to be much less efficient than thermal nucleation.

6   SQM in Cosmic Rays

De Rújula and Glashow [118] argued that unusual meteor-events, earthquakes, etched tracks in old mica, in meteorites and in cosmic-ray detectors might be used for observation of quark nuggets hitting the Earth or its atmosphere. In particular they were interested in the possibility of detecting a galactic dark matter halo of nuggets, where typical velocities would be a few hundred kilometers per second, given by the depth of the gravitational potential. Even if nuggets only survived from the Big Bang in small numbers, or were spread in our galaxy by secondary processes such as strange star collisions, there could be a potentially observable flux of nuggets hitting the Earth. The only data actually investigated in their paper came from a negative search for tracks in ancient mica, and corresponded to a lower nugget flux limit of $8 \times 10^{-19}\,\mathrm{cm}^{-2}\,\mathrm{s}^{-1}\,\mathrm{sr}^{-1}$, for nuggets with $A > 1.4 \times 10^{14}$ (smaller nuggets would be trapped in layers above the mica samples studied). This can also be expressed as an excluded range of $1.4 \times 10^{14} < A < 8 \times 10^{23}\rho_{24}v_{250}$, where $v \equiv 250\mathrm{km\,s}^{-1}v_{250}$ and $\rho \equiv 10^{-24}\mathrm{g\,cm}^{-3}\rho_{24}$ are the typical speeds and mass density of nuggets in the galactic halo. The speed is given by the depth of the gravitational potential of our galaxy, whereas $\rho_{24} \approx 1$ corresponds to the density of dark matter. In these units the number of nuggets hitting the Earth per cm^2 per second per steradian is $6.0 \times 10^5 A^{-1} \rho_{24} v_{250}$.

Later investigations have improved these flux limits somewhat. These Earth-based flux-limits [119,120] are shown in Figure 7. It is seen that quark nuggets with $3 \times 10^7 < A < 5 \times 10^{25}$ seem incapable of explaining the dark halo around our galaxy, but a low flux either left over from the Big Bang or arising from collision of strange stars cannot be ruled out. If the strange matter hypothesis is valid, one should indeed expect a significant background flux from stellar collisions, since several pulsars are members of binary systems, where the two components are ultimately going to collide. If such collisions spread as little as $0.1 M_\odot$ of non-relativistic strangelets with baryon number A, a single collision will lead to a flux of $10^{-6} A^{-1} v_{250}$ cm^{-2}s^{-1}sterad^{-1}, assuming strangelets to be spread homogeneously in a halo of radius 10 kpc.

Such a flux-level is below the sensitivity of present experiments, but Madsen [121] suggested that neutron stars and their stellar "parents" may be used as alternative large surface area, long integration time detectors. The reason is simple. The presence of a single quark nugget in the interior of a neutron star is sufficient to initiate a transformation of the star into a strange star [2,68,107]. The time-scale for the transformation is short, between seconds and minutes [107,108,111,109,110], so observed pulsars would have been converted long ago, if their stellar progenitors ever captured a quark nugget, or if the neutron stars themselves absorbed one after formation.

The rate at which quark nuggets hit the surface of a star depends on the phase space distribution of nuggets relative to the star. For an infinite bath of positive energy nuggets with an isotropic, monoenergetic distribution function, the number accretion rate is given by

$$F = 1.39 \times 10^{30}\,\text{s}^{-1}\,A^{-1} M R \rho_{24} v_{250}^{-1} \left[1 + 0.164 v_{250}^2 R M^{-1}\right], \qquad (94)$$

where M and R denote the stellar mass and radius in solar units. For the Sun the second term in parenthesis (the geometrical term) contributes only slightly to the accretion rate, and the contribution is even less important for more massive stars and for compact objects like white dwarfs and neutron stars (in contrast, the geometrical term dominates for accretion onto the Earth). In the following I therefore only take the first term (gravitational) into account.

To convert a neutron star into strange matter a quark nugget should not only hit a supernova progenitor but also be caught in the core. Similarly, nuggets hitting a neutron star after its creation have to penetrate the outer layers and reach the neutron drip region. These issues were discussed in [121].

A main sequence star is capable of capturing quark nuggets with baryon numbers below A_{STOP}, where

$$A_{\text{STOP}} = 5.0 \times 10^{31} M^{-1.8}. \qquad (95)$$

This works for non-relativistic nuggets, which are basically braked by inertia, i. e. they are slowed down by electrostatic scatterings after plowing through a column of mass similar to their own, and afterwards settle in the stellar core. In particular it is valid for nuggets moving with virial speed in our galactic halo. Relativistic nuggets, like those reported in some cosmic ray observations, may be destroyed after collisions with nuclei in the stellar atmospheres, and so the limit can not be used immediately, but it is worth noticing, that even a tiny fraction of a nugget surviving such an event and settling in the star is sufficient to convert the neutron star to a strange star.

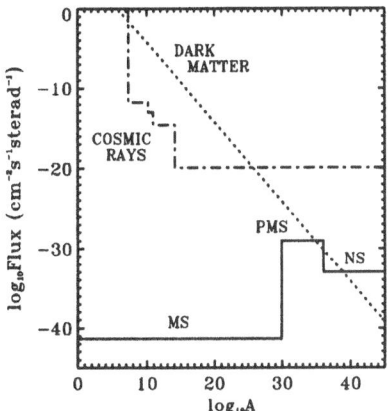

Fig. 7: Astrophysical flux-limits [121] compared to the flux expected for a galactic halo of nuggets being the dark matter [118], and to the experimental results for cosmic rays [119,120]. The three horizontal parts of the solid curve correspond to capture in main sequence supernova progenitors, post main sequence stars, and neutron stars younger than the Vela pulsar (10^4 years).

For nuggets with $A < A_{\text{STOP}}$ the sensitivity of main sequence stars as detectors is remarkable, as it is given by the limit of one nugget hitting the surface of the supernova progenitor in its main sequence lifetime! Converted into a flux, $\mathcal{F}$, of nuggets hitting the Earth per cm^2 per sec per steradian, it corresponds to

$$\mathcal{F} = 4 \times 10^{-42} M^{0.1} v_{250}^2. \qquad (96)$$

As can be seen from Figure 7, this is a factor of 10^{20}–10^{40} more sensitive than ordinary experiments!

If it is possible to prove that some neutron stars are indeed neutron stars rather than strange stars, the sensitivity of the astrophysical detectors rules out quark nuggets as being the dark matter for baryon numbers in the range $A < 10^{34-38}$. And it questions the whole idea of stable strange quark matter, since it seems impossible to avoid polluting the interstellar medium with nuggets from strange star collisions or supernova explosions at fluxes many orders of magnitude above the limit measurable in this way.

If on the other hand SQM is stable, then all neutron stars are likely to be strange stars, again because some pollution can not be avoided.

The Sun would in this way accrete $3.7 \times 10^{-20} \rho_{24} v_{250}^{-1} M_{\odot}$/year, or a total of $10^{-10} \rho_{24} v_{250}^{-1} M_{\odot}$ in its total lifetime on the main sequence. Very low-mass nuggets collected near the solar center in this manner might have an impact on the energy production [122], but the effect is negligible unless the electrostatic barrier at the nugget surface is much smaller than expected, or unless very special circumstances allow nuggets to catalyze nuclear reactions [21].

The Sun will develop into a white dwarf in about 6×10^{9} years. As just mentioned, the Sun would accrete a core of $10^{-10} \rho_{24} v_{250}^{-1} M_{\odot}$ in its total lifetime on the main sequence. Such accretion is too small to lead to a strange dwarf distinguishable from an ordinary white dwarf as suggested in [72,73]. However, higher concentrations could occur if quark nuggets were somehow mixed into the gas cloud from which the star originally formed. Whether this is likely to happen depends strongly on assumptions regarding the velocity distribution of the nuggets formed, and the possibility of interactions with the gas [123].

Most of the discussion above dealt with halo nuggets moving at non-relativistic velocities. Relativistic nuggets are not as easily detected using neutron stars, since they may be destroyed in collisions with nuclei in the star. On the other hand two relativistic candidate events with charge $Z = 14$ and mass $A \approx 370$ were found in a balloon experiment by Saito *et al.* [32]. This corresponds to a rather high flux, and it is not quite clear how to produce such nuggets, though spallation of larger nuggets originating from strange star collisions may be involved [124].

Quark nuggets have also been suggested as candidates for the Centauro cosmic-ray events [2,125,126]. Centauro primaries may have a flux as high as $10^{-14} \mathrm{cm}^{-2} \mathrm{s}^{-1}$ and $A \approx 10^{3}$. Since Centauro primaries move at relativistic speeds they are destroyed by inelastic collisions when hitting a star, so the flux-limits given above cannot directly be used to rule out quark nuggets as Centauro primaries. However the mechanism producing the primaries must be tuned so that it only produces relativistic quark nuggets in order not to conflict with the flux-limits in Figure 7 for non-relativistic nuggets.

7   Conclusion and Outlook

The possible stability of strange quark matter is a fundamentally exciting idea. Should it turn out to be true, many textbooks in nuclear, particle and astrophysics will need revisions, but our daily lives will not be changed dramatically, apart from possible technological applications such as energy production and disposal of radioactive waste [14,127]. There are two main reasons why stability of SQM is possible without drastic consequences. The first reason is that stability requires a certain minimum strangeness content, so ordinary nuclei do not decay into strangelets. The second reason is the positive electrostatic potential of the quark phase in a strangelet, which means that you could walk around with a lump of SQM in your pocket without being swallowed.

While heavy-ion collisions is the way to look for small (meta)stable strangelets, astrophysics gives a possibility for testing larger (and therefore more stable) SQM-systems. Direct cosmic ray searches is an obvious way to look for intermediate baryon numbers in the form of relativistic or non-relativistic lumps produced in strange star collisions, and for leftovers from the Big Bang. The latter can only exist for very high baryon numbers (cf. Section 4), whereas a galactic background of the former seems unavoidable if the strange matter hypothesis is correct.

Strange stars may be the most promising place to look for SQM, but as explained in Section 5 it is actually hard to find clear-cut ways of distinguishing strange stars from neutron stars, unless one finds an object of very low mass. Pulsar rotation properties at present seem to provide the best clue, in particular after the finding that young strange stars in contrast to neutron stars are not braked by gravitational wave emission due to r-mode instabilities.

If SQM is only metastable, heavy-ion physicists may still have a chance of finding it; the cosmological quark-hadron phase transition may still lead to inhomogeneities of importance for Big Bang nucleosynthesis (without quark nuggets left over); and neutron stars may still have strange matter cores.

In any case the confirmation or disproof of the existence of (meta)stable strange quark matter via experiments and astrophysical tests makes it possible to limit strong interaction parameters that are otherwise difficult to probe. This in itself is a good reason for continued studies of the physics and astrophysics of strange quark matter.

Acknowledgments

This work was supported in part by the Theoretical Astrophysics Center under the Danish National Research Foundation. I take this opportunity to thank the "strangers" among my present and former PhD-students, Michael Olesen, Dan Jensen, Michael Christiansen, and Gregers Neergaard for collaboration and many enlightening discussions.

References

1. A. R. Bodmer, Phys. Rev. D 4, 1601 (1971).
2. E. Witten, Phys. Rev. D 30, 272 (1984).
3. *Strange Quark Matter in Physics and Astrophysics*, edited by J. Madsen and P. Haensel (Nucl. Phys. B (Proc. Suppl.), 24B, 1991).

4. C. Alcock and A. Olinto, Annu. Rev. Nucl. Part. Sci. **38**, 161 (1988).

5. B. S. Kumar, in *Physics and Astrophysics of Quark-Gluon Plasma*, edited by B. Sinha, Y. P. Viyogi, and S. Raha (World Scientific, Singapore, 1994), pp. 63–74.

6. B. S. Kumar, in *Strangeness and Quark Matter*, edited by G. Vassiliadis, A. D. Panagiotou, S. Kumar, and J. Madsen (World Scientific, Singapore, 1995), pp. 318–332.

7. B. S. Kumar, Nucl. Phys. A **590**, 29c (1995).

8. J. Madsen, in *Physics and Astrophysics of Quark-Gluon Plasma*, edited by B. Sinha, Y. P. Viyogi, and S. Raha (World Scientific, Singapore, 1994), pp. 186–205.

9. J. Madsen, in *Strangeness and Quark Matter*, edited by G. Vassiliadis, A. D. Panagiotou, S. Kumar, and J. Madsen (World Scientific, Singapore, 1995), pp. 191–205.

10. J. Madsen, in *Strangeness in Hadronic Matter*, Vol. 340 of *AIP Conf. Proc.*, edited by J. Rafelski (AIP, New York, 1995), pp. 32–45.

11. P. Papazoglou *et al.*, in *Strangeness and Quark Matter*, edited by G. Vassiliadis, A. D. Panagiotou, S. Kumar, and J. Madsen (World Scientific, Singapore, 1995), pp. 206–219.

12. C. Greiner and J. Schaffner-Bielich, preprint nucl-th/9801062.

13. S. Kabana for the NA52 collaboration, J. Phys. G **23**, 2135 (1997).

14. G. L. Shaw, M. Shin, R. H. Dalitz, and M. Desai, Nature **337**, 436 (1989).

15. S. L. Shapiro and S. A. Teukolsky, *Black Holes, White Dwarfs, and Neutron Stars* (John Wiley & Sons, New York, 1983).

16. E. Farhi and R. L. Jaffe, Phys. Rev. D **30**, 2379 (1984).

17. A. Chodos *et al.*, Phys. Rev. D **9**, 3471 (1974).

18. T. A. DeGrand, R. L. Jaffe, K. Johnson, and J. Kiskis, Phys. Rev. D **12**, 2060 (1975).

19. D. Vasak, W. Greiner, and L. Neise, Phys. Rev. C **34**, 1307 (1986).

20. C. Greiner, D.-H. Rischke, H. Stöcker, and P. Koch, Phys. Rev. D **38**, 2797 (1988).

21. K. Takahashi and R. N. Boyd, Astrophys. J. **327**, 1009 (1988).

22. E. P. Gilson and R. L. Jaffe, Phys. Rev. Lett. **71**, 332 (1993).

23. J. Madsen, Phys. Rev. D **50**, 3328 (1994).

24. J. Schaffner-Bielich, C. Greiner, A. Diener, and H. Stöcker, Phys. Rev. C **55**, 3038 (1997).

25. M. S. Desai, H. J. Crawford, and G. L. Shaw, Phys. Rev. D **47**, 2063 (1993).

26. H. J. Crawford, M. S. Desai, and G. L. Shaw, Phys. Rev. D **48**, 4474 (1993).

27. M. S. Berger and R. L. Jaffe, Phys. Rev. C **35**, 213 (1987).

28. J. Madsen, Phys. Rev. Lett. **70**, 391 (1993).

29. J. Madsen, Phys. Rev. D **47**, 5156 (1993).

30. H. Heiselberg, Phys. Rev. D **48**, 1418 (1993).

31. R. Balian and C. Bloch, Ann. Phys. **60**, 401 (1970).

32. T. Saito, Y. Hatano, Y. Fukada, and H. Oda, Phys. Rev. Lett. **65**, 2094 (1990).

33. H. Heiselberg, J. Madsen, and K. Riisager, Phys. Scr. **34**, 556 (1986).

34. P. Koch, in *Strange Quark Matter in Physics and Astrophysics*, edited by J. Madsen and P. Haensel (Nucl. Phys. B (Proc. Suppl.), **24B**, 1991), pp. 255–259.

35. J. Madsen, Phys. Rev. D **47**, 325 (1993).

36. H. Heiselberg, Phys. Scr. **46**, 485 (1992).
37. D. M. Jensen and J. Madsen, Phys. Rev. D (Rapid Communication) **53**, R4719 (1996).
38. Y. B. He, C. S. Gao, X. Q. Li, and W. Q. Chao, Phys. Rev. C **53**, 1903 (1996).
39. M. G. Mustafa and A. Ansari, Phys. Rev. D **53**, 5136 (1996); Erratum *ibid* **54**, 4694 (1996).
40. M. G. Mustafa and A. Ansari, Phys. Rev. C **55**, 2005 (1997).
41. J. H. Applegate and C. J. Hogan, Phys. Rev. D **31**, 3037 (1985).
42. H. Kurki-Suonio, Phys. Rev. D **37**, 2104 (1988).
43. K. Sumiyoshi, T. Kajino, C. R. Alcock, and G. J. Mathews, Phys. Rev. D **42**, 3963 (1990).
44. K. Jedamzik and G. M. Fuller, Astrophys. J. **423**, 33 (1994).
45. C. Alcock and E. Farhi, Phys. Rev. D **32**, 1273 (1985).
46. J. Madsen, H. Heiselberg, and K. Riisager, Phys. Rev. D **34**, 2947 (1986).
47. K. Sumiyoshi and T. Kajino, in *Strange Quark Matter in Physics and Astrophysics*, edited by J. Madsen and P. Haensel (Nucl. Phys. B (Proc. Suppl.), **24B**, 1991), pp. 80–83.
48. P. Bhattacharjee, J. Alam, B. Sinha, and S. Raha, Phys. Rev. D **48**, 4630 (1993).
49. C. Alcock and A. Olinto, Phys. Rev. D **39**, 1233 (1989).
50. J. Madsen and M. L. Olesen, Phys. Rev. D **43**, 1069 (1991); Erratum *ibid* **44**, 566 (1991).
51. M. L. Olesen and J. Madsen, Phys. Rev. D **47**, 2313 (1993).
52. H.-C. Liu and G. L. Shaw, Phys. Rev. D **30**, 1137 (1984).
53. C. Greiner, P. Koch, and H. Stöcker, Phys. Rev. Lett. **58**, 1825 (1987).
54. C. Greiner and H. Stöcker, Phys. Rev. D **44**, 3517 (1991).
55. H. W. Barz, B. L. Friman, J. Knoll, and H. Schulz, in *Strange Quark Matter in Physics and Astrophysics*, edited by J. Madsen and P. Haensel (Nucl. Phys. B (Proc. Suppl.), **24B**, 1991), pp. 211–220.
56. H. J. Crawford, M. S. Desai, and G. L. Shaw, Phys. Rev. D **45**, 857 (1992).
57. C. Spieles *et al.*, Phys. Rev. Lett. **76**, 1776 (1996).
58. C. H. Lee and H. K. Lee, Phys. Rev. D **44**, 398 (1991).
59. S. J. Cho, K. S. Lee, and U. Heinz, Phys. Rev. D **50**, 4771 (1994).
60. J. Madsen and K. Riisager, Phys. Lett. **158B**, 208 (1985).
61. M. L. Olesen, Master Thesis, University of Aarhus, 1990.
62. R. Schaeffer, P. Delbourgo-Salvador, and J. Audouze, Nature **317**, 407 (1985).
63. K. Jedamzik, G. M. Fuller, and G. J. Mathews, Astrophys. J. **422**, 423 (1994).
64. G. Baym and S. A. Chin, Nucl. Phys. A **262**, 527 (1976).
65. G. Chapline and M. Nauenberg, Nature **264**, 23 (1976).
66. B. Freedman and L. McLerran, Phys. Rev. D **17**, 1109 (1978).
67. P. Haensel, J. L. Zdunik, and R. Schaeffer, Astron. Astrophys. **160**, 121 (1986).
68. C. Alcock, E. Farhi, and A. Olinto, Astrophys. J. **310**, 261 (1986).
69. C. Alcock, in *Strange Quark Matter in Physics and Astrophysics*, edited by J. Madsen and P. Haensel (Nucl. Phys. B (Proc. Suppl.), **24B**, 1991), pp. 93–102.
70. T. Chmaj, P. Haensel, and W. Słomiński, in *Strange Quark Matter in Physics and Astrophysics*, edited by J. Madsen and P. Haensel (Nucl. Phys. B (Proc. Suppl.), **24B**, 1991), pp. 40–44.
71. C. Kettner, F. Weber, M. K. Weigel, and N. K. Glendenning, Phys. Rev. D **51**, 1440 (1995).

72. N. K. Glendenning, C. Kettner, and F. Weber, Phys. Rev. Lett. **74**, 3519 (1995).
73. N. K. Glendenning, C. Kettner, and F. Weber, Astrophys. J. **450**, 253 (1995).
74. O. G. Benvenuto and J. E. Horvath, Mon. Not. R. Ast. Soc. **247**, 584 (1990).
75. F. C. Michel, Phys. Rev. Lett. **60**, 677 (1988).
76. N. K. Glendenning, Phys. Rev. D **46**, 1274 (1992).
77. H. Heiselberg, C. J. Pethick, and E. F. Staubo, Phys. Rev. Lett. **70**, 1355 (1993).
78. N. K. Glendenning and S. Pei, Phys. Rev. C **52**, 2250 (1995).
79. M. B. Christiansen and N. K. Glendenning, Phys. Rev. C **56**, 2858 (1997).
80. J. M. Lattimer, C. J. Pethick, M. Prakash, and P. Haensel, Phys. Rev. Lett. **66**, 2701 (1991).
81. C. J. Pethick, Rev. Mod. Phys. **64**, 1133 (1992).
82. C. Schaab, B. Hermann, F. Weber, and M. K. Weigel, Astrophys. J. **480**, L111 (1997).
83. Y. F. Huang and T. Lu, Astron. Astrophys. **325**, 189 (1997).
84. T. Lu, preprint astro-ph/9807052.
85. M. A. Alpar, Phys. Rev. Lett. **58**, 2152 (1987).
86. N. K. Glendenning and F. Weber, Astrophys. J. **400**, 647 (1992).
87. B. Datta, P. K. Sahu, J. D. Anand, and A. Goyal, Phys. Lett. B **283**, 313 (1992).
88. Q. D. Wang and T. Lu, Phys. Lett. B **148**, 211 (1984).
89. R. F. Sawyer, Phys. Lett. B **233**, 412 (1989).
90. C. Cutler, L. Lindblom, and R. J. Splinter, Astrophys. J. **363**, 603 (1990).
91. J. Madsen, Phys. Rev. D **46**, 3290 (1992).
92. J. L. Zdunik, in *Strange Quark Matter in Physics and Astrophysics*, edited by J. Madsen and P. Haensel (Nucl. Phys. B (Proc. Suppl.), **24B**, 1991), pp. 119–124.
93. M. Colpi and J. C. Miller, Astrophys. J. **388**, 513 (1992).
94. N. Andersson, Astrophys. J. in press (1998). Preprint gr-qc/9706075.
95. J. L. Friedman and S. M. Morsink, Astrophys. J. in press (1998). Preprint gr-qc/9706073.
96. L. Lindblom, B. J. Owen, and S. M. Morsink, Phys. Rev. Lett. **80**, 4843 (1998).
97. J. Madsen, Phys. Rev. Lett. submitted (1998). Preprint astro-ph/9806032.
98. H. Heiselberg and C. J. Pethick, Phys. Rev. D **48**, 2916 (1993).
99. P. Haensel and A. J. Jerzak, Acta Phys. Pol. B **20**, 141 (1989).
100. C. Alcock, E. Farhi, and A. Olinto, Phys. Rev. Lett. **57**, 2088 (1986).
101. J. E. Horvath, H. Vucetich, and O. G. Benvenuto, Mon. Not. R. Ast. Soc. **262**, 506 (1993).
102. K. S. Cheng and Z. G. Dai, Phys. Rev. Lett. **80**, 18 (1998).
103. K. S. Cheng, Z. G. Dai, D. M. Wei, and T. Lu, Science **280**, 407 (1998).
104. P. Haensel, B. Paczyński, and P. Amsterdamski, Astrophys. J. **375**, 209 (1991).
105. X.-D. Li, Z.-G. Dai, and Z.-R. Wang, Astron. Astrophys. **303**, L1 (1995).
106. J. Madsen, Astron. Astrophys. **318**, 466 (1997).
107. G. Baym *et al.*, Phys. Lett. B **160**, 181 (1985).
108. A. V. Olinto, Phys. Lett. B **192**, 71 (1987).
109. H. Heiselberg, G. Baym, and C. J. Pethick, in *Strange Quark Matter in Physics and Astrophysics*, edited by J. Madsen and P. Haensel (Nucl. Phys. B (Proc. Suppl.), **24B**, 1991), pp. 144–147.

110. M. L. Olesen and J. Madsen, in *Strange Quark Matter in Physics and Astrophysics*, edited by J. Madsen and P. Haensel (Nucl. Phys. B (Proc. Suppl.), **24B**, 1991), pp. 170–174.

111. A. Olinto, in *Strange Quark Matter in Physics and Astrophysics*, edited by J. Madsen and P. Haensel (Nucl. Phys. B (Proc. Suppl.), **24B**, 1991), pp. 103–109.

112. J. E. Horvath and O. G. Benvenuto, Phys. Lett. B **213**, 516 (1988).

113. M. L. Olesen and J. Madsen, Phys. Rev. D **49**, 2698 (1994).

114. M. I. Krivoruchenko and B. V. Martemyanov, Astrophys. J. **378**, 628 (1991).

115. J. E. Horvath, O. G. Benvenuto, and H. Vucetich, Phys. Rev. D **45**, 3865 (1992).

116. K. Iida and K. Sato, Prog. Theor. Phys. **98**, 277 (1997).

117. K. Iida, Prog. Theor. Phys. **98**, 739 (1997).

118. A. De Rújula and S. L. Glashow, Nature **312**, 734 (1984).

119. P. B. Price, Phys. Rev. D **38**, 3813 (1988).

120. D. M. Lowder, in *Strange Quark Matter in Physics and Astrophysics*, edited by J. Madsen and P. Haensel (Nucl. Phys. B (Proc. Suppl.), **24B**, 1991), pp. 177–183.

121. J. Madsen, Phys. Rev. Lett. **61**, 2909 (1988).

122. M. Jändel, Z. Phys. C **40**, 599 (1988).

123. R. R. Caldwell and J. L. Friedman, Phys. Lett. **264B**, 143 (1991).

124. R. N. Boyd and T. Saito, Phys. Lett. B **298**, 6 (1993).

125. J. D. Bjorken and L. D. McLerran, Phys. Rev. D **20**, 2353 (1979).

126. F. Halzen and H. C. Liu, Phys. Rev. D **32**, 1716 (1985).

127. M. S. Desai and G. L. Shaw, in *Strange Quark Matter in Physics and Astrophysics*, edited by J. Madsen and P. Haensel (Nucl. Phys. B (Proc. Suppl.), **24B**, 1991), pp. 207–210.

Out of Equilibrium Thermal Field Theories - Elimination of Pinching Singularities

Ivan Dadić[1]

Ruder Bošković Institute, Zagreb, Croatia

Abstract. We analyze ill-defined pinch singularities characteristic of out of equilibrium thermal field theories. We identify two mechanisms that eliminate pinching even at the single self-energy insertion approximation to the propagator: the first is based on the vanishing of phase space at the singular point (threshold effect). It is effective in QED with a massive electron and a massless photon. In massless QCD, this mechanism fails, but the pinches cancel owing to the second mechanism, i.e., owing to the spinor/tensor structure of the single self-energy insertion contribution to the propagator. The constraints imposed on distribution functions are very reasonable.The same mechanism eliminates pinching from the resummed Schwinger-Dyson series.

1 Introduction

Out of equilibrium thermal field theories have recently attracted much interest. From the experimental point of view, various aspects of heavy-ion collisions and the related hot QCD plasma are of considerable interest, in particular the supposedly gluon-dominated stage.

Contrary to the equilibrium case [1,2] where pinch, collinear, and infrared problems have been successfully controlled [3–6], out of equilibrium theory [7–9] has suffered from them to these days. However, progress has been made in this field, too.

Weldon [10] has observed that the out of equilibrium pinch singularity does not cancel; hence it spoils analyticity and causality. The problem gets worse with more than one self-energy insertions.

Bedaque has argued that in out of equilibrium theory the time extension should be finite. Thus, the time integration limits from $-\infty$ to $+\infty$, which are responsible for the appearance of pinches, have to be abandoned as unphysical [11].

Le Bellac and Mabilat [12] have shown that pinching singularity gives a contribution of order $g^2 \delta n$, where δn is a deviation from equilibrium. They have also found that collinear singularities cancel in scalar theory, and in QCD using physical gauges, but not in the case of covariant gauges. Niégawa [13] has found that the pinchlike term contains a divergent part that cancels collinear singularities in the covariant gauge.

Altherr and Seibert have found that in massive $g^2 \phi^3$ theory pinch singularity does not occur owing to the kinematical constraint [14]. This result is restricted to the case of one-loop self-energies.

Altherr has suggested a regularization method in which the propagator is modified by the width γ which is an arbitrary function of momentum to be calculated in a self-consistent way. In $g^2\phi^4$ theory, for small deviations from equilibrium, γ was found to be just the usual equilibrium damping rate [15].

This recipe has been justified in the resummed Schwinger-Dyson series in various problems with pinching [16–20].

Baier, Dirks, and Redlich [16] have calculated the $\pi - \rho$ self-energy contribution to the pion propagator, regulating pinch contributions by the damping rate. In subsequent papers with Schiff [17,18] they have calculated the quark propagator within the HTL approximation [21–23]; in the resummed Schwinger-Dyson series, the pinch is naturally regulated by $Im\Sigma_R$.

Carrington, Defu, and Thoma [19] have found that no pinch singularities appear in the HTL approximation to the resummed photon propagator .

Niégawa [20] has introduced the notion of renormalized particle-number density. He has found that, in the appropriately redefined calculation scheme, the amplitudes and reaction rates are free from pinch singularities.

By pinching singularity we understand the contour passing between two infinitely close poles:

$$\int \frac{dx}{(x + i\epsilon)(x - i\epsilon)},$$ (1)

where $x = q^2 - m^2$. It is controlled by some parameter, e.g., ϵ. For finite ϵ, the expression is regular. However, when ϵ tends to zero, the integration path is "pinched" between the two poles, and the expression is ill-defined. Integration gives an ϵ^{-1} contribution plus regular terms. By performing a simple decomposition of $(x \pm i\epsilon)^{-1}$ into $PP(1/x) \mp i\pi\delta(x)$, one obtains the related ill-defined δ^2 expression.

The following expression, which is similar to (1), corresponds to the resumed Schwinger-Dyson series:

$$\int dx \frac{\omega(x)}{(x - \Sigma_R(x) + i\epsilon)(x - \Sigma_R^*(x) - i\epsilon)},$$ (2)

where $\omega(x)$ and $\bar{\omega}(x)$ (which appears in (3)) are, respectively, proportional to $\Omega(x)$ and $\bar{\Omega}(x)$, where $\Omega(x)$, $\Sigma_R(x)$, and $\bar{\Omega}(x)$ are the components of the self-energy matrix to be defined in Sec. III.

In expression (2), pinching is absent [16–20] if $Im\Sigma_R(x_o) \neq 0$ at a value of x_o satisfying $x_o - Re\Sigma_R(x_o) = 0$.

The expression corresponding to the single self-energy insertion approximation to the propagator is similar to (2):

$$\int dx \frac{\bar{\omega}(x)}{(x + i\epsilon)(x - i\epsilon)}.$$ (3)

One can rewrite the integral as

$$\int \frac{dx}{2} \left(\frac{1}{x + i\epsilon} + \frac{1}{x - i\epsilon} \right) \frac{\bar{\omega}(x)}{x}.$$ (4)

If it happens that

$$\lim_{x \to 0} \frac{\bar{\omega}(x)}{x} = K < \infty, \tag{5}$$

then the integral (4) decomposes into two pieces that, although possibly divergent, do not suffer from pinching.

There are two cases in which the function $\bar{\omega}(x)$ is even identically zero in the vicinity of the $x = 0$ point: in thermal equilibrium, because of detailed balance relations; in massive $g^2\phi^3$ theory out of equilibrium, owing to the mass shell condition [14]. The latter mechanism also works in out of equilibrium QED if a small photon mass m_γ is introduced. However, this elimination of pinching can be misleading: the domain of x, where $\bar{\omega}(x) = 0$, shrinks to a point as $m_\gamma \to 0$. We shall show that the elimination of pinching also occurs in the $m_\gamma = 0$ case.

In this paper we identify two mechanisms leading to relation (5). They are based on the observation that in the pinchlike contribution loop particles have to be on mass shell.

The first mechanism is effective in out of equilibrium QED: in the pinchlike contribution to the electron propagator, phase space vanishes linearly as $x \to 0$. In the pinchlike contribution to the photon propagator, the domain of integration is shifted to infinity as $x \to 0$. For distributions disappearing rapidly enough at large energies, the contribution again vanishes linearly in the $x \to 0$ limit. This mechanism is also valid in QCD in the cases with massive quarks.

In out of equilibrium massless QCD, phase space does not vanish, but there is an alternative mechanism: the spinor/tensor structure in all cases leads to relation (5).

Also, in out of equilibrium massless QCD, introduction of a small gluon mass does not help. In this case, processes like $q\bar{q} \to g$ are kinematically allowed, the spinor/tensor structure is modified, and $\bar{\Omega}$ does not vanish in the $x \to 0$ limit.

In a few cases, none of the mentioned mechanisms works and one has to sum the Schwinger-Dyson series. This is the case of the $\pi - \rho$ loop in the π self-energy . Even in the limit of zero pion mass, $\bar{\omega}(x)$ vanishes only as $|x|^{1/2}$ and relation (5) is not fulfilled. A similar problem appears in electroweak interactions involving decays of Z and W bosons, decay of Higgs particles, etc. Another important case is massless $g^2\phi^3$ theory. In contrast to massless QCD, massless $g^2\phi^3$ theory contains no spin factor to provide a q^2 factor necessary to obtain (5).

The densities are restricted only mildly: they should be cut off at high energies, at least as $|k_o|^{-3-\delta}$, in order to obtain a finite total particle density; for nonzero k_o, they should be finite; for k_o near zero, they should not diverge more rapidly than $|k_o|^{-1}$, the electron (positron) distribution should have a finite derivative. Further restrictions may come from Slavnov-Taylor identities [24–26], but they are not crucial for our analysis.

The paper is organized as follows.

In Sec. II we analyze the Schwinger-Dyson equation in the Keldysh representation, solve it formally, and identify pinchlike expressions. For one-loop self-energy insertions, we find that the Keldysh component $(\bar{\Omega}(q^2))$ of the self-energy is responsible for pinches. We further find that the nonzero Keldysh component requires loop particles to be on shell.

In Sec. III we analyze functions such as Ω, $\bar{\Omega}$, and $Im\Sigma_R$, and investigate their threshold properties.

In Sec. IV we show that the electron and photon propagators, calculated in the single self-energy insertion approximation, are free from pinching.

In Sec. V we analyze pinchlike expressions in the $q - \bar{q}$, $g - g$, and ghost-ghost contributions to the gluon propagator, the quark propagator and the ghost propagator in the single self-energy insertion approximation. We find that, in all the cases, the spinor/tensor factor F contains a factor q^2 that is sufficient to eliminate pinching.

In Sec. VI we briefly discuss the elimination of pinching in the resummed Schwinger-Dyson equation.

In Sec. VII we briefly recollect the main results of the paper.

2   Propagators and the Schwinger-Dyson Equation

We start [27,28] by defining out of equilibrium thermal propagators for bosons, in the case when we can ignore the variations of slow variables in Wigner functions [12,29]:

$$D = \begin{pmatrix} D_{11} & D_{12} \\ D_{21} & D_{22} \end{pmatrix}, \tag{6}$$

$$D_{11}(k) = D_{22}^*(k) = \frac{i}{k^2 - m^2 + 2i\epsilon|k_o|} + 2\pi \sinh^2 \theta \delta(k^2 - m^2), \tag{7}$$

$$D_{12}(k) = -2\pi\delta(k^2 - m^2)(\cosh^2 \theta \Theta(k_o) + \sinh^2 \theta \Theta(-k_o)), \tag{8}$$

$$D_{21}(k) = -2\pi\delta(k^2 - m^2)(\cosh^2 \theta \Theta(-k_o) + \sinh^2 \theta \Theta(k_o)). \tag{9}$$

For particles with additional degrees of freedom, relations (6)-(9) are provided with extra factors $(\not{k} + m)$ for spin 1/2, $(g_{\mu\nu} - (1 - a)k_\mu k_\nu/(k^2 \pm 2i\epsilon k_o))$ for vector particle, etc., and similarly for internal degrees of freedom. To keep the discussion as general as possible, we show these factors explicitly only when necessary. The propagator defined by relations (6)-(9) satisfies the important condition

$$0 = D_{11} + D_{12} + D_{21} + D_{22}. \tag{10}$$

In the case of equilibrium, we have

$$\sinh^2 \theta(k_o) = n_B(k_o) = \frac{1}{\exp \beta|k_o| - 1}. \tag{11}$$

To obtain the corresponding relations for fermions, we only need to make the substitution

$$\sinh^2 \theta(k_o) \rightarrow -\sin^2 \bar{\theta}(k_o). \tag{12}$$

In the case of equilibrium, for fermions we have

$$\sin^2 \bar{\theta}_{F,F}(k_o) = n_{F,F}(k_o) = \frac{1}{\exp \beta(|k_o| \mp \mu) + 1}. \tag{13}$$

Out of equilibrium, $n_B(k_o)$ and $n_F(k_o)$ will be some given functions of k_o.

To transform into the Keldysh form, one defines the matrix Q [28] as

$$Q = \frac{1}{\sqrt{2}} \begin{pmatrix} 1 & 1 \\ -1 & 1 \end{pmatrix}. \tag{14}$$

Now

$$\begin{pmatrix} 0 & D_R \\ D_A & D_K \end{pmatrix} = QDQ^{-1}, \tag{15}$$

$$D_R(k) = -(D_{11} + D_{21}) = \frac{-i}{k^2 - m^2 + 2i\epsilon k_o}, \tag{16}$$

$$D_A(k) = -(D_{11} + D_{12}) = \frac{-i}{k^2 - m^2 - 2i\epsilon k_o} = -D_R^*(k) = D_R(-k), \tag{17}$$

$$D_K(k) = D_{11} + D_{22} = 2\pi\delta(k^2 - m^2)(1 + 2\sinh^2 \theta). \tag{18}$$

We need D_K expressed through D_R and D_A:

$$D_K = h(k_o)(D_R - D_A), \quad h(k_o) = -\epsilon(k_o)(1 + 2\sinh^2 \theta). \tag{19}$$

Again for fermions, D_K is equal to

$$D_K(k) = D_{11} + D_{22} = 2\pi\delta(k^2 - m^2)(1 - 2\sin^2 \bar{\theta}). \tag{20}$$

The proper self-energy

$$\Sigma = \begin{pmatrix} \Sigma_{11} & \Sigma_{12} \\ \Sigma_{21} & \Sigma_{22} \end{pmatrix} \tag{21}$$

satisfies the condition

$$0 = \Sigma_{11} - \Sigma_{12} - \Sigma_{21} + \Sigma_{22}. \tag{22}$$

It is also transformed into the Keldysh form (in Niemi's paper there is a misprint using Q^{-1} instead of Q):

$$\begin{pmatrix} \Omega & \Sigma_A \\ \Sigma_R & 0 \end{pmatrix} = Q\Sigma Q^{-1}, \tag{23}$$

$$\Sigma_R = -(\Sigma_{11} - \Sigma_{21}), \tag{24}$$

$$\Sigma_A = -(\Sigma_{11} - \Sigma_{12}), \tag{25}$$

$$\Omega = \Sigma_{11} + \Sigma_{22}. \tag{26}$$

We also find

$$\Sigma_A = \Sigma_R^*. \tag{27}$$

The calculation of the Σ matrix gives (propagators $S(k)$ and $G(k)$ in the self-energy matrix and in the Schwinger-Dyson equation are also given by (6) to (19), with the spin indices suppressed to keep the discussion as general as possible):

$$\Sigma_R = -i\frac{1}{2}g^2 \quad \int \frac{d^4k}{(2\pi)^4} \big(D_R(k)S_A(k-q) \\ + (D_A(k) - D_K(k))(S_R(k-q) - S_K(k-q))\big), \tag{28}$$

$$\Sigma_A = -i\frac{1}{2}g^2 \quad \int \frac{d^4k}{(2\pi)^4} \big(D_A(k)S_R(k-q) \\ + (D_R(k) - D_K(k))(S_A(k-q) - S_K(k-q))\big), \tag{29}$$

$$\Omega = i\frac{1}{2}g^2 \quad \int \frac{d^4k}{(2\pi)^4} \big((D_R(k) \\ + D_A(k))(S_A(k-q) + S_R(k-q)) + D_K(k)S_K(k-q)\big). \tag{30}$$

A simple exercise with the help of (47) will convince us that only on-shell loop-particle momenta contribute to Ω. The Schwinger-Dyson equation

$$\mathcal{G} = G + iG\Sigma\mathcal{G}, \tag{31}$$

can be written in the Keldysh form as

$$\begin{pmatrix} 0 & \mathcal{G}_R \\ \mathcal{G}_A & \mathcal{G}_K \end{pmatrix} = \begin{pmatrix} 0 & G_R \\ G_A & G_K \end{pmatrix} \\ + i\begin{pmatrix} 0 & G_R\Sigma_R\mathcal{G}_R \\ G_A\Sigma_A\mathcal{G}_A & G_A\Omega\mathcal{G}_R + G_K\Sigma_R\mathcal{G}_R + G_A\Sigma_A\mathcal{G}_K \end{pmatrix} \tag{32}$$

By expanding (32), we deduce the contribution from the single self-energy insertion to be of the form

$$\mathcal{G}_R \approx G_R + iG_R\Sigma_R G_R, \quad \mathcal{G}_A \approx G_A + iG_A\Sigma_A G_A, \tag{33}$$

which is evidently well defined, and the Keldysh component suspected for pinching:

$$\mathcal{G}_K \approx G_K + iG_A\Omega G_R + iG_K\Sigma_R G_R + iG_A\Sigma_A G_K. \tag{34}$$

It is easy to obtain a solution [28] for $\mathcal{G}_R$ and $\mathcal{G}_A$ using the form (32). One observes that the equations for $\mathcal{G}_R$ and $\mathcal{G}_A$ are simple and the solution is straightforward:

$$\mathcal{G}_R = \frac{1}{G_R^{-1} - i\Sigma_R} = -\mathcal{G}_A^*. \tag{35}$$

To calculate $\mathcal{G}_K$, we can use the solution (35) for $\mathcal{G}_R$ and $\mathcal{G}_A$:

$$\mathcal{G}_K = \mathcal{G}_A(G_A^{-1}G_K G_R^{-1} + i\Omega)\mathcal{G}_R. \tag{36}$$

Now we eliminate G_K with the help of (19):

$$\mathcal{G}_K = \mathcal{G}_A \left(h(q_o)(G_A^{-1} - G_R^{-1}) + i\Omega \right) \mathcal{G}_R. \tag{37}$$

The first term in (37) is not always zero, but it does not contain pinching singularities! The second term in (37) is potentially ill-defined (or pinchlike). The pinchlike contribution appears only in this equation; thus it is the key for the whole problem of pinch singularities. In the one-loop approximation, it requires loop particles to be on mass shell. This will be sufficient to remove ill-defined expressions in all studied cases.

We start with (34). After substituting (19) into (34), we obtain the regular term plus the pinchlike contribution:

$$\mathcal{G}_K \approx \mathcal{G}_{Kr} + \mathcal{G}_{Kp}, \tag{38}$$

$$\mathcal{G}_{Kr} = h(q_o) \left(G_R - G_A + iG_R\Sigma_R G_R - iG_A\Sigma_A G_A \right), \tag{39}$$

$$\mathcal{G}_{Kp} = iG_A\bar{\Omega}G_R, \quad \bar{\Omega} = \Omega - h(q_o)(\Sigma_R - \Sigma_A). \tag{40}$$

For equilibrium densities, we have $\Sigma_{21} = e^{-\beta q_o}\Sigma_{12}$, and expression (40) vanishes identically. This is also true for fermions.

Expression (40) is the only one suspected of pinch singularities at the single self-energy insertion level. The function $\bar{\Omega}$ in (40) belongs to the type of functions characterized by the fact that both loop particles have to be on mass shell. It is analyzed in detail in Secs. III and IV (for threshold effect) and in Sec. V (for spin effect). With the help of this analysis we show that relation (40) transforms into

$$\mathcal{G}_{Kp} = -i\frac{K(q^2, m^2, q_o)}{2} \left(\frac{1}{q^2 - m^2 + 2i\epsilon q_o} + \frac{1}{q^2 - m^2 - 2i\epsilon q_o} \right), \tag{41}$$

where $K(q^2, m^2, q_o)$ is $\bar{\Omega}/(q^2 - m^2)$ multiplied by spinor/tensor factors included in the definition of $G_{R,A}$. The finiteness of the limit

$$\lim_{q^2 \to m^2 \neq 0} K(q^2, m^2, q_o) = K_{\mp}(q_o) < \infty \tag{42}$$

is important for cancellation of pinches. The index $\mp$ indicates that the limiting value m^2 is approached from either below or above, and these two values are generally different. To isolate the potentially divergent terms, we express the function $K(q^2, m^2, q_o)$ in terms of functions that are symmetric $(K_1(q^2, m^2, q_o))$ and antisymmetric $(K_2(q^2, m^2, q_o))$ around the value $q^2 = m^2$.

$$K(q^2, m^2, q_o) = (K_1(q^2, m^2, q_o) + \epsilon(q^2 - m^2)K_2(q^2, m^2, q_o)). \qquad (43)$$

These functions are given by

$$K_{1,2}(q^2, m^2, q_o) = \frac{1}{2}(K(q^2, m^2, q_o) \pm K(2m^2 - q^2, m^2, q_o)). \qquad (44)$$

Locally (around the value $q^2 = m^2$) this functions are related to the limits $K\pm(q_o)$ by

$$K_{1,2}(q^2, m^2, q_o) = \frac{1}{2}(K_+(q_o) \pm K_-(q_o)). \qquad (45)$$

As a consequence, the right-hand side of expression (41) behaves locally as

$$\mathcal{G}_{Kp}(q^2, m^2, q_o) \approx -\frac{i}{2}(K_1(q_o) + \epsilon(q^2 - m^2)K_2(q_o))$$
$$\left(\frac{1}{q^2 - m^2 + 2i\epsilon q_o} + \frac{1}{q^2 - m^2 - 2i\epsilon q_o} \right), \qquad (46)$$

and the term proportional to K_2 is capable of producing logarithmic singularity. Furthermore, we were unable to eliminate pinches related to the double, triple, etc., self-energy insertion contributions to the propagator. However, their sum Σ_2^∞ is free from pinching under the assumption that the resumed Schwinger-Dyson series is also free from pinching.

3   Threshold Factor

In this section we analyze the phase space of the loop integral with both loop particles on mass shell. Special care is devoted to the behavior of this integral near thresholds. In this analysis the densities are constrained only mildly: they are supposed to be finite and smooth, with a possible exception at zero energy. We also assume that the total density of particles is finite. The expressions are written for all particles being bosons, and spins are not specified; change to fermions is elementary.

To obtain the integrals over the products of $D_{R,A}$ and $S_{R,A}$, we start with a useful relation [35]:

$$\int dk_o f(k)\left(\frac{1}{(k^2 - m_D^2 + i\epsilon)((k - q)^2 - m_S^2 + i\lambda\epsilon)}\right.$$
$$\left. + \frac{1}{(k^2 - m_D^2 - i\epsilon)((k - q)^2 - m_S^2 - i\lambda\epsilon)}\right)$$

$$= -(2\pi)^2 \frac{\int dk_o f(k)}{4\omega_D \omega_S} \left(\delta(k_o - \omega_D)\delta(k_o - q_o + \lambda\omega_S)\right.$$
$$\left. + \delta(k_o + \omega_D)\delta(k_o - q_o - \lambda\omega_S)\right). \tag{47}$$

where $f(k,q)$ is some polynomial, $\omega_D = (k^2 + m_D^2)^{1/2}$, and $\omega_S == ((k - vecq)^2 + m_S^2)^{1/2}$. Similar relations could be obtained for higher powers of $D_{R,A}$ and $S_{R,A}$. For example, for the nth power of $k^2 - m_D^2 + i\lambda k_o \epsilon$, the real part of the integral will be obtained by substituting $\delta^{(n)}(k^2 - m_D^2)(-1)^{(n)}$ instead of $\delta(k^2 - m_D^2)$. Now we easily calculate $Re\Sigma_R$ as

$$Re\Sigma_R = \frac{-g^2}{2(2\pi)^3} \int d^4k \mathcal{P} \left(\frac{\epsilon(k_o)\delta(k^2 - m_D^2)h_D(k_o)}{((k-q)^2 - m_S^2)}\right.$$
$$\left. + \frac{\epsilon(k_o - q_o)\delta((k-q)^2 - m_S^2)h_S(k_o - q_o)}{(k^2 - m_D^2)}\right) F. \tag{48}$$

F is the factor dependent on spin and internal degrees of freedom.

As we assume that the zero-temperature renormalization has already been performed, the zero-temperature part is in fact eliminated by counter terms and only the thermal part remains:

$$Re\Sigma_{R\,th} = \frac{g^2}{(2\pi)^3} \int d^4k \mathcal{P} \left(\frac{\delta(k^2 - m_D^2)\sinh_D^2(k_o)}{((k-q)^2 - m_S^2)}\right.$$
$$\left. + \frac{\delta((k-q)^2 - m_S^2)\sinh_S^2(k_o - q_o)}{(k^2 - m_D^2)}\right) F. \tag{49}$$

Now, starting from (24) to (26), we calculate Ω and $Im\Sigma_R$.

$$\Omega = 2iIm\Sigma_{11} = 2\frac{ig^2}{2} \int \frac{d^4k}{(2\pi)^4} 4\pi^2 \delta(k^2 - m_D^2)\delta((k-q)^2 - m_S^2)$$
$$N_\Omega(k_o, k_o - q_o)F, \tag{50}$$

where

$$N_\Omega(k_o, k_o - q_o) = \frac{1}{2}(-\epsilon(k_o(k_o - q_o)) + (1 + 2\sinh^2 \theta_D(k_o))$$
$$(1 + 2\sinh^2 \theta_S(k_o - q_o))), \tag{51}$$

$$Im\Sigma_R = \frac{g^2}{2} \int \frac{d^4k}{(2\pi)^4} 4\pi^2 \delta(k^2 - m_D^2)\delta((k-q)^2 - m_S^2)N_R(k_o, k_o - q_o)F, \tag{52}$$

and

$$N_R(k_o, k_o - q_o) = (\sinh^2 \theta_D(k_o)\epsilon(k_o - q_o) + \sinh^2 \theta_S(k_o - q_o)\epsilon(-k_o)$$

$$+\Theta(-k_o)\Theta(k_o - q_o) - \Theta(k_o)\Theta(q_o - k_o)). \tag{53}$$

It is useful to define $N_{\tilde{\Omega}}(k_o, k_o - q_o)$ as

$$N_{\tilde{\Omega}}(k_o, k_o - q_o) = N_{\Omega}(k_o, k_o - q_o) - h(q_o)N_R(k_o, k_o - q_o). \quad (54)$$

After integrating over δ's, one obtains expressions of the general form

$$\mathcal{I} = \frac{1}{4|q|} \int \frac{|k_o|dk_o}{|k|} d\phi N(k_o, k_o - q_o) F(q_o, |q|, k_o, |k|, qk, ...)\Theta(1 - z_o^2), \quad (55)$$

where $|k| = (k_o^2 - m_D^2)^{1/2}$,

$$qk = |q||k|z_o, \quad (56)$$

$$z_o = \frac{q^2 + k^2 - (q - k)^2}{2|k||q|}. \quad (57)$$

$\phi\epsilon(0, \pi)$ is the angle between k_T and x axes.

Let us start with the $q^2 > 0$ case. Solution of $\Theta(1 - z_o^2)$ gives the integration limits

$$k_{o\ 1,2} = \frac{1}{2q^2} \left(q_o(q^2 + m_D^2 - m_S^2) \mp |q|((q^2 - q_{+tr}^2)(q^2 - q_{-tr}^2))^{1/2}\right), \quad (58)$$

or

$$|k|_{1,2} = \frac{1}{2q^2} \left(|q|(q^2 + m_D^2 - m_S^2) \mp q_o((q^2 - q_{+tr}^2)(q^2 - q_{-tr}^2))^{1/2}\right), \quad (59)$$

$$q_{\pm tr} = |m_D \pm m_S|. \quad (60)$$

Assume now that $q_{tr} \neq 0$. In this case, at the threshold, the limits shrink to the value

$$k_{o\ tr} = \frac{q_o(q_{tr}^2 + m_D^2 - m_S^2)}{2q_{tr}^2}, \quad |k|_{tr} = \frac{|q|(q_{tr}^2 + m_D^2 - m_S^2)}{2q_{tr}^2}. \quad (61)$$

Near the threshold, it is convenient to change the integration variable by $dk_o|k_o|/|k| = d|veck|$. Now, for $|q^2 - q_{tr}^2|$ small enough, we have

$$k_o = \epsilon(k_{o,tr}(|k|^2 + m_D^2)^{1/2}$$

We define the coefficient c_1 by

$$c_1 = \frac{1}{4|q|} \int d\phi N(k_{o\ tr}, k_{o\ tr} - q_o) F(q_o, |q|, k_{o\ tr}, |k|_{tr}, qk_{tr}, ...). \quad (62)$$

Now the expression (55) can be approximated by

$$\mathcal{I} \approx c_1(|k|_2 - |k|_1)$$

$$\approx c_1(\Theta(q^2 - q_{+tr}^2) + \Theta(-q^2 + q_{-tr}^2)) \frac{q_o((q^2 - q_{+tr}^2)(q^2 - q_{-tr}^2))^{1/2}}{q^2}. \quad (63)$$

Relation (63) is the key to further discussion of the threshold effect.

We obtain this also for higher dimension (D=6, for example).

Relation (63) put some limits on the behavior of density functions: they should not tend to infinity at any value of $q_o \neq 0$; near $q_o = 0$, owing to the presence of the factor q_o, they should not rise more rapidly than q_o^{-1}.

Owing to (60) and (63), the function $\mathcal{I}(q^2, m_D^2, m_S^2)$ has the following properties important for cancellation of pinches.

It vanishes between the thresholds, i.e., the domain $(m_D - m_S)^2 < q^2 < (m_D + m_S)^2$ is forbidden ($\mathcal{I} = 0$). If it happens, that the bare mass m^2 belongs to this domain, the single self-energy insertion will be free of pinching. In this case also multiple (double, triple, etc.) self-energy insertions will be free of pinching.

It is (in principle) different from zero in the allowed domain $q^2 < (m_D - m_S)^2$ and $(m_D + m_S)^2 < q^2$. In this case one cannot get rid of pinching. An exception to this rule are occasional zeros owing to the specific form of densities.

The behavior at the boundaries (i.e., in the allowed region near the threshold) depends on the masses m_D and m_S and there are a few possibilities.

If both masses are nonzero and different $(0 \neq m_D \neq m_S \neq 0)$, then there are two thresholds and $\mathcal{I}$ behaves as $(q^2 - q_{\pm tr}^2)^{1/2}$ in the allowed region near the threshold $q_{\pm tr}^2$. For $m^2 m = q_{tr}^2$, the power 1/2 is not large enough to suppress pinching.

If one of the masses is zero $(m_D \neq 0, m_S = 0$ or $m_D = 0, m_S \neq 0)$, then (63) gives that the thresholds are identical (i.e., the forbidden domain shrinks to zero) and one obtains the $(q^2 - m_D^2)^1$ behavior near m_D^2. This case (for $m^2 = m_D^2$) is promising. Elimination of pinching in electron propagator, considered in Sec.IV, is one of important examples.

If the masses are equal but different from zero $(m_D = m_S \neq 0)$, then there are two thresholds with different behavior. The function $\mathcal{I}$ behaves as $(q^2 - q_{+tr}^2)^{1/2}$ in the allowed region near the threshold $q_{+tr}^2 = 4m_D^2$, and this behavior cannot eliminate pinching in the supposed case $m^2 = 4m_D^2$.

However, at the other threshold, namely at $q_{-tr}^2 = 0$, the physical region is determined by $q^2 < 0$ and the above discussion does not apply. In fact, the integration limits (58) or (59) are valid, but the region between $k_{o\,1}$ and $k_{o\,2}$ is now excluded from integration. One has to integrate over the domain $(-\infty, k_{o\,1}) \bigcup (k_{o\,2}, +\infty)$. This leads to the limitation in the high-energy behavior of the density functions. Important example of such behavior, elimination of pinching in photon propagator (m_γ), is discussed in Sec.IV.

If both masses vanish $(m_D = m_S = 0)$, the thresholds coincide, there is no forbidden region and no threshold behavior. The behavior depends on the spin of the particles involved. For scalars, the leading term in the expansion of $\mathcal{I}$ does not vanish. Pinching is not eliminated.

The case of vanishing masses ($m_D = m_S = 0$) for particles with spin exhibits a peculiar behavior. In all studied examples (see the Sec. V for details), $\mathcal{I}$ behaves as q^2 as $q^2 \to 0$, which promises elimination of pinching.

4   Pinch Singularities in QED

4.1   Pinch Singularities in the Electron Propagator

In this subsection we apply the results of preceding section to cancel the pinching singularity appearing in a single self-energy insertion approximation to the electron propagator. To do so, we have to substitute $m_D = m$, $m_S = 0$, $\sinh^2 \Theta_D(k_o) \to -n_e(k_o)$, $\sinh^2 \Theta_S(k_o - q_o) \to n_\gamma(k_o - q_o)$, and $h(k_o) = -\epsilon(k_o)(1 - 2n_e(k_o))$, where n_e and n_γ are given non-equilibrium distributions of electrons and photons in relations (51), (53),(54), and (19). The thresholds are now identical

$$q^2_{\pm tr} = m^2, \tag{64}$$

and the integration limits are

$$k_{o\,1,2} = \frac{1}{2q^2}\left(q_o(q^2 + m^2) \mp |q|((q^2 - m^2))\right) \tag{65}$$

or

$$|k|_2 - |k|_1 = \frac{q_o}{q^2}(q^2 - m^2)). \tag{66}$$

At threshold the limits shrink to the value

$$k_{o\,tr} = q_o, \qquad |k|_{tr} = |q|. \tag{67}$$

Then, with the help of (62), we define

$$\mathcal{K}(q^2, m^2, q_o) = \frac{(\not{k} + m)\not{\bar{P}}(\not{k} + m)}{(q^2 - m^2)}$$

$$\approx \frac{1}{16\pi^2|q|(q^2 - m^2)} \int d\phi N_\Omega(k_{o\,tr} = q_o, k_{o\,tr} - q_o = 0)$$

$$(\not{k} + m)\mathcal{F}(q_o, |q|, k_{o\,tr}, |k|_{tr}, qk_{tr}, ...)$$

$$(\not{k} + m)(|k|_2 - |k|_1). \tag{68}$$

The trace factor $\mathcal{F}$ is calculated with loop particles on mass shell:

$$\mathcal{F}_{e\gamma} = \left(g_{\mu\nu} - (1-a)\frac{(k-q)_\mu(k-q)_\nu}{(k-q)^2 \pm 2i(k_o - q_o)\epsilon}\right)\gamma^\mu(\not{k} + m)\gamma^\nu$$

$$= \left(-2\not{k} + 4m - (1-a)(-\not{k} + m - \frac{(\not{k} - \not{q})(-k^2)q^2}{(k-q)^2 \pm 2i(k_o - q_o)\epsilon}\right). \tag{69}$$

In calculating the term proportional to $(1 - a)$, we have to use the trick

$$((k - q)^2 \pm i\epsilon)^{-2} = \lim_{m_\gamma \to 0} \left[\frac{\partial}{\partial m_\gamma^2} ((k - q)^2 \pm i\epsilon) - m_\gamma^2)^{-1} \right]. \quad (70)$$

For $q^2 \neq 0$, we can decompose the vector k as

$$k = \frac{(k.q)}{q^2} q + \frac{(k.\tilde{q})}{\tilde{q}^2} \tilde{q} + k_T = (q - \frac{q_o}{|q|} \tilde{q}) \frac{-m_\gamma^2 + m^2 + q^2}{2q^2} + \frac{k_o}{|q|} \tilde{q} + k_T, \quad (71)$$

where, in the heat-bath frame with the z axis oriented along the vector q, we have

$$q = (q_o, 0, 0, |q|), \quad \tilde{q} = (|q|, 0, 0, q_o), \quad q\tilde{q} = 0, \quad \tilde{q}^2 = -q^2, \quad (72)$$

The transverse component of k, k_T vanishes after integration over ϕ.
Finally, we obtain

$$(\slashed{k} + m)\slashed{F}(\slashed{k} + m) = 2m(q^2 + m^2 + 2m\slashed{k})$$

$$+(q^2 - m^2)\left(-\frac{q^2 - m^2}{q^2} \slashed{k} + (-\frac{q_o(q^2 + m^2)}{q^2|q|} + 2\frac{k_o}{|q|})\slashed{\tilde{q}} + 2k_T \right.$$

$$-(1 - a)\frac{(q^2 - m^2)}{2q^2}(-\slashed{k} + \frac{q_o}{|q|}\slashed{\tilde{q}}). \quad (73)$$

Now we can study the limit

$$\slashed{K}(q_o) = \lim_{q^2 \to m^2} \slashed{K}(q^2, m^2, q_o)$$

$$= (\slashed{k} + m)\frac{q_o}{2\pi|q|m^2} N_{\tilde{\Omega}}(k_{o\ tr}, k_{o\ tr} - q_o). \quad (74)$$

It is easy to find that $\slashed{K}(q_o)$ is finite provided that $m^2 \neq 0$ and $N_{\tilde{\Omega}}(q_o, 0) < \infty$. The last condition is easy to investigate using the limiting procedure:

$$N_{\tilde{\Omega}}(q_o, 0) = \lim_{k_o \to q_o} N_{\tilde{\Omega}}(k_o, k_o - q_o)$$

$$= \lim_{k_o \to q_o} 2n_\gamma(k_o - q_o)(n_e(q_o) - n_e(k_o))$$

$$+ \lim_{k_o \to q_o} ((n_e(q_o) - n_e(k_o) - (\epsilon(q_o)\epsilon(k_o - q_o)$$

$$(n_e(q_o) + n_e(k_o) - 2n_e(q_o)n_e(k_o))). \quad (75)$$

One should observe here that the integration limits imply that the limit $k_o \to q_o$ is taken from below for $q^2 > m^2$, and from above for $q^2 < m^2$. The two limits lead to different values of $N_{\tilde{\Omega}}(q_o, 0)$. Only the first term in (75) can give rise to problems. We rewrite it as $\lim_{k_o \to 0} \left(2k_o n_\gamma(k_o) \frac{\partial n_e(k_o + q_o)}{\partial k_o} \right)$.

As relation (74) should be valid at any q_o, we can integrate over q_o to find that the photon distribution should not grow more rapidly than $|k_o|^{-1}$ as k_o approaches zero, while the derivative of the electron distribution $n_e(q_o)$ should be finite at any q_o:

$$\lim_{k_o \to 0} k_o n_\gamma(k_o) < \infty, \tag{76}$$

$$\left| \frac{\partial n_e(q_o)}{\partial q_o} \right| < \infty. \tag{77}$$

Under the very reasonable conditions (76) and (77) the electron propagator is free from pinches.

It is interesting to observe the discontinuity of $K(q^2, m^2, q_o)$ at the point $q^2 = m^2$. This feature will be repeated in massless QCD.

It is worth observing that $K(q_o)$ is gauge independent, at least within the class of covariant gauges.

4.2   Pinch Singularities in the Photon Propagator

To consider the pinching singularity appearing in a single self-energy insertion approximation to the photon propagator, we have to make the substitutions $m_D = m = m_S$, $\sinh^2 \Theta_D(k_o) \to -n_e(k_o)$, $\sinh^2 \Theta_S(k_o - q_o) \to -n_e(k_o - q_o)$, and $h(k_o) = -\epsilon(k_o)(1 + 2n_\gamma(k_o))$. There are two thresholds, but only $q_{1,tr}^2 = 0$ and the domain where $q^2 < 0$ are relevant to a massless photon. The integration limits are given by the same expression (65), but now we have to integrate over the domain $(-\infty, k_{o\,1}) \bigcup (k_{o\,2}, +\infty)$. As $q^2 \to -0$, we find $(k_{o\,1} \to -\infty)$ and $(k_{o\,2} \to +\infty)$. The integration domain is still infinite but is shifted toward $\pm\infty$ where one expects that the particle distribution vanishes:

$$K_{\mu\nu}(q^2, q^o) = \left(g_{\mu\rho} - (1-a)\frac{q_\mu q_\rho}{q^2 - 2iq_o\epsilon} \right) \frac{\bar{\Omega}^{\rho\sigma}}{q^2} \left(g_{\sigma\nu} - (1-a)\frac{q_\sigma q_\nu}{q^2 + 2iq_o\epsilon} \right)$$

$$= \frac{1}{16\pi^2 |q| q^2} \left(\int_{-\infty}^{k_{o1}} + \int_{k_{o2}}^{\infty} \right) \frac{k_o dk_o}{|k|}$$

$$\int d\phi N_\Omega(k_o, k_o - q_o) \left(g_{\mu\rho} - (1-a)\frac{q_\mu q_\rho}{q^2 - 2iq_o\epsilon} \right)$$

$$F^{\rho\sigma}(q_o, |q|, k_o, |k|, qk, \ldots) \left(g_{\sigma\nu} - (1-a)\frac{q_\sigma q_\nu}{q^2 + 2iq_o\epsilon} \right). \tag{78}$$

To calculate $F^{\mu\nu}$ for the $e - \bar{e}$ loop, we parametrize the loop momentum k by introducing an intermediary variable l perpendicular to q. m is the mass of loop particles:

$$k = \alpha q + l, \quad q.l = 0, \quad k^2 = (k - q)^2 = m^2,$$
$$l^2 = m^2 - \alpha^2 q^2,$$
$$\alpha = \frac{k^2 + q^2 - (k-q)^2}{2q^2}. \tag{79}$$

At the end of the calculation we eliminate l in favor of k. After all possible singular denominators are canceled, one can set $\alpha = 1/2$.

$$F_{e\bar{e}}^{\mu\nu} = -Tr(\not{k} + m)\gamma^\mu(\not{k} - \not{q} + m)\gamma^\nu$$

$$= (2q^2 g^{\mu\nu} - 2q^\mu q^\nu + 8l^\mu l^\nu)$$

$$= \left(\frac{4m^2 q_o^2}{q^2}A^{\mu\nu}(q)\right.$$

$$+\frac{q^2}{q^2}\left((4k_o(k_o - q_o) - 4m^2 - q^2)A^{\mu\nu}(q)\right.$$

$$\left.\left. + (-8(k_o - \frac{q_o}{2})^2 + 2q^2)B^{\mu\nu}(q)\right)\right). \tag{80}$$

Using relation (156) we obtain

$$K_{\mu\nu}(q^2, q_o) = \frac{1}{16\pi^2|q|q^2}\left(\int_{-\infty}^{k_{o1}} + \int_{k_{o2}}^\infty\right)\frac{k_o dk_o}{|k|}\int d\phi N_{\tilde{\Omega}}(k_o, k_o - q_o)$$

$$\left(\frac{4m^2 q_o^2}{q^2}A_{\mu\nu}(q)\right.$$

$$+\frac{q^2}{q^2}\left((4k_o(k_o - q_o) - 4m^2 - q^2)A_{\mu\nu}(q)\right.$$

$$\left.\left.+(-8(k_o - \frac{q_o}{2})^2 + 2q^2)B_{\mu\nu}(q)\right)\right). \tag{81}$$

In the integration over k_o the terms proportional to $(k_o^2 q^2)^n$ dominate and $\lim_{q^2 \to 0}|K_{\mu\nu}(q^2, q_o)| < \infty$ if

$$\left|\frac{1}{16\pi^2|q|q^2}\left(\int_{-\infty}^{k_{o1}} + \int_{k_{o2}}^\infty\right)\frac{k_o dk_o}{|k|}(\alpha + \beta k_o^2 q^2)\cdot\right.$$

$$\left.\int d\phi N_{\tilde{\Omega}}(k_o, k_o - q_o)\right| < \infty. \tag{82}$$

Here $N_{\tilde{\Omega}}(k_o, k_o - q_o)$ is given by

$$N_{\tilde{\Omega}}(k_o, k_o - q_o) = -2n_e(k_o - q_o)(-n_\gamma(q_o) - n_e(k_o))$$

$$-n_\gamma(q_o) - n_e(k_o)$$
$$-\epsilon(q_o)\epsilon(k_o - q_o)(-n_\gamma(q_o)$$
$$+n_e(k_o) + 2n_\gamma(q_o)n_e(k_o)). \tag{83}$$

Assuming that the distributions obey the inverse-power law at large energies $n_e(k_o) \propto |k_o|^{-\delta_e}$ and $n_{\bar{e}}(k_o) \propto |k_o|^{-\delta_{\bar{e}}}$, we find that the terms linear in densities dominate. Thus, for $n = 0, 1$, one finds

$$\frac{-1}{q^2}\left(\int_{-\infty}^{k_{o\,1}} + \int_{k_{o\,2}}^{+\infty}\right)\frac{|k_o|dk_o}{|k|}|k_o|^{2n-\delta}(-q^2)^n$$

$$\propto (\delta - 1 - 2n)^{-1}(|q|m)^{1+2n-\delta}(-q^2)^{(\delta-3)/2}. \tag{84}$$

It follows that (82) is finite (in fact, it vanishes) if $\delta_e, \delta_{\bar{e}} > 3$. This is exactly the condition

$$\int d^3 k n_{e,\bar{e}}(k_o) < \infty. \tag{85}$$

Thus the pinching singularity is canceled in the photon propagator under the condition that the electron and positron distributions should be such that the total number of particles is finite.

A:so, in the photon propagator, the quantity $\lim_{q^2 \to 0} K_{\mu\nu}(q^2, q_o)$ does not depend on the gauge parameter.

Expression (84) is not valid for $m = 0$.

5   Pinch Singularities in Massless QCD

In this section we consider the case of massless QCD. Pinching singularities, related to massive quarks, are eliminated by the methods used in the preceding section.

In self-energy insertions related to gluon, quark, and ghost propagators, the masses in the loop as well as the masses of the propagated particles are zero. Thus, the methods of the preceding section do not produce the expected result. Attention is turned to the spin degrees of freedom, i.e., to the function F of the integrand in (50) to (55). In the calculation of F it has been anticipated that the loop particles have to be on mass shell. In this case, F provides an extra q^2 factor in all the cases considered, in which not all particles are scalars. This q^2 factor suffices for the elimination of pinching singularities.

The integration limits are now

$$k_{o\,1,2} = \frac{1}{2}\left(q_o \mp |q|\right). \tag{86}$$

The difference $|k|_2 - |k|_1$ is finite and there is no threshold effect.

It is worth observing that for $q^2 > 0$, we have to integrate between k_{o1} and k_{o2}, whereas for $q^2 < 0$, the integration domain is $(-\infty, k_{o\,1}) \cup (k_{o\,2}, +\infty)$. This leads to two limits, $\lim_{q^2 \to \pm 0} K(q^2, q_o) = K_{\pm}(q_o)$, in all cases of massless QCD.

By inspection of the final results (89),(90), and (91), we find that the case $q^2 < 0$ requires integrability of the function $k_o^2 N_{\Omega}(k_o, k_o - q_o)$ leading to the condition (85) on the quark, gluon, and ghost distribution functions.

By using (79), we again introduce the intermediary variable l perpendicular to q; now we have to set $m = 0$.

5.1   Self-Energy Insertions Contributing to the Gluon Propagator

The function $K_{\mu\nu}(q^2, q_o)$ related to the gluon propagator is the sum

$$K_{\mu\nu}(q^2, q_o) = \Sigma_i K_{q_i \bar{q}_i massive, \mu\nu}(q^2, q_o)$$

$$+\Sigma_i K_{q_i \bar{q}_i massless, \mu\nu}(q^2, q_o) + K_{ghgh, \mu\nu}(q^2, q_o) + K_{gg, \mu\nu}(q^2, q_o), \quad (87)$$

where the terms in the sum are defined as

$$K_{\mu\nu}(q^2, q_o) = (g_{\mu\rho} - (1-a)D_{R\mu\rho}(q))\frac{\bar{\Omega}^{\rho\sigma}}{q^2}(g_{\sigma\nu} - (1-a)D_{A\sigma\nu}(q)). \quad (88)$$

Pinching singularities, related to massive quarks, are eliminated by the methods used in the preceding section. The tensor F related to the massless quark-antiquark contribution to the gluon self-energy is

$$F_{q\bar{q}}^{\mu\nu} = -\frac{\delta_{ab}}{6}Tr\not{k}\gamma^\mu(\not{k} - \not{q})\gamma^\nu$$

$$= \frac{\delta_{ab}}{6}(2q^2 g^{\mu\nu} - 2q^\mu q^\nu + 8l^\mu l^\nu)$$

$$= \frac{\delta_{ab}}{6}\left[(\frac{q^2}{q^2}\left((4k_o(k_o - q_o) - q^2)A^{\mu\nu}(q)\right.\right.$$

$$\left.\left. + (-8(k_o - \frac{q_o}{2})^2 + 2q^2)B^{\mu\nu}(q)\right) + O^{\mu\nu}(k_T)\right]. \quad (89)$$

As $F_{\mu\nu}$ contains only A and B projectors, relation (156) guarantees that the result does not depend on the gauge parameter.

Relation (89) contains only terms proportional to q^2, and

$$\lim_{q^2 \to 0} K_{\mu\nu}(q^2, q_o)$$

is finite.

For the ghost-ghost contribution to the gluon self-energy, the tensor F is given by

$$F_{ghgh}^{\mu\nu} = -\delta_{ab}N_c k^\mu(k - q)^\nu$$

$$= -\delta_{ab}N_c\left(-\frac{q^\mu q^\nu}{4} + l^\mu l^\nu + \frac{q^\mu l^\nu - l^\mu q^\nu}{2}\right)$$

$$= -\delta_{ab}N_c\frac{q^2}{q^2}\left(\frac{4k_o(k_o - q_o) + q^2}{8}A^{\mu\nu}(q)\right.$$

$$\left. - (k_o - \frac{q_o}{2})^2 B^{\mu\nu}(q) - \frac{q^2}{4}D^{\mu\nu}(q) + O^{\mu\nu}(k_T)\right). \quad (90)$$

The antisymmetric part vanishes after integration, so we have left it out from the final result in (90).

The tensor F for the gluon-gluon contribution to the gluon self-energy is

$$F_{gg}^{\mu\nu} = \frac{\delta_{ab}N_c}{2}\left(g^{\mu\sigma}(q+k)^\tau - g^{\sigma\tau}(2k-q)^\mu + g^{\tau\mu}(k-2q)^\sigma\right)$$

$$\left(g_{\sigma\sigma'} - (1-a)\frac{(k-q)_\sigma(k-q)_{\sigma'}}{(k-q)^2 \pm 2i(k_o - q_o)\epsilon}\right)$$

$$\left(g^{\nu\sigma'}(q+k)^{\tau'} - g^{\sigma'\tau'}(2k-q)^\nu + g^{\tau'\nu}(k-2q)^{\sigma'}\right)$$

$$\left(g_{\tau\tau'} - (1-a)\frac{k_\tau k_{\tau'}}{k^2 \pm 2ik_o\epsilon}\right)$$

$$= \frac{\delta_{ab}N_c}{2}\left(4q^2 g_{\mu\nu} - \frac{9}{2}q_\mu q_\nu + 10 l_\mu l_\nu - (1-a)(-5q_\mu q_\nu + 3q^2 g_{\mu\nu})\right)$$

$$-\frac{1-a}{k^2 \pm 2ik_o\epsilon}\left(-\frac{q^2}{4}q_\mu q_\nu + 5q^2 l_\mu l_\nu + q^2(l_\mu q_\nu + l_\nu q_\mu) + \frac{q^4}{4}g_{\mu\nu}\right)$$

$$-\frac{1-a}{(k-q)^2 \pm 2i(k_o - q_o)\epsilon}\left(-\frac{q^2}{4}q_\mu q_\nu + 5q^2 l_\mu l_\nu - q^2(l_\mu q_\nu + q_\mu l_\nu) + \frac{q^4}{4}g_{\mu\nu}\right)$$

$$+(1-a)^2\left(-2q_\mu q_\nu\right.$$

$$+\left(\frac{(1-a)^2}{k^2 \pm 2ik_o\epsilon} - \frac{(1-a)^2}{(k-q)^2 \pm 2i(k_o - q_o)\epsilon}\right)2q^2(q_\mu l_\nu + l_\mu q_\nu)$$

$$+\left.\frac{(1-a)^2}{(k^2 \pm 2ik_o\epsilon)((k-q)^2 \pm 2i(k_o - q_o)\epsilon)}4q^4 l_\mu l_\nu\right)$$

$$\rightarrow \frac{\delta_{ab}N_c q^2}{2}\left(\frac{1}{q^2}\left((10(k_o - \frac{q_o}{2})^2 + \frac{3}{2}q^2)A^{\mu\nu}(q)\right.\right.$$

$$+(-10(k_o - \frac{q_o}{2})^2 + 4q^2)B^{\mu\nu}(q) - \frac{q^2}{2}D^{\mu\nu}(q)\right)$$

$$-(1-a)\left(\frac{1}{2}A^{\mu\nu} - B^{\mu\nu} - \frac{q_o}{|q|}C^{\mu\nu}\right)$$

$$+(1-a)^2\left(-\frac{q^2}{q^2}A^{\mu\nu} + 2\frac{q_o^2}{q^2}B^{\mu\nu} - 2\frac{q_o}{|q|}C^{\mu\nu} - 2D^{\mu\nu}\right)$$

$$+O^{\mu\nu}(k_T)\Big). \tag{91}$$

Expressions (89), (90), and (91) have been obtained by substitution of (152), (154), and (155) and, finally, by eliminating the variable l in favor of k. The tensor $O^{\mu\nu}(k_T)$ is linear in k, thus it vanishes after integration over ϕ.

We note here that, in the Feynman gauge ($a = 1$), the operator C is absent from the gluon self-energy! Consequently, the relation originating from Slavnov-Taylor identities (proved in [24–26] for equilibrium densities), $\Pi_C^2 =$

$(q^2 - \Pi_L)\Pi_D$, is fulfilled at $a = 0$ only if $\Pi_D = 0$. Thus the contributions to π_D from the ghost-ghost and gluon-gluon self-energies mutually cancel, imposing restrictions on the densities related to unphysical degrees of freedom. As it does not interfere with the cancellation of pinches, the problem of unphysical degrees of freedom will be discussed elsewhere.

Finally, we need (156) in all three cases.

Expressions (89), (90), and (91) for the ghost-ghost, quark-antiquark, and gluon-gluon contributions to the gluon self-energy contain only terms proportional to q^2. The function $K_{\mu\nu}(q^2, q_o)$ approaches the finite value $K_{\mu\nu}(\pm, q_o)$.

Thus we have shown that the single self-energy contribution to the gluon propagator is free from pinching under the condition (85) .

5.2   Quark-Gluon Self-Energy Contribution to the Quark Propagator

The K spinor for the quark-gluon contribution to the massless quark propagator is defined as

$$K(q^2, q_o) = \not{q} \frac{\not{\mathcal{Q}}}{q^2} \not{q}. \tag{92}$$

In the self-energy of a massless quark coupled to a gluon the spin factor F is given by

$$
\begin{aligned}
F_{qg} &= \delta_{ab} \frac{N_c^2 - 1}{2N_c} \left(g_{\mu\nu} - (1-a) \frac{(k-q)_\mu (k-q)_\nu}{(k-q)^2 \pm 2i(k_o - q_o)\epsilon} \right) \gamma^\mu \not{k} \gamma^\nu \\
&= \delta_{ab} \frac{N_c^2 - 1}{2N_c} \left(-2\not{k} - (1-a)(-\not{k} - \frac{(\not{k} - \not{q})(q^2 - k^2)}{(k-q)^2 \pm 2i(k_o - q_o)\epsilon} \right) \\
&\to \delta_{ab} \frac{N_c^2 - 1}{2N_c} \left(-\not{k} + \frac{q_o}{|q|}\not{q} - 2\frac{k_o}{|q|}\not{q} - 2\not{k}_T - \frac{1-a}{2}(-\not{k} - \frac{q_o}{|q|}\not{q}) \right) . \tag{93}
\end{aligned}
$$

For our further discussion, we need the product

$$
\begin{aligned}
\not{q} F_{qg} \not{q} &= \delta_{ab} \frac{N_c^2 - 1}{2N_c} q^2 \left(-\not{k} - \frac{q_o}{|q|}\not{q} + 2\frac{k_o}{|q|}\not{q} + 2\not{k}_T \right. \\
&\left. - \frac{1-a}{2}(-\not{k} + \frac{q_o}{|q|}\not{q}) \right), \tag{94}
\end{aligned}
$$

which contains the damping factor q^2. The term $\not{k}_T$ will be integrated out.

By inserting (94) into (92), we obtain (42) free from pinches.

To calculate $K(q_o)$, we need the limit

$$\lim_{q^2 \to 0} \frac{\not{q} F_{qg} \not{q}}{q^2} = \delta_{ab} \frac{N_c^2 - 1}{2N_c} \frac{2(k_o - q_o)}{q_o} \not{q}, \tag{95}$$

From (95) we conclude that $K(q_o)$ does not depend on the gauge parameter.

Omitting details, we observe that pinching is absent from the quark propagator, also in the Coulomb gauge.

5.3 Ghost-Gluon Self-Energy Contribution to the Ghost Propagator

The K factor is defined as

$$K(q^2, q_o) = \frac{\bar{\Omega}}{q^2}. \tag{96}$$

The F factor for the ghost-gluon contribution to the ghost self-energy is

$$F_{ghg} = \delta_{ab} N_c k^\mu q^\nu \left(g_{\mu\nu} - (1-a)\frac{(k_\mu - q_\mu)(k_\nu - q_\nu)}{(k-q)^2 \pm 2i(k_o - q_o)\epsilon} \right)$$

$$\rightarrow \delta_{ab} N_c \frac{q^2}{2}. \tag{97}$$

The factor q^2 ensures the absence of pinch singularity and a well-defined perturbative result.

5.4 Scalar -Photon Self-Energy Contribution to the Scalar Propagator

Although the massless scalar boson interacting with a photon is not part of massless QCD , it is treated using the same methods.

The K factor is defined as

$$K(q^2, q_o) = \frac{\bar{\Omega}}{q^2}. \tag{98}$$

The F factor for the massless scalar-photon contribution to the scalar self-energy,

$$F_{s\gamma} = (q+k)^\mu (q+k)^\nu \left(g_{\mu\nu} - (1-a)\frac{(k-q)_\mu (k-q)_\nu}{(k-q)^2 \pm 2i(k_o - q_o)\epsilon} \right)$$

$$= q^2 \left(2 - (1-a)\frac{q^2 - k^2}{(k-q)^2 \pm 2i(k_o - q_o)\epsilon} \right)$$

$$\rightarrow 2q^2, \tag{99}$$

clearly exhibits the q^2 damping factor!

6 Pinch Singularities in Schwinger-Dyson Series

For the resummed Schwinger-Dyson series, instead of the zero-temperature renormalized mass m_G used in preceding section, we have to use the thermal mass defined through the solution $q_o(|q|)$ of equation (100) for fixed $|q|$:

$$q^2 - m_G^2 - Re\Sigma_R(q_o, |q|) = 0, \quad m_G^2(|q|) = q_o^2(|q|) - q^2. \tag{100}$$

This mass is now compared with the thresholds of the functions Ω and $Im\Sigma_R$ and one obtains a classification of possible situations appearing in the resummed Schwinger-Dyson series.

With increasing thermal mass, the same loop particles will appear in different cases, contrary to the case of single self-energy insertion .

Now, with the help of (101), one defines Z_m, the factor correction to the renormalization constant due to the presence of matter out of equilibrium.

$$q^2 - m_G^2 - Re\Sigma_R(q_o, |\boldsymbol{q}|) = Z_m^{-1}(q_o, |\boldsymbol{q}|)(q^2 - m_G^2(|\boldsymbol{q}|)). \tag{101}$$

However, the full renormalization programme, although parallel to the equilibrium one, deserves a more detailed analysis and remains outside of the scope of this paper.

In the case of particles with spin, (35) will include projection operators and (100) will be modified accordingly. This will not change arguments of this section!

6.1   Schwinger-Dyson Series for Scalars

We start with a scalar (pseudo-scalar) propagated particle. Owing to the analysis in Sec. III threshold effect determines the Keldysh component (37) in the form

$$\mathcal{G}_K = 0 \qquad\qquad\qquad q_{-tr}^2 < q^2 < q_{+tr}^2$$

$$= i\frac{1}{G_A^{-1} - i\Sigma_A}\Omega\frac{1}{G_R^{-1} - i\Sigma_R}, \quad q^2 < q_{-tr}^2 \ \ or \ \ q^2 > q_{+tr}^2. \tag{102}$$

The threshold for $Im\Sigma_R$ depends on the masses of involved propagators. While $\mathcal{G}_{A,R}$ is the resummed propagator and its mass is shifted to $m_{\mathcal{G}}(|\boldsymbol{q}|)$, m_D and m_S are zero-temperature renormalized. Now we have three possible cases, each of them free from pinches.

I) If the thermal mass belongs to the kinematically forbidden region,

$$|m_S - m_D| < m_{\mathcal{G}}(|\boldsymbol{q}|) < m_D + m_S, \tag{103}$$

there are no pinches.

II) If the thermal mass belongs to the kinematically allowed region,

$$m_{\mathcal{G}}(|\boldsymbol{q}|) > m_D + m_S, \tag{104}$$

or

$$m_{\mathcal{G}}(|\boldsymbol{q}|) < |m_S - m_D|, \tag{105}$$

the first term in (37) vanishes again, and the two poles are separated by finite $2|Im\Sigma_R|$ and there are no pinches.

The function

$$K_{\Omega/Im\Sigma_R} = \frac{\Omega}{Im\Sigma_R} \tag{106}$$

is non-singular and poles can be separated by

$$\mathcal{G}_K = -\frac{K_{\Omega/Im\Sigma_R}}{2}\left(\frac{1}{q^2 - \Sigma_R + 2iq_o\epsilon} - \frac{1}{q^2 - \Sigma_R^* - 2iq_o\epsilon}\right). \tag{107}$$

In this case, in order to remove pinching, we have had to assume that the Ω and $Im\Sigma_R$ are different from zero whenever it is kinematically allowed. As our densities are almost completely arbitrary, and $Im\Sigma_R$ is not a positive definite quantity, we cannot exclude the case that at some q^2, $Im\Sigma_R$ has occasional zero which is not zero of Ω. If it happens that this is just the point where $q^2 = m_{\mathcal{G}}(|q|)$ we shall have a new sort of pinch. Such a situation, which we could not prove to be excluded, should be treated separately, with, possibly, no better way out than performing the the calculation of two-loop contributions)!

III) As the thermal mass varies with (input) particle densities, and with momentum, it is necessary to treat also the boundary cases i.e. when one of thresholds coincides with the mass shell. Specially, as the energy of the propagated particle rises, it should be expected that the thermal mass converges to its zero-temperature limit. The solution is found in the threshold effect and in the spin effect.

Among the boundary cases there is one which exhibits a behaviour typical of the allowed region ($m_S = m_D = 0$, $m_{\mathcal{G}}(|q|) = 0$, all particles are scalars). There is no threshold and no threshold behaviour. Consequently, for scalar particles the poles will be separated for both $0 < q^2$ and $q^2 < 0$, and there is no special behaviour as $q^2 \to 0$.

Case of massless loop particles with spin coupled to massless scalar (examples: $ghost \to ghost + gluon$, $scalar \to scalar + photon$) is described by the same expression (107), but now there is a special point $q^2 = 0$ where Ω and $Im\Sigma_R$ vanish. At $|q|$, for which $m_{\mathcal{G}}(|q|) \neq 0$, the $K_{\Omega/Im\Sigma_R}$ is well defined even at $q^2 = 0$ and one can still use (107).

However we have to analyze the possibility that for some $|q|$, $m_G^2 + Re\Sigma_R \to 0$ as $(q^2)^n$. For $n < 1$, the expression is integrable, $n = 1$ we obtain the case treated below in (108). If there is a case such that $n > 1$, a problem will arise!

Case ($m_D \neq 0$, $m_S = 0$), thresholds are identical ($q^2_{-tr} = q^2_{+tr} = m_D^2$) and the effect is linear in ($q^2 - q^2_{tr}$). The behaviour near the threshold is obtained by an expansion around singular point and retaining first non-zero contributions:

$$\mathcal{G}_K \approx \frac{-4q_o h(q_o)\epsilon - iC_\Omega x}{(Z_m^{-1}x - iC_{Im\Sigma_R}x + 2iq_o\epsilon)(Z_m^{-1}x + iC_{Im\Sigma_R}x - 2iq_o\epsilon)}, \tag{108}$$

where

$$x = q^2 - m_{\mathcal{G}}^2(|q|) \equiv q^2 - m_{tr}^2, \tag{109}$$

and

$$\Omega = C_\Omega x + O(x^2), \tag{110}$$

$$Im\Sigma_R = C_{Im\Sigma_R} x + O(x^2). \tag{111}$$

For small x, one obtains

$$
\begin{aligned}
\mathcal{G}_K \approx\; &-2Z_m h(q_o)\pi\epsilon(q_o)\delta(x) \\
&- \frac{iC_\Omega}{2(Z_m^{-2} + C_{Im\Sigma_R}^2)}\left[\frac{1}{x + 2iq_o\epsilon} + \frac{1}{x - 2iq_o\epsilon}\right] \\
&-2Z_m C_{Im\Sigma_R}\,\pi\epsilon(q_o)\delta(x)].
\end{aligned}
\tag{112}
$$

The result is well defined whenever $Z_m^{-1} \neq 0$.

Case ($m_S \neq m_D \neq 0$, $m_{\mathcal{G}}^2(|q|) = (m_D \pm m_S)^2$) and ($m_S = m_D \neq 0$, $m_{\mathcal{G}}^2(|q|) = 4m_D^2$), the threshold effect is proportional to the square root of $(q^2 - q_{tr}^2)$ at the corresponding threshold. The behaviour near the threshold is determined by

$$\mathcal{G}_K \approx -i\frac{\bar{C}_\Omega|x|^{1/2}}{(Z_m^{-1}x - i\bar{C}_R|x|^{1/2} + 2i\epsilon q_o)(Z_m^{-1}x + i\bar{C}_R|x|^{1/2} - 2i\epsilon q_o)}, \tag{113}$$

$$\Omega = |x|^{1/2}(\bar{C}_\Omega + O(x)), \tag{114}$$

$$Im\Sigma_R = |x|^{1/2}(\bar{C}_{Im\Sigma_R} + O(x)). \tag{115}$$

In this case, the $|x|^{1/2}$ dominates over x for $|x|$ small, the resulting $|x|^{-1/2}$ is integrable. The small $|x|$ behaviour is

$$\mathcal{G}_K \approx -i\frac{\bar{C}_\Omega}{\bar{C}_{Im\Sigma_R}^2|x|^{1/2}}. \tag{116}$$

The elimination of pinches is non-perturbative as we cannot put $\bar{C}_{Im\Sigma_R}$ equal to zero in (116).

In the next two subsections we analyze the resummed Schwinger-Dyson series for the vector particle and for spinors. The discussion is to a large extent parallel to that given above, so we do not repeat it.

6.2   Schwinger-Dyson Series for Vector Bosons

The gluon self-energy is the sum of quark-antiquark, ghost-ghost and gluon-gluon contributions. These contributions are evaluated with the help of (89), (90) and (91). After inserting the gluon self-energy into (50) and (52) for imaginary parts and calculating the real part with the help of (49), one can proceed by projecting the contributions into the invariant tensor subspaces. We shall not write it explicitly because, for our purpose, it is enough to have an overall q^2 factor in Ω and $Im\Sigma_R$.

The self-energy equals

$$\Pi(q) = \Pi_T \ _R(q)A_{\mu\nu}(q) + \Pi_L \ _R(q)B_R \ _{\mu\nu}(q)$$
$$+\Pi_D \ _R(q)D_R \ _{\mu\nu}(q) + \Pi_C \ _R(q)C_R \ _{\mu\nu}(q). \tag{117}$$

In the presence of the tensor $C_R \ _{\mu\nu}$ the inversion of the Schwinger-Dyson equation requires [26] construction of two operators $T_{\mu\nu}^{\pm}$ as

$$T_R^+ \ _{\mu\nu}(q) = \sigma_R(q)B_R \ _{\mu\nu}(q) + (1 - \sigma_R(q))D_R \ _{\mu\nu}(q)$$

$$+ \left(\frac{\sigma_R^2(q) - \sigma_R(q)}{a} \right)^{1/2} \left(aB_R \ _{\mu\nu}(q)C_R \ _{\mu\nu}(q) \right.$$

$$\left. + C_R \ _{\mu\nu}(q)B_R \ _{\mu\nu}(q) \right), \tag{118}$$

$$T_R^- \ _{\mu\nu}(q) = (1 - \sigma_R(q))B_R \ _{\mu\nu}(q) + \sigma_R(q)D_R \ _{\mu\nu}(q)$$

$$- \left(\frac{\sigma_R^2(q) - \sigma_R(q)}{a} \right)^{1/2} \left(aB_R \ _{\mu\nu}(q)C_R \ _{\mu\nu}(q) \right.$$

$$\left. + C_R \ _{\mu\nu}(q)B_R \ _{\mu\nu}(q) \right), \tag{119}$$

$$T_{R,A}^{\pm \ 2} = T_{R,A}^{\pm}, \ T_{R,A}^+ T_{R,A}^- = T_{R,A}^- T_{R,A}^+ = 0. \tag{120}$$

Here $\sigma_R(q)$ is the solution of the equation

$$\frac{\Pi_C \ _R a^{1/2}}{a\Pi_D \ _R - \Pi_L \ _R} = \frac{(\sigma_R^2(q) - \sigma_R(q))^{1/2}}{1 - 2\sigma_R(q)}. \tag{121}$$

With $\Pi_R^{\pm}$ defined as

$$\Pi_R^{\pm}(q) = \frac{1}{2}(a\Pi_D \ _R + \Pi_L \ _R \pm \frac{(a\Pi_D \ _R - \Pi_L \ _R)(1 + 2\sigma_R(q))}{1 - 2\sigma_R(q)}), \tag{122}$$

one obtains

$$\mathcal{G}_R \ _{\mu\nu}(q) = \left(A_{\mu\nu}(q) + B_R \ _{\mu\nu}(q) + aD_R \ _{\mu\nu}(q) \right)$$

$$\frac{iA_{\mu\nu}(q)}{q^2 - \Pi_T(q) + 2iq_o\epsilon} + \frac{iT_R^+ \ _{\mu\nu}(q)}{q^2 - \Pi_R^+(q) + 2iq_o\epsilon} + \frac{iT_R^- \ _{\mu\nu}(q)}{q^2 - \Pi_R^-(q) + 2iq_o\epsilon}. \tag{123}$$

Similarly, one obtains the Keldysh component

$$\mathcal{G}_K \ _{\mu\nu}(q) = \left(-\frac{iA_{\mu\rho}(q)}{q^2 - \Pi_T^*(q) - 2iq_o\epsilon} \right.$$

$$-\frac{iT_{R\ \mu\rho}^{+*}(q)}{q^2 - \Pi_R^{+*}(q) - 2iq_o\epsilon}$$

$$-\frac{iT_{R\ \mu\rho}^{-*}(q)}{q^2 - \Pi_R^{-*}(q) - 2iq_o\epsilon}\Bigg)$$

$$\cdot\left(A^{\rho\tau*}(q) + B_R^{*\ \rho\tau}(q) + aD_R^{*\ \rho\tau}(q)\right)$$

$$\cdot\left(-4q_oh(q_o)\epsilon g_\sigma^\tau - i\Omega_{\tau\eta}\left(A^{\eta\sigma}(q) + B_R^{\eta\sigma}(q) + aD_R^{\eta\sigma}(q)\right)\right)$$

$$\cdot\left(\frac{iA_{\sigma\nu}(q)}{q^2 - \Pi_T(q) + 2iq_o\epsilon}\right.$$

$$+\frac{iT_{R\ \sigma\nu}^{+}(q)}{q^2 - \Pi_R^{+}(q) + 2iq_o\epsilon}$$

$$+\frac{iT_{R\ \sigma\nu}^{-}(q)}{q^2 - \Pi_R^{-}(q) + 2iq_o\epsilon}\Bigg). \tag{124}$$

6.3 Schwinger-Dyson Series for Spin 1/2 Particles

The self-energy contribution can be written as

$$\Omega(q) = U_\Omega \slashed{q} + \slashed{L}_\Omega + W_\Omega, \tag{125}$$

$$\Sigma_R(q) = U_R \slashed{q} + \slashed{L}_R + W_R, \tag{126}$$

where U, L_μ and W are functions of q, and L_μ satisfies $Lq = o$. For isotropic densities L_μ is a linear combination of q_μ and U_μ. By multiplication we find $i\Sigma_R G_R$ to have a similar form

$$i\Sigma_R(q)G_R = (t_R + u_R\slashed{q} + v_R\slashed{L}_R + w_R\slashed{L}_R\slashed{q})/(q^2 - m_G^2 + 2iq_o\epsilon). \tag{127}$$

With the help of the projection operators

$$P_{\pm R}^2 = P_{\pm R}, \ P_{+R}P_{-R} = P_{-R}P_{+R} = 0$$

,

$$P_{\pm R} = \frac{1}{2}(1 \pm \frac{u_R\slashed{q} + v_R\slashed{L}_R + w_R\slashed{L}_R\slashed{q}}{\eta}), \tag{128}$$

$$\eta_R(q) = (u_R^2 q^2 + v_R^2 L_R^2 - w_R^2 L_R^2 q^2)^{1/2},$$
$$a_{\pm R} = (t_R \pm \eta_R)/(q^2 - m_G^2 + 2iq_o\epsilon), \tag{129}$$

we find

$$i\Sigma_R(q)G_R = a_{+R}P_{+R} + a_{-R}P_{-R}, \tag{130}$$

$$G_R = -i \frac{P_{+R} + P_{-R}}{q^2 - m_G^2 + 2iq_o\epsilon}. \tag{131}$$

Finally,

$$G_R = -i(\not{q} + m) \left(\frac{1}{q^2 - m_G^2 - \Sigma_{+R} + 2iq_o\epsilon} P_{+R} \right.$$
$$\left. + \frac{1}{q^2 - m_G^2 - \Sigma_{-R} + 2iq_o\epsilon} P_{-R} \right),$$

$$\Sigma_{\pm R} = (t_R \pm \eta_R), \tag{132}$$

$$G_K = \left(\frac{1}{q^2 - m_G^2 - \Sigma_{+R}^* - 2iq_o\epsilon} P_{+R}^* + \frac{1}{q^2 - m_G^2 - \Sigma_{-R}^* - 2iq_o\epsilon} P_{-R}^* \right)$$

$$(-4q_o h(q_o)\epsilon(\not{q} + m) - i(\not{q} + m)\Omega(\not{q} + m))$$

$$\left(\frac{1}{q^2 - m_G^2 - \Sigma_{+R} + 2iq_o\epsilon} P_{+R} \right.$$
$$\left. + \frac{1}{q^2 - m_G^2 - \Sigma_{-R} + 2iq_o\epsilon} P_{-R} \right). \tag{133}$$

Complex conjugation does not affect γ matrices.

In this case we have two modes with physical masses (which can be different) defined by

$$q^2 - m_G^2 - Re\Sigma_{\pm R}(q_o, |q|) = 0, \quad m_{\pm G}^2(|q|) = q_{\pm o}^2(|q|) - q^2. \tag{134}$$

Analogously, one defines $Z_{\pm m}$ (for each mode separately!)

$$q^2 - m_G^2 - Re\Sigma_{\pm R}(q_o, |q|) = Z_{\pm m}^{-1}(q_o, |q|)(q^2 - m_{\pm G}^2(|q|)). \tag{135}$$

To eliminate pinch singularities, our general discussion applies also to fermions. In the product of terms with the same physical mass, the cases when it belongs to either the forbidden or the allowed region are evidently free from pinches.

For a massive fermion coupled to a massless boson (electron+photon, massive quark+gluon) at the threshold (for the impulses at which the thermal mass is equal to the bare mass), all the imaginary parts in Ω and $Im\Sigma_R(U_\Omega, V_\Omega, W_\Omega, Im\ t, Im\ u, Im\ w)$ vanish linearly with $q^2 - m_G^2$ owing to the threshold effect. The "ϵ" term in (133) is well defined, and properties of Ω (with the natural assumption that $Z_m^{-1} \neq 0$) guarantee that expression (133) remains well-defined also at the boundary.

A similar case is with massless fermion coupled to a massless boson (massless quark + gluon) in the limiting case where also the physical mass tends to zero ($m_G = m = 0$). We have that $W_{\Omega,R} = o$ and thus $u = v = 0$. From (133) we find that it is enough if $\not{q}\Omega\not{q}$ vanishes linearly with $q^2 \to 0$, and this is guaranteed by relation (94).

7 Conclusion

Studying the out of equilibrium Schwinger-Dyson equation, we have found that ill-defined pinchlike expressions appear exclusively in the Keldysh component ($\mathcal{G}_K$) of the resumed propagator (37), or in the single self-energy insertion approximation to it (40). This component does not vanish only in the expressions with the Keldysh component (26) (Ω or $\bar{\Omega}$ for the single self-energy approximation) of the self-energy matrix. This then requires that loop particles be on mass shell. This is the crucial point to eliminate pinch singularities.

We have identified two basic mechanisms for the elimination of pinching: the threshold and the spin effects.

For a massive electron and a massless photon (or quark and gluon) it is the threshold effect in the phase space integration that produces, respectively, the critical $q^2 - m^2$ or q^2 damping factors.

In the case of a massless quark, ghost, and gluon, this mechanism fails, but the spinor/tensor structure of the self-energy provides an extra q^2 damping factor.

We have found that, in QED, the pinching singularities appearing in the single self-energy insertion approximation to the electron and the photon propagators are absent under very reasonable conditions: the distribution function should be finite, exceptionally the photon distribution is allowed to diverge as k_o^{-1} as $k_o \to 0$; the derivative of the electron distribution should be finite; the total density of electrons should be finite.

For QCD, identical conditions are imposed on the distribution of massive quarks and the distribution of gluons; the distributions of massless quarks and ghosts (observe here that in the covariant gauge, the ghost distribution is not required to be identically zero) should be integrable functions; they are limited by the finiteness of the total density.

In the preceding sections we have shown that all pinchlike expressions appearing in QED and QCD (with massless and massive quarks!) at the single self-energy insertion level do transform into well-defined expressions. Many other theories behave in such a way. However, there are important exceptions: all theories in which lowest-order processes are kinematically allowed do not acquire well-defined expressions at this level. These are electroweak interactions, processes involving Higgs and two light particles, a ρ meson and two π mesons, Z, W, and other heavy particles decaying into a pair of light particles, etc. The second important exception is massless $g^2\phi^3$ theory. This theory, in contrast to massless QCD, contains no spin factors to provide (5). In these cases, one has to resort to the resumed Schwinger-Dyson series.

Indeed in the resummed Schwinger-Dyson series all the problems are solved.

But the case in which pinch singularity is not eliminated in the single self-energy insertion approximation [16] shows a peculiar properties in also in the resummed Schwinger-Dyson series: at threshold, Keldish component is enhanced near the threshold and dominates over the other low order contributions!

The main result of the present paper is the cancellation of pinching singularities at the single self-energy insertion level in QED- and QCD-like theories. This, together with the reported [12,13] cancellation of collinear singularities, allows the extraction of useful physical information contained in the imaginary parts of the two-loop diagrams. This is not the case with three-loop diagrams, because some of them contain double self-energy insertions. In this case, one again has to resort to the sophistication of resumed propagators.

Appendix

We start [1] by defining a heat-bath four-velocity U_μ, normalized to unity, and define the orthogonal projector

$$\Delta_{\mu\nu} = g_{\mu\nu} - U_\mu U_\nu. \tag{136}$$

We further define spacelike vectors in the heat-bath frame:

$$\kappa_\mu = \Delta_{\mu\nu} q^\nu, \qquad \kappa_\mu \kappa^\mu = \kappa^2 = -q^2. \tag{137}$$

There are four independent symmetric tensors (we distinguish retarded from advanced tensors by the usual modification of the $i\epsilon$ prescription) A, B, and D (which are mutually orthogonal projectors), and C:

$$A_{\mu\nu}(q) = \Delta_{\mu\nu} - \frac{\kappa_\mu \kappa_\nu}{\kappa^2}, \tag{138}$$

$$B_{R\ \mu\nu}(q) = U_\mu U_\nu + \frac{\kappa_\mu \kappa_\nu}{\kappa^2} - \frac{q_\mu q_\nu}{(q^2 + 2iq_o\epsilon)}, \tag{139}$$

$$C_{R\ \mu\nu}(q) = \frac{(-\kappa^2)^{1/2}}{U.q} \left(\frac{(U.q)^2}{\kappa^2} U_\mu U_\nu - \frac{\kappa_\mu \kappa_\nu}{\kappa^2} + \frac{q_o^2 + q^2}{q^2} \frac{q_\mu q_\nu}{q^2 + 2iq_o\epsilon} \right), \tag{140}$$

$$D_{R\ \mu\nu}(q) = \frac{q_\mu q_\nu}{q^2 + 2iq_o\epsilon}. \tag{141}$$

In addition to the known multiplication [1,26] properties

$$A(q)A(q) = A(q), \qquad B_{R,A}(q)B_{R,A}(q) = B_{R,A}(q), \tag{142}$$

$$C_{R,A}(q)C_{R,A}(q) = -(B_{R,A}(q) + D_{R,A}(q)), \quad D_{R,A}(q)D_{R,A}(q) = D_{R,A}(q), \tag{143}$$

$$A(q)B(q) = B(q)A(q) = 0, \quad A(q)C(q) = C(q)A(q) = 0,$$

$$A(q)D(q) = D(q)A(q) = 0, \quad B(q)D(q) = D(q)B(q) = 0, \quad (144)$$

$$(B_{R,A}(q)C_{R,A}(q))_{\mu\nu} = (C_{R,A}(q)D_{R,A}(q))_{\mu\nu} = \frac{\tilde{q}_\mu q_\nu}{q^2 \pm 2iq_o\epsilon}, \quad (145)$$

$$(C_{R,A}(q)B_{R,A}(q))_{\mu\nu} = (D_{R,A}(q)C_{R,A}(q))_{\mu\nu} = \frac{q_\mu \tilde{q}_\nu}{q^2 \pm 2iq_o\epsilon}, \quad (146)$$

we need mixed products

$$B_{R,A}(q)B_{A,R}(q) = \frac{1}{2}(B_R(q) + B_A(q)), \quad (147)$$

$$C_{R,A}(q)C_{A,R}(q) = -\frac{1}{2}(B_R(q) + B_A(q) + D_R(q) + D_A(q)), \quad (148)$$

$$D_{R,A}(q)D_{A,R}(q) = \frac{1}{2}(D_R(q) + D_A(q)), \quad (149)$$

$$(B_{R,A}(q)C_{A,R}(q))_{\mu\nu} = (C_{R,A}(q)D_{A,R}(q))_{\mu\nu}$$
$$= \frac{1}{2}\left(\frac{\tilde{q}_\mu q_\nu}{q^2 + 2iq_o\epsilon} + \frac{\tilde{q}_\mu q_\nu}{q^2 - 2iq_o\epsilon}\right), \quad (150)$$

$$(C_{R,A}(q)B_{A,R}(q))_{\mu\nu} = (D_{R,A}(q)C_{A,R}(q))_{\mu\nu}$$
$$= \frac{1}{2}\left(\frac{q_\mu \tilde{q}_\nu}{q^2 + 2iq_o\epsilon} + \frac{q_\mu \tilde{q}_\nu}{q^2 - 2iq_o\epsilon}\right). \quad (151)$$

By calculating the traces of the tensors $l^\mu l^\nu$, $q^\mu l^\nu + l^\mu q^\nu$, and $q^\mu q^\nu$ with projectors, we find

$$l^\mu l^\nu = m^2 \frac{q_o^2}{2q^2} A^{\mu\nu}(q) + \frac{q^2}{q^2}\left(\frac{4l^2 - q_o^2}{8} A^{\mu\nu}(q) - l_o^2 B^{\mu\nu}(q)\right) + O^{\mu\nu}(k_T), \quad (152)$$

$$q^\mu l^\nu + l^\mu q^\nu = -\frac{q^2 l_o}{|q|} C^{\mu\nu}(q) + O^{\mu\nu}(k_T), \quad (153)$$

$$q^\mu q^\nu = q^2 D^{\mu\nu}(q), \quad (154)$$

$$g^{\mu\nu} = (A^{\mu\nu}(q) + B^{\mu\nu}(q) + D^{\mu\nu}(q)). \quad (155)$$

The tensor $O^{\mu\nu}(k_T)$ is linear in k, and vanishes after integration over ϕ. One should observe that (152) to (155) are valid for an arbitrary (but the same for B and D) R/A prescription, so we do not indicate it.

Using the multiplication rules one obtains

$$(g_{\mu\rho} - (1-a)D_{R\mu\rho}(q))$$
$$(f_a A^{\rho\sigma} + f_b B^{\rho\sigma} + f_c C^{\rho\sigma} + f_d D^{\rho\sigma}(g_{\sigma\nu} - (1-a)D_{A\sigma\nu}(q))$$

$$= \frac{1}{2}\left(f_a A_R^{\rho\sigma} + f_b B_R^{\rho\sigma} - (1-a)f_c C_R^{\rho\sigma} + (1-a)^2 f_d D_R^{\rho\sigma} \right)$$

$$+ \frac{1}{2}\left(f_a A_A^{\rho\sigma} + f_b B_A^{\rho\sigma} - (1-a)f_c C_A^{\rho\sigma} + (1-a)^2 f_d D_A^{\rho\sigma} \right). \tag{156}$$

It is important to observe that, owing to the properties of the mixed products (147) to (151), the R/A assignment of F does not influence the final result!

References

1. N. P. Landsman and Ch. G. van Weert, Phys. Rep. 145, 141 (1987).
2. M. Le Bellac, "Thermal Field Theory", (Cambridge University Press, Cambridge, 1996).
3. R. Baier, B. Pire, and D. Schiff, *Phys. Rep.D38, 2814 (1988)*.
4. J. Cleymans and I. Dadić, Z. Phys. C42, 133 (1989).
5. T. Altherr, P. Aurenche, and T. Becherrawy, Nucl. Phys. B315, 436 (1989).
6. T. Grandou, M. Le Bellac, and J. -L. Meunier, Z. Phys. C43, 575 (1989).
7. J. Schwinger, J. Math. Phys. 2, 407 (1961).
8. L. V. Keldysh, ZH. Eksp. Teor. Fiz.47, 1515 (1964) [Sov. Phys.-JETP 20, 1018 (1965)].
9. J. Rammer and H. Smith, Rev. Mod. Phys. 58, 323 (1986).
10. H. A. Weldon, Phys. Rep.D45, 352 (1992).
11. P. F. Bedaque, Phys. Lett. B344, 23 (1995).
12. M. Le Bellac and H. Mabilat, *Phys. Rep.D55, 3215 (1997)*.
13. A. Niégawa, Preprint HEP-PH/9711477, OCU-PHYS-168,1997.
14. T. Altherr and D. Seibert, Phys. Lett. B333, 149 (1994).
15. T. Altherr, Phys. Lett. B341, 325 (1995).
16. R. Baier, M. Dirks, and K. Redlich, Phys. Rep.D55, 4344 (1997).
17. R. Baier, M. Dirks, K. Redlich, and D. Schiff, Preprint HEP-PH/9704262.
18. R. Baier, M. Dirks, and K. Redlich, Contribution to the XXXLII Cracow School (to appear in Acta. Phys. Pol.), Preprint HEP-PH/9711213.
19. M. E. Carrington, H. Defu, and M. H. Thoma, Preprint HEP-PH/9708363.
20. A. Niégawa, Preprint HEP-TH/9709140, OCU-PHYS-166.
21. R. D. Pisarski, *Phys. Rev. Lett63, 1129 (1989)*.
22. E. Braaten and R. D. Pisarski, Nucl. Phys. B337, 569 (1990).
23. J. Frenkel and J. C. Taylor, Nucl. Phys. B334, 199 (1990).
24. R. Kobes, G. Kunstatter, and K. W. Mak, Z. Phys. C45, 129 (1989).
25. R. Kobes, G. Kunstatter, and A. Rebhan, Phys. Rev. Lett64, 2992 (1990).
26. H. A. Weldon, preprint HEP-PH/9701279.
27. K.-C. Chou, Z.-B. Su, B.-L. Hao, and L. Yu, Phys. Rep. 118, 1 (1985).
28. A. J. Niemi, Phys. Lett. B203, 425 (1988).
29. J-P, Blaizot, E. Iancu, and J-Y. Ollitrault, in Quark-Gluon Plasma II, edited by R. Hwa (World Scientific, Singapore, 1995).
30. E. Calzetta and B. L. Hu, *Phys. Rep.D37, 2878 (1988)*.
31. J. F. Donoghue and B. R. Holstein, Phys. Rep.D28, 340 (1983) and D29, 3004 (1984).

32. J. F. Donoghue, B. R. Holstein, and R. W. Robinet, Ann. Phys. 164, 233 (1985).
33. W. Keil and R. L. Kobes, Physica A158, 47 (1989).
34. M. Le Bellac and D. Poizat, Z. Phys. C47, 125 (1990).
35. R. L. Kobes and G. W. Semenoff, Nucl. Phys. B260, 714 (1985).

Exact Conservation of Quantum Numbers in the Statistical Description of High Energy Particle Reactions

Esko Suhonen[1], Antti Keränen[1], and Jean Cleymans[2]

[1] Department of Physical Sciences, University of Oulu, FIN-90571 Oulu, Finland
[2] Department of Physics, University of Cape Town, Rondebosch 7701, South Africa

Abstract. Relativistic heavy ion collisions are studied taking the exact conservation of baryon number, strangeness and charge explicitly into account.

1 Introduction

In high energy particle collisions, the interaction volume is often very small and large deviations can occur from the thermodynamic limit depending on the beam energy. In some cases it may be easy to produce a pion since one needs only 140 MeV for its rest mass, however, to produce an anti-proton one needs at least twice the rest mass, 1.88 GeV and not simply 0.94 GeV, since they can only be produced in pairs. The probability to produce anti-protons therefore cannot simply follow the same law (e.g. a Boltzmann distribution) as the production of pions. Fortunately, statistical mechanics provides us with the tools to take into account constraints like baryon number or strangeness conservation. This will be presented in these lecture notes.

Basically, if the system is large and hot, the corrections are negligible. In $Pb - Pb$ collisions at the CERN/SPS collider they are negligible because the system is large enough but for $p - p$ collisions at the same energy these corrections are large and must be taken into account because the final system is too small.

The main statistical concepts are presented in section 2. In chapter 3, a statistical method for taking the exact strangeness conservation into account is presented. This is applied in section 3.1 to describe the particle production as observed at the GSI in Ni+Ni collisions. The exact treatment of quantum numbers is extended to include the strangeness, baryon number and charge in section 4. A comparison of numerical results with the AGS E802 data is given in section 4.1. Finally in section 5, the generalization of the method to arbitrary internal symmetry is reviewed.

2   Quantum Statistical Concepts in Brief

Throughout this paper we follow the usual convention of natural units, so the speed of light, Planck constant and Boltzmann constant have values $c \equiv 1$, $\hbar \equiv 1$ and $k \equiv 1$, respectively.

All the physical information about a collection of particles is contained in a density operator, $\hat{\rho}$. The average of an observable A in this *statistical ensemble* is calculated as $\langle A \rangle = \mathrm{tr}(\hat{\rho}\hat{A})$, were $\hat{A}$ is a Hermitian operator corresponding to the observable. Using the density operator, one defines the entropy of the system considered as $S = -\mathrm{tr}(\hat{\rho}\ln\hat{\rho})$. In any system in nature, the entropy is known to tend to its maximum, so one has to find a representation of the density operator satisfying this condition. In thermodynamical equilibrium, the average occupations of different quantum states do not change in time, so $\partial\hat{\rho}/\partial t = 0$. The density operator satisfies the equation of motion of the form $i\partial\hat{\rho}/\partial t = -[\hat{\rho}, \hat{H}]$, where $\hat{H}$ is the Hamiltonian of the system. Thus, the thermodynamical, stationary density operator is diagonal in the basis formed by Hamiltonian eigenstates.

The choice of constraints used in maximizing the entropy defines the type of the statistical ensemble obtained. The closed system with fixed energy, E, volume, V, and number of particles i, N_i, is a *microcanonical* ensemble. System in heat bath (the *ensemble average* of energy $\langle E \rangle = \mathrm{tr}(\hat{\rho}\hat{H})$, V and N_i are conserved) corresponds to *canonical* ensemble. Further, if we let the particle number N_i fluctuate such that the average $\mathrm{tr}(\hat{\rho}\hat{N}_i)$ is conserved, we obtain a *grand canonical* ensemble.

Maximization of entropy using the canonical boundary conditions leads to the density operator of the form $\hat{\rho} = e^{-\beta\hat{H}}/Z$, where β is the inverse of temperature T, and Z is the canonical *partition function*,

$$Z = \mathrm{tr}\, e^{-\beta\hat{H}} = \sum_i e^{-\beta E_i(N)}. \tag{1}$$

Here i labels the different quantum states in the system and $E_i(N)$ is the eigenvalue of the N particle Hamiltonian. In the last step the trace is expressed in the basis of Hamiltonian eigenstates. Once knowing the correct partition function, one is able to calculate the thermodynamical quantities describing the system. For example, the average energy is $\langle E \rangle = T^2 \partial \ln Z/\partial T$.

Choosing the grand canonical boundary conditions yields

$$\hat{\rho}_G = e^{-\beta(\hat{H}-\mu_i\hat{N}_i)}/Z_G.$$

Here we have employed the grand canonical partition function,

$$Z_G(T, \{\lambda_i\}, V) = \mathrm{tr}\, e^{-\beta(\hat{H}-\mu_i\hat{N}_i)} = \prod_i \sum_{N_i} \lambda_i^{N_i} Z_{N_i}, \tag{2}$$

where $\lambda_i = e^{\beta\mu_i}$ is the *fugacity* of the particle species i and Z_{N_i} is the N particle canonical partition function of the species i. Chemical potentials μ_i

take care of particle number conservation in an average sense. In this work, we are mainly interested in the mean particle numbers in the grand canonical system, so we employ frequently the equation

$$\langle N_i \rangle = \lambda_i \frac{\partial \ln Z_G}{\partial \lambda_i}. \tag{3}$$

In properly quantized, finite volume system the partition function is often rather difficult to compute. In this work we never know the exact geometry of the system, so we settle for a finite volume, V, sample of the infinite volume system. Thus, the summation over discrete quantum states in partition function is changed to simple phase space integration over continuum. The one particle canonical partition function of particle i is now given by

$$Z_i^1 = g_i \frac{V}{2\pi^2} \int_0^\infty dp\, p^2 e^{-\beta\sqrt{p^2 + m_i^2}}, \tag{4}$$

where g_i is the spin degeneration factor, and m_i is the mass of the particle i. Using the previous result and taking care of the correct occupation of quantum states, the grand canonical partition function can be written in the form

$$\ln Z_G(T, \{\lambda_i\}, V) = \sum_i g_i \frac{V}{2\pi^2} \int_0^\infty dp\, p^2 \ln\left[1 + \eta_i \lambda_i e^{-\beta\sqrt{p^2 + m_i^2}}\right]^{\eta_i}, \tag{5}$$

where η is the statistics factor: $\eta_i = 1$ for fermions and $\eta_i = -1$ for bosons. Now we can write the mean particle number as

$$\langle N_i \rangle = g_i \frac{V}{2\pi^2} \int_0^\infty dp\, p^2 \left[\lambda_i^{-1} e^{\beta\sqrt{p^2 + m_i^2}} + \eta_i\right]^{-1}. \tag{6}$$

In a rare gas (i.e. Boltzmann) limit, which is mostly applied here, we just put the statistics factors $\eta_i = 0$ in particle numbers formula, or let the possible occupation of one particle states be only one to obtain

$$\ln Z_G = \sum_i \lambda_i Z_i^1. \tag{7}$$

In the relativistic, multispecies gas, where the conservation of number of distinct particles is not the main interest, we associate the chemical potentials to conserved quantum numbers. Given that baryon number B, strangeness S and electric charge Q are conserved averagely in a relativistic hadron gas, the particle numbers (6) are

$$\langle N_i \rangle = g_i \frac{V}{2\pi^2} \int_0^\infty dp\, p^2 \left[(\lambda_B^{B_i} \lambda_S^{S_i} \lambda_Q^{Q_i})^{-1} e^{\beta\sqrt{p^2 + m_i^2}} + \eta_i\right]^{-1}, \tag{8}$$

where B_i, S_i and Q_i are the quantum numbers of individual particle species. The average net quantum number, say baryon number, is a sum over particle numbers weighted by the chosen quantum number,

$$\langle N_B \rangle = \sum_i B_i \langle N_i \rangle. \tag{9}$$

3 Exact Strangeness Conservation

Let us consider first a gas composed of neutral kaons and antikaons and request that the overall strangeness be zero. The canonical (with respect to strangeness) partition function is given by

$$Z_{S=0} = \prod_{i=0}^{\infty} \left(\sum_{n_i=0}^{\infty} \frac{1}{n_i!} e^{-\beta \varepsilon_i n_i} \right) \left(\sum_{\bar{n}_i=0}^{\infty} \frac{1}{\bar{n}_i!} e^{-\beta \varepsilon_i \bar{n}_i} \right) \delta_{n_{K^0}, n_{\overline{K}^0}}, \qquad (10)$$

where ε_i is the energy and n_i the number of K^0's in the state denoted by i. The Kronecker delta ensures that the overall strangeness is zero: $n_{K^0} - n_{\overline{K}^0} = \sum_{i=0}^{\infty}(n_i - \bar{n}_i) = 0$. By including the $1/n_i!$ and $1/\bar{n}_i!$ one gets the sums over all *distinct* states. Replacing the sums over single particle levels by the Boltzmannian momentum integrals and using the Dirac representation

$$\delta(n - m) = \frac{1}{2\pi} \int_0^{2\pi} d\alpha e^{i(n-m)\alpha} \qquad (11)$$

of delta function, one obtains the following result

$$Z_{S=0} = \frac{1}{2\pi} \int_0^{2\pi} d\alpha \exp\left[\frac{V}{2\pi^2} \int_0^{\infty} dp\, p^2 e^{-\beta \varepsilon_{K^0} + i\alpha} \right]$$
$$\times \exp\left[\frac{V}{2\pi^2} \int_0^{\infty} dp\, p^2 e^{-\beta \varepsilon_{\overline{K}^0} - i\alpha} \right], \qquad (12)$$

where $\varepsilon_{K^0} = \sqrt{p^2 + m_K^2}$. Applying the notation of single particle partition function (4) gives

$$Z_{S=0} = \frac{1}{2\pi} \int_0^{2\pi} d\alpha \exp\left(Z_{K^0}^1 e^{i\alpha} + Z_{\overline{K}^0}^1 e^{-i\alpha} \right). \qquad (13)$$

By expanding the exponentials in power series it is easy to perform the α-integration to obtain

$$Z_{S=0} = \sum_{p=0}^{\infty} \frac{1}{(p!)^2} (Z_{K^0}^1)^p (Z_{\overline{K}^0}^1)^p. \qquad (14)$$

This series converges to a modified Bessel function

$$Z_{S=0} = I_0(2\sqrt{Z_{K^0}^1 Z_{\overline{K}^0}^1}), \qquad (15)$$

which is generally defined by

$$I_\nu(x) = \sum_{l=0}^{\infty} \frac{1}{l!} \frac{1}{(l+\nu)!} \left(\frac{x}{2} \right)^{2l+\nu}. \qquad (16)$$

The generalization of the calculation to a gas containing any particles i carrying strangeness $S_i = 0, \pm 1$ is straightforward. For this case one obtains

$$Z_{S=0} = \frac{1}{2\pi} \int_0^{2\pi} d\alpha \exp\left(\mathcal{N}_1 e^{i\alpha} + \mathcal{N}_{-1} e^{-i\alpha} + \mathcal{N}_0\right), \qquad (17)$$

where $\mathcal{N}_x$ stands for the sum of single particle partition functions of particles having strangeness $S_i = x$:

$$\begin{cases} \mathcal{N}_1 = Z_{K^+}^1 + Z_{K^0}^1 + \ldots + Z_{\bar{\Lambda}}^1 + Z_{\bar{\Sigma}^+}^1 + Z_{\bar{\Sigma}^0}^1 + \ldots \\[2mm] \mathcal{N}_{-1} = Z_{K^-}^1 + Z_{\bar{K}^0}^1 + \ldots + Z_{\Lambda}^1 + Z_{\Sigma^+}^1 + Z_{\Sigma^0}^1 + \ldots \\[2mm] \mathcal{N}_0 = Z_{\pi^+}^1 + Z_{\pi^0}^1 + Z_{\pi^-}^1 + Z_{\eta}^1 + \ldots + Z_p^1 + Z_n^1 + \ldots \end{cases} \qquad (18)$$

In this case, where the strangeness is being treated canonically and the baryon number and electric charge grand canonically, the Boltzmannian one species partition function is

$$Z_i^1 = \lambda_B^{B_i} \lambda_Q^{Q_i} g_i \frac{V}{2\pi^2} \int_0^\infty dp\, p^2 e^{\beta \sqrt{p^2 + m_i^2}}. \qquad (19)$$

Performing the power expansion of the exponentials and the integration in eq. (17) we are left with

$$Z_{S=0} = Z_0 \sum_{p=0}^\infty \frac{1}{p!^2} (\mathcal{N}_1 \mathcal{N}_{-1})^p = Z_0 I_0(2\sqrt{\mathcal{N}_1 \mathcal{N}_{-1}}), \qquad (20)$$

where Z_0 is the grand canonical partition function for particles having strangeness zero. To calculate the average abundances of particles, we substitute the fictitious strangeness fugacity and derive

$$\langle N_i \rangle = \lambda_i \frac{\partial \ln Z_{S=0}}{\partial \lambda_i} \bigg|_{\lambda_S = 1}, \qquad (21)$$

which gives

$$\langle N_i \rangle = Z_i^1 \left(\sqrt{\frac{\mathcal{N}_1}{\mathcal{N}_{-1}}} \right)^{S_i} \frac{I_{S_i}(2\sqrt{\mathcal{N}_1 \mathcal{N}_{-1}})}{I_0(2\sqrt{\mathcal{N}_1 \mathcal{N}_{-1}})}. \qquad (22)$$

Each term in the sum of eq. (20) is the product of a strangeness plus one and a strangeness minus one particle and one sees the exact strangeness conservation explicitly at work. Due to the strict conservation, the number of strange particles increases nonlinearly with volume, which is illustrated in Fig. (1). The method used above and the expression of the partition function $Z_{S=0}$ indicate that to impose the strict strangeness conservation, one projects

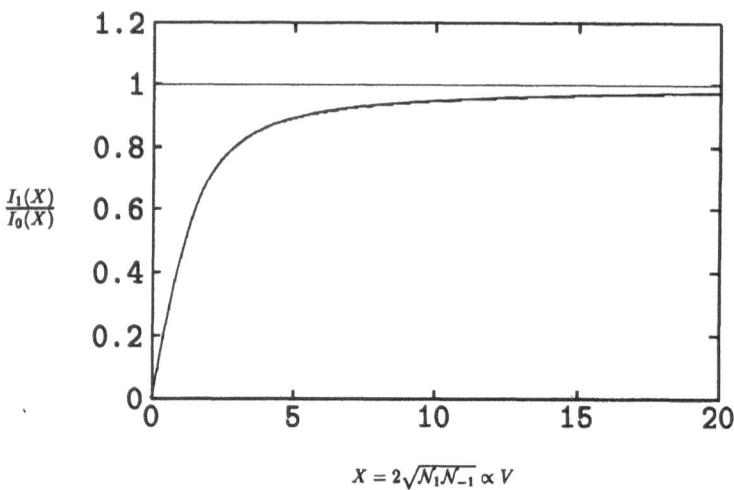

$$X = 2\sqrt{N_1 N_{-1}} \propto V$$

Fig. 1: Nonlinear volume dependence of strange particle abundances in canonical treatment (see eq. 22) compared to grand canonical case ($X \equiv 1$).

the grand canonical partition function $Z_G(T, \lambda_B, \lambda_Q, \lambda_S)$ onto the state with strangeness S,

$$Z_S = \frac{1}{2\pi} \int_0^{2\pi} d\alpha e^{-iS\alpha} Z_G(T, \lambda_B, \lambda_Q, \lambda_S), \qquad (23)$$

where the fugacity factor λ_S has been replaced by

$$\lambda_S = e^{i\alpha}. \qquad (24)$$

The partition function for a gas containing particles with strangeness $S_i = 0, \pm 1, \pm 2$ is given by

$$Z_S = \frac{Z_0}{2\pi} \int_0^{2\pi} d\alpha e^{-iS\alpha} \exp\Big[\mathcal{N}_1 e^{i\alpha} + \mathcal{N}_{-1} e^{-i\alpha} \qquad (25)$$

$$+ \mathcal{N}_{\Xi^0}(\lambda_B e^{-i2\alpha} + \lambda_B^{-1} e^{i2\alpha}) + \mathcal{N}_{\Xi^-}(\lambda_B \lambda_Q^{-1} e^{-i2\alpha} + \lambda_B^{-1} \lambda_Q e^{i2\alpha})\Big],$$

where the sums $\mathcal{N}_{\Xi^0}$ and $\mathcal{N}_{\Xi^-}$ include also the heavier resonances carrying the same quantum numbers, as the $\mathcal{N}_{\pm 1}$ do. Using the generating function for modified Bessel functions I_ν defined by

$$\exp\Big\{ \frac{\rho}{2}\Big(t + \frac{1}{t}\Big) \Big\} = \sum_{\nu=-\infty}^{\infty} I_\nu(\rho) t^\nu \qquad (26)$$

and expanding again the exponentials in power series we obtain

$$Z_S = \frac{Z_0}{2\pi} \int_0^{2\pi} d\alpha e^{-iS\alpha} \sum_{m=-\infty}^{\infty} I_m(2\mathcal{N}_{\Xi^0}) \lambda_B^m e^{-i2m\alpha} \qquad (27)$$

$$\times \sum_{n=-\infty}^{\infty} I_n(2\mathcal{N}_{\Xi^-})\lambda_B^n \lambda_Q^{-n} e^{-i2n\alpha} \sum_{p=0}^{\infty} \frac{1}{p!} \mathcal{N}_1^p e^{ip\alpha} \sum_{q=0}^{\infty} \frac{1}{q!} \mathcal{N}_{-1}^q e^{-iq\alpha}.$$

Carrying out the integration and rearranging the summations the result can be expressed in terms of I_ν -functions:

$$Z_S = Z_0 \sum_{m=-\infty}^{\infty} I_m(2\mathcal{N}_{\Xi^0})\lambda_B^m \sum_{n=-\infty}^{\infty} I_n(2\mathcal{N}_{\Xi^-})\lambda_B^n \lambda_Q^{-n}$$

$$\times \sum_{p=0}^{\infty} \frac{1}{p!} \frac{1}{(-S+p-2m-2n)!}$$

$$\times \left(\sqrt{\mathcal{N}_1 \mathcal{N}_{-1}}\right)^{-S+2p-2m-2n} \left(\frac{\mathcal{N}_1}{\mathcal{N}_{-1}}\right)^{\frac{S}{2}+m+n}$$

$$= Z_0 \sum_{m=-\infty}^{\infty} I_m(2\mathcal{N}_{\Xi^0})\lambda_B^m \sum_{n=-\infty}^{\infty} I_n(2\mathcal{N}_{\Xi^-})\lambda_B^n \lambda_Q^{-n}$$

$$\times \left(\sqrt{\frac{\mathcal{N}_1}{\mathcal{N}_{-1}}}\right)^{S+2m+2n} I_{S+2m+2n}(2\sqrt{\mathcal{N}_1 \mathcal{N}_{-1}}). \tag{28}$$

Using the same techniques the result can be generalized to the case, where the Ω -like hadrons ($S_i = \pm 3$) are included as well:

$$Z_S = Z_0 \sum_{m=-\infty}^{\infty} I_m(2\mathcal{N}_{\Xi^0})\lambda_B^m \sum_{n=-\infty}^{\infty} I_n(2\mathcal{N}_{\Xi^-})\lambda_B^n \lambda_Q^{-n}$$

$$\times \sum_{l=-\infty}^{\infty} I_l(2\mathcal{N}_{\Omega^-})\lambda_B^l \lambda_Q^{-l}$$

$$\times \left(\sqrt{\frac{\mathcal{N}_1}{\mathcal{N}_{-1}}}\right)^{S+3l+2m+2n} I_{S+2m+2n+3l}(2\sqrt{\mathcal{N}_1 \mathcal{N}_{-1}}). \tag{29}$$

The mean number of hadrons i with strangeness S_i in the system becomes

$$\langle N_i \rangle = Z_i^1 \frac{Z_{S-S_i}}{Z_S} \left(\sqrt{\frac{\mathcal{N}_1}{\mathcal{N}_{-1}}}\right)^{-S_i}. \tag{30}$$

The modified Bessel functions decrease quickly with increasing absolute value of their indecies, so the numerical evaluation of mean particle numbers is not cumbersome.

3.1   Application of Z_S

In a recent paper [1] we have analysed the particle production in GSI SIS Ni+Ni experiments. We addressed especially the abundance of kaons who can not be described by hadronic gas model in its standard form. Although the size of the Ni system is relatively large the corrections due to exact strangeness conservation turned out to be crucial at low temperatures, $T < 100$ MeV, involved. At these temperatures the width of resonances had to be taken into account. A summary of our results for particle ratios is presented in table 1.

Table 1: Particle ratios given by present model compared to experimental results. The best fit value, $\mu_B = 0.72$ GeV, for the baryon chemical potential is used.

Ratio	Model				Data	
	$R = 4.2$ fm		$R = 3$ fm			
	$T = 65$ MeV	75 MeV	$T = 65$ MeV	75 MeV	ratio	ref.
K^+/K^-	25.7	22.4	23.9	21.1	21 ± 9	[2–4]
K^+/π^+	0.0071	0.0339	0.0027	0.0132	0.0074 ± 0.0021	[3,6,9]
ϕ/K^-	0.103	0.082	0.276	0.212	0.1	[7]
π^+/π^-	0.893	0.895	0.894	0.898	0.89	[8]
η/π^0	0.008	0.015	0.008	0.015	0.037 ± 0.002	[5]
π^+/p	0.225	0.247	0.224	0.246	0.195 ± 0.020	[9,8]
π^0/B	0.104	0.108	0.104	0.107	0.125 ± 0.007	[5]
d/p	0.129	0.188	0.129	0.188	0.26	[9]

The measured hadronic ratios with corresponding errorbars are described as bands in the (T, μ_B) plane as shown in Fig. 1. The intervals of temperature and of chemical potential

$$T = 70 \pm 10 \text{MeV}$$
$$\mu_B = 720 \pm 30 \text{MeV}$$

give a good fit to the data. The freeze-out radius of $R \simeq 4$ fm was extracted from the volume dependence of the ratios K^+/π^+ and ϕ/K^-.

4   Exact Baryon, Charge and Strangeness Conservation

In the case of three exactly conserved, additive quantum numbers we start from the single particle partition function of particle i,

$$Z_i = \sum_j g_j \frac{V}{2\pi^2} \int_0^\infty dp\, p^2 e^{-\beta \epsilon_j} \delta_{B_j, B_i} \delta_{Q_j, Q_i} \delta_{S_j, S_i}. \tag{31}$$

Making use of the integral representation for δ -functions and the overall conservation constraints

$$\sum B_i = B, \quad \sum Q_i = Q, \quad \sum S_i = S, \tag{32}$$

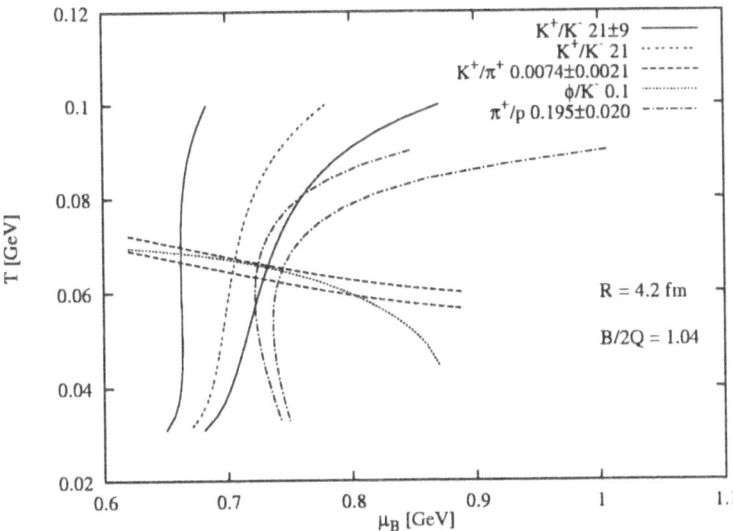

Fig. 2: Curves in the (μ_B, T) plane corresponding to the hadronic ratios indicated. The interaction volume corresponds to the radius of 4.2 fm, and the isospin asymmetry is $B/2Q = 1.04$.

the resulting integral corresponds to a projection of the grand canonical partition function onto the desired values of B, Q and S:

$$Z_{B,Q,S} = \frac{1}{(2\pi)^3} \int_0^{2\pi} d\psi e^{-iB\psi} \int_0^{2\pi} d\phi e^{-iQ\phi} \int_0^{2\pi} d\alpha e^{-iS\alpha}$$
$$\times Z_G(T, \lambda_B, \lambda_Q, \lambda_S). \tag{33}$$

Here the fugacity factors have been replaced by

$$\lambda_B = e^{i\psi}, \quad \lambda_Q = e^{i\phi}, \quad \lambda_S = e^{i\alpha}. \tag{34}$$

As the contributions always come pairwise for particles and antiparticles, the fugacity factors will give rise to the cosine of the corresponding angle. It is useful to group the particles according to their quantum numbers. Leaving out charm, bottom and $S_i \geq 2$ particles we are left with ten categories which will be labeled by their particle content. For instance, $\mathcal{N}_{K^0}$ stands for the sum of one particle partition functions of neutral strange and antistrange mesons (K^0, $\overline{K}^0$, K^{*0}, $\overline{K^*}^0$, ...), $\mathcal{N}_{K^c}$ stands for charged strange mesons (K^+, K^-, K^{*+}, K^{*-}, ...), and $\mathcal{N}_\Lambda$ stands for all neutral hyperons and antihyperons. With these notations the partition function can be rewritten in the form

$$Z_{B,Q,S} = \frac{1}{(2\pi)^3} \int_0^{2\pi} d\psi e^{-iB\psi} \int_0^{2\pi} d\phi e^{-iQ\phi} \int_0^{2\pi} d\alpha e^{-iS\alpha} \tag{35}$$

$$
\times \exp\{2\mathcal{N}_n \cos\psi + 2\mathcal{N}_{\pi^c}\cos\phi + 2\mathcal{N}_{K^0}\cos\alpha + 2\mathcal{N}_{K^c}\cos(\phi+\alpha)
$$
$$
+2\mathcal{N}_p\cos(\psi+\phi) + 2\mathcal{N}_{\Delta^-}\cos(\psi-\phi) + 2\mathcal{N}_{\Delta^{++}}\cos(\psi+2\phi)
$$
$$
+2\mathcal{N}_\Lambda \cos(\psi-\alpha)
$$
$$
+2\mathcal{N}_{\Sigma^+}\cos(\psi+\phi-\alpha) + 2\mathcal{N}_{\Sigma^-}\cos(\psi-\phi-\alpha)\}.
$$

The integration above can not be done directly due to cosine terms of multiple angles. To circumvent this difficulty, we introduce a new angle whenever more than one appears. For example, in the term involving $\mathcal{N}_p$ we introduce an intermediate angle ξ in the following way

$$
1 = \int_0^{2\pi} d\xi \delta(\psi+\phi-\xi) = \sum_{\nu=-\infty}^{\infty} \frac{1}{2\pi}\int_0^\pi d\xi e^{i\nu(\psi+\phi-\xi)}. \tag{36}
$$

The application of the integral representation of the modified Bessel function,

$$
I_n(z) = \frac{1}{\pi}\int_0^\pi d\omega\, e^{z\cos\omega}\cos n\omega, \tag{37}
$$

allows one to write the partition function in the form

$$
Z_{B,Q,S}(T,V) = Z_0 \left(\prod_{\nu=1}^{7}\sum_{n_\nu=-\infty}^{\infty}\right)
$$
$$
\times I_{-B+n_2+n_3+n_4+n_5+n_6+n_7}(2\mathcal{N}_n)
$$
$$
\times I_{-Q+n_1+n_2-n_3+n_5-n_6+2n_7}(2\mathcal{N}_{\pi^c}) \tag{38}
$$
$$
\times I_{-S+n_1-n_4-n_5-n_6}(2\mathcal{N}_{K^0})
$$
$$
\times I_{n_1}(2\mathcal{N}_{K^c})I_{n_2}(2\mathcal{N}_p)I_{n_3}(2\mathcal{N}_{\Delta^-})
$$
$$
\times I_{n_4}(2\mathcal{N}_\Lambda)I_{n_5}(2\mathcal{N}_{\Sigma^+})I_{n_6}(2\mathcal{N}_{\Sigma^-})I_{n_7}(2\mathcal{N}_{\Delta^{++}}).
$$

The differentiation of eq. (33) for particle abundances yields the result

$$
\langle N_i \rangle = \frac{Z_{B-B_i,Q-Q_i,S-S_i}}{Z_{B,Q,S}} Z_i^1. \tag{39}
$$

The evaluation of the canonical partition function with three simultaneously conserved quantum numbers becomes numerically very time consuming for large values of B. So far, for systems with $B > 20$ we have been forced to resort to the grand canonical treatment.

4.1   Application of $Z_{B,Q,S}$

In order to compare our numerical results with the E802 experimental data shown in Table 2, we estimate the baryon number and charge of the experimental system via geometrical considerations. Letting R_P and R_T be the radius of a projectile and target nucleus respectively, we assume that the radii are directly proportional to the cubic roots of the mass numbers A_P and A_T of the interacting nuclei. In the case of central collisions the interaction region is taken to be a cylinder of radius R_P, length $2\sqrt{R_T^2 - R_P^2}$ plus, two remaining spherical segments at the ends of the cylinder with R_T as the radius. We find the total number of participating nucleons (or, baryon number, B) and the total charge Q to be

$$B = A_P + A_T \left\{ 1 - \left[1 - \left(\frac{A_P}{A_T} \right)^{\frac{2}{3}} \right]^{\frac{3}{2}} \right\} \tag{40}$$

$$Q = Z_P + Z_T \left\{ 1 - \left[1 - \left(\frac{A_P}{A_T} \right)^{\frac{2}{3}} \right]^{\frac{3}{2}} \right\} \tag{41}$$

respectively.

The comparison of our numerical results with the E802 data (Table 2) of the K^+/π^+ and K^-/π^- ratios as functions of the size of the system or, equivalently, B are shown in Fig. 3. In the upper figure we observe that the theoretical curve for $B/2Q = 5/6$ approximates the K^+/π^+ data best. The theoretical curves lie systematically above the data but drop closer as $B/2Q$ decreases towards the collision value. The effect of the isospin asymmetry of the system is seen also in the K^-/π^- data comparison. As the ratio $B/2Q$ approaches the collision value the theoretical curves begin to approximate the data more closely.

Table 2: Experimental results reported by the E802 collaboration. B and Q are calculated using equations 40 and 41.

Collision	K^+/π^+	Ref.	K^-/π^-	Ref.	B	Q
$p + {}^4Be_9$	7.8±0.4%	[11,12]	2.0±0.2%	[11]	3.9	2.3
$p + {}^{13}Al_{27}$	9.9±0.5%	[12]			5.4	3.1
$p + {}^{29}Cu_{64}$	10.8±0.6%	[12]			6.9	3.7
$p + {}^{79}Au_{197}$	12.5±0.6%	[11,12]	2.8±0.3%	[11]	9.7	4.5
${}^{14}Si_{28} + {}^{79}Au_{197}$	18.2±0.9%	[11]	3.2±0.3%	[11]	102.7	44.0
	19.2±3%	[13]	3.6±0.8%	[13]		

5   Generalization of the Projection Method

In this section, we review the projection method generalized to arbitrary internal symmetry of the system in addition to $U(1)$ of strangeness and $U(1) \times U(1) \times U(1)$ of baryon number, electric charge and strangeness. For complete

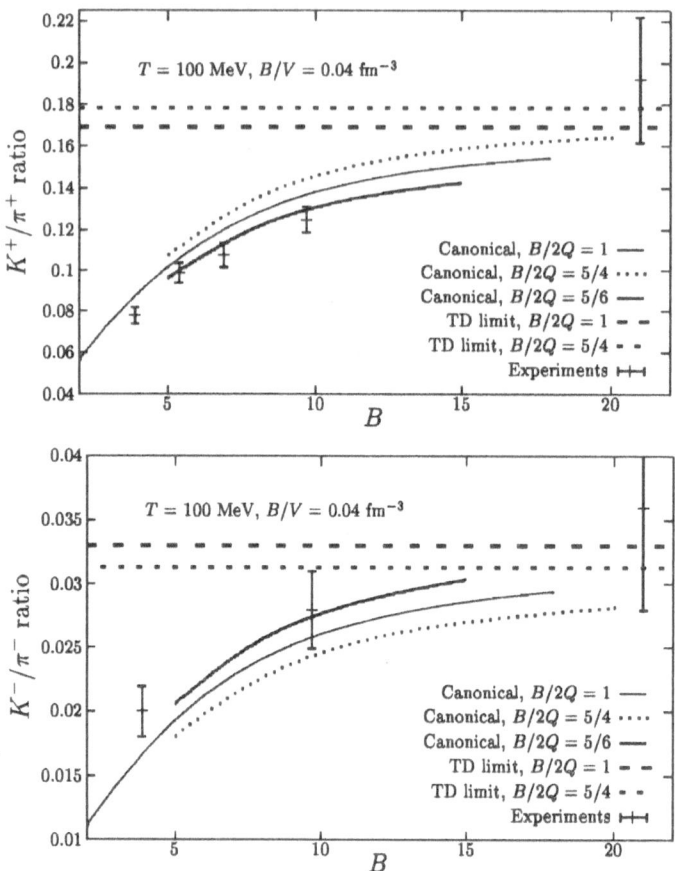

Fig. 3: Thermal model expectations for the production ratios K^+/π^+ and K^-/π^- at a temperature of 100 MeV and a baryon density of 0.04 fm^{-3} compared to experimental results from the Brookhaven AGS. The experimental ratios from $Si-Au$ collisions ($B \sim 103$) is moved to $B = 21$ for the sake of convenience.

derivation, see the original texts of Turko, Redlich and Hagedorn [14]. The general method is suitable for non-abelian symmetries, such as $SU(2)$ of isospin [15] or angular momentum [16], and $SU(3)$ of color [17] as well.

If the system is exactly symmetric under the operations of internal symmetry group G, the corresponding group generators Q_k have the same eigenstates as the Hamiltonian. Thus

$$[\hat{H}, Q_k] = 0, \quad k = 1, \ldots, n, \qquad (42)$$

where n is the number of parameters in the group. Let us define the *generating function* $\hat{Z}$ as $\hat{Z} = \text{tr}[U(g)e^{-\beta \hat{H}}]$, where $U(g)$ is an unitary representation of the group. With the aid of irreducible presentations of U, this decomposes to

$$\hat{Z} = \sum_{\nu} \frac{\hat{\chi}^{\nu}(g)}{d(\nu)} Z_{\nu}. \tag{43}$$

Here we used a character $\hat{\chi}^{\nu}(g)$ and the dimension $d(\nu)$ of an irreducible presentation $U^{\nu}(g)$, and the corresponding canonical partition function Z_{ν}. Using the orthogonality of characters,

$$\int d\mu(g)\overline{\hat{\chi}^{\nu}(g)}\hat{\chi}^{\nu'}(g) = \delta_{\nu\nu'}, \tag{44}$$

we may compute the canonical partition function once we know the generating function:

$$Z_{\nu} = d(\nu) \int d\mu(g)\overline{\hat{\chi}^{\nu}(g)}\hat{Z}. \tag{45}$$

Further investigation of the generating function reveals that

$$\hat{Z} = \mathrm{tr}\exp\left(-\beta\hat{H} + i\sum_{k=1}^{r} Q_k\gamma_k\right)$$

$$= \prod_{j=1}^{\infty}\prod_{\rho=1}^{d(\nu)}\sum_{n}\exp\left[n\left(-\beta E_j + i\sum_{k=1}^{r} q_k^{(\rho)}\gamma_k\right)\right]. \tag{46}$$

In the last step, we have expressed the trace in the basis of n-particle Hamiltonian eigenstates. The $q_k^{(\rho)}$ are the conserved charges, and the γ_k are the variables of the Cartan subgroup of the group G of rank r. Eq. (46) resembles the grand canonical partition function, and is actually obtained from it by the Wick rotation: $\beta\mu_i \rightarrow -i\gamma_i$.

As an example, let us choose the internal symmetry of the system correspond to $U(1)_{q_1} \times \cdots \times U(1)_{q_r}$, where the q_i are the conserved charges. The character of $U(1)_{q_i}$ is $e^{iq_i\gamma_i}$, so the character of the direct product group is $\exp(i\sum_{i=1}^{r} q_i\gamma_i)$. The canonical partition function respecting the exact conservation of charges q_i has now the form

$$Z_{q_1,\ldots,q_r}(T,V) = \frac{1}{(2\pi)^r}\int_0^{2\pi} d\gamma_1 \cdots \int_0^{2\pi} d\gamma_r$$

$$\times \exp\left[-i\sum_{i=1}^{r} q_i\gamma_i\right]\hat{Z}(T,V,\gamma_1,\cdots,\gamma_r). \tag{47}$$

The special cases, Z_S and $Z_{B,Q,S}$ for a Boltzmannian hadron resonance gas are considered in previous sections.

6   Summary

The particle abundances have been computed in the canonical formalism using the formulation for the exact conservation of baryon number, strangeness and charge in the thermal model of particle production. A good agreement with the experimental data of GSI Ni+Ni collisions and of E802 collaboration in $p - A$ collisions was reported.

The good agreement with chemical equilibrium does not mean that the particle spectra should follow exactly a Boltzmann distribution since the momenta of particles can be severely affected by flow. As an example, a model with Bjorken expansion in the longitudinal direction will still have its particle ratios determined by Boltzmann factors even though the longitudinal distribution is nowhere near a Boltzmann distribution [18].

References

1. J. Cleymans, D. Elliott, A. Keränen and E. Suhonen, Phys. Rev. **C57** (1998) 3319

2. KaoS Collaboration, P. Barth et al., Phys. Rev. Lett. **78** (1997) 4007

3. KaoS Collaboration, H. Oeschler, "Hadrons in Dense Matter and Hadrosynthesis", these proceedings.

4. FOPI Collaboration, Y. Leifels : "FOPI Results - Strangeness in 4π." Talk presented at the International Workshop : "Hadrons in Dense Matter.", GSI, Darmstadt, July 2-4, 1997

5. TAPS Collaboration, M. Appenheimer et al., GSI 97-1, page 58; R. Averbeck, "Hadronische Materie bei SIS-Energien - Eine Thermodynamische Analyse." (unpublished), presented at the DPG Frühjahrstagung, Göttingen, February 1997

6. FOPI Collaboration, D. Best et al., Nucl. Phys. **A625** (1997) 367

7. FOPI Collaboration, N. Herrmann, Nucl. Phys. **A610** (1996) 49c

8. FOPI Collaboration, D. Pelte et al. , Z. Phys. **A359** (1997) 55

9. KaoS Collaboration, C. Müntz et al. , Z. Phys. **A352** (1995) 175; Z. Phys. A357 (1997) 399

10. J. Cleymans, A. Keränen, M. Marais and E. Suhonen, Phys. Rev. **C56** (1997) 2474

11. T. Abbot et al (E802 Collaboration), Phys. Rev. Lett. **66** (1991) 1567

12. T. Abbot et al (E802 Collaboration), Phys. Rev. **D45** (1992) 3906

13. T. Abbot et al (E802 Collaboration), Phys. Rev. Lett. **64** (1990) 847

14. K. Redlich and L. Turko, Z. Phys. **C5** (1980) 201; L. Turko, Phys. Lett. **B104** (1981) 153; R. Hagedorn and K. Redlich, Z. Phys. **C27** (1985) 541

15. B. Müller and J. Rafelski, Phys. Lett. **B116** (1982) 274

16. W. Blümel, P. Koch and U. Heinz, Z. Phys. **C63** (1994) 637

17. Th. Elze and W. Greiner, Phys. Rev. **A33** (1986) 1879

18. J. Cleymans, 3rd International Conference on "Physics and Astrophysics of Quark-Gluon Plasma", Jaipur, India, March 17 - 21, 1997 to be published, nucl-th/9704046.

Springer
and the
environment

At Springer we firmly believe that an
international science publisher has a
special obligation to the environment,
and our corporate policies consistently
reflect this conviction.

We also expect our business partners –
paper mills, printers, packaging
manufacturers, etc. – to commit
themselves to using materials and
production processes that do not harm
the environment. The paper in this
book is made from low- or no-chlorine
pulp and is acid free, in conformance
with international standards for paper
permanency.

Lecture Notes in Physics

For information about Vols. 1–478
please contact your bookseller or Springer-Verlag

Monographs
For information about Vols. 1–10
please contact your bookseller or Springer-Verlag